Atlas of Selected
Oil and Gas
Reservoir Rocks
From North America

Atlas of Selected Oil and Gas Reservoir Rocks From North America

A COLOR GUIDEBOOK
TO THE PETROLOGY OF SELECTED
OIL AND GAS RESERVOIR ROCKS
FROM THE UNITED STATES
AND CANADA

Edwin W. Biederman, Jr.
The Pennsylvania State University

A WILEY-INTERSCIENCE PUBLICATION
JOHN WILEY & SONS
New York · Chichester · Brisbane · Toronto · Singapore

Copyright © 1986 by John Wiley & Sons, Inc.

All rights reserved. Published simultaneously in Canada.

Reproduction or translation of any part of this work beyond that permitted by Section 107 or 108 of the 1976 United States Copyright Act without the permission of the copyright owner is unlawful. Requests for permission or further information should be addressed to the Permissions Department, John Wiley & Sons, Inc.

Library of Congress Cataloging in Publication Data:

Biederman, Edwin W.
    Atlas of selected oil and gas reservoir rocks from North America.

    Includes bibliographical references and index.
    1. Rocks, Sedimentary—North America—Pictorial works. 2. Thin sections (Geology)—Pictorial works.
3. Petroleum—Geology—North America—Pictorial works.
4. Gas, Natural—Geology—North America—Pictorial works. I. Title.
QE471.B485 1986   553.2.8'097   85-29467
ISBN 0-471-81666-3

Printed in the United States of America

10 9 8 7 6 5 4 3 2 1

*To my wife Margaret-Jane and to Sphene
for their patience and to my children
Robert, Mary, Jane, and James for their
generosity in making this book possible*

Preface

Scattered throughout the literature on petroleum exploration and recovery are numerous photomicrographs of thin sections of oil and gas reservoir rocks; however, most of them are in black and white. This is not what the petrologist actually sees. So much depends on the colors produced under the crossed nicols of the petrographic microscope that if photomicrographs are not reproduced in color, they are of limited value. It was with this in mind that I began to take photomicrographs of reservoir rocks as they appeared in thin section, and it is the collection resulting from 10 years of such work that is presented in this book.

Oil companies take cores from oil and gas reservoirs, analyze them for engineering purposes, and then store them for future reference. Because of their bulk and weight, storage space soon becomes a problem and gradually they are disposed of to the point where only the 1-in. plugs on which the engineering measurements were made are saved. Although these cores retrieved from thousands of feet or even miles down in the earth's crust are very expensive to obtain, they are frequently discarded or lost.

Research projects tend to focus on specific fields or particular productive horizons. The result of this selection process is that the cores from relatively uninteresting reservoirs receive little attention. Furthermore, the time and effort required to sample long rock cores on a foot-by-foot basis and to have these

Preface

samples thin-sectioned is no small matter; however, the next step is the most time consuming of all. This involves the careful examination of the thin sections with the petrographic microscope by an experienced petrologist. It is an unfortunate fact that relatively few sedimentary petrologists have access to thin-section collections from a number of different oil and gas reservoir rocks. Yet, it is these same rocks that are the subject of a worldwide search that is costing the oil industry billions of dollars. Furthermore, once the primary and secondary crops of oil and gas are harvested, it is becoming very worthwhile to spend even more dollars in attempting to use tertiary recovery methods. This book is designed to give the petrologist a broader view of the range of reservoir rock types using North American examples.

Clearly, there is also a much wider audience that can benefit a great deal from studying this type of a collection. The petroleum geologist, the geophysicist, and particularly the petroleum engineer all need to have a good idea of what the actual target looks like. The student majoring in petroleum engineering may also wish to have this volume on his shelf so that he will be one step ahead when he reports to his first job assignment.

Each chapter that discusses a specific reservoir rock is designed to provide a brief discussion of the geologic background, the mineralogy and grainsize (where appropriate), color photomicrographs, engineering data, summaries of experimental tertiary recovery projects where available, recent sediment equivalents, and exploration potential. It must be recognized that in many cases some of the subjects in this outline cannot be addressed because the information is not available. In other instances, topics such as the geochemistries of the crude oils are added; hence, the outline is designed to be a flexible guide that expands and contracts along with the amount of information.

In the best of all possible worlds, one would have access to all kinds of supporting geologic and geophysical data; and it would be available uniformly for all the fields described. Unfortunately, this is far from the case. In a fair number of instances, the literature has filled the gaps; however, a number of the producing fields discussed are not covered elsewhere in any detail, and the thin sections are the only samples left from which evidence can be obtained.

Perhaps one result of this initial effort may be to inspire other workers to study each core that becomes available and to record their observations before these costly samples become lost. Others may be stimulated to dig deeper and ask more penetrating questions concerning the events that produced the various rock types illustrated. Some will undoubtedly thumb through the pages, glance at the pictures, and wonder about the artistic caprice of the Creator, which produced the juxtaposition of shape and color. If this work can serve in some small way both the artistic soul in man and his longing for understanding, it will have been worth the effort.

EDWIN W. BIEDERMAN, JR.

University Park, Pennsylvania
March 1986

Acknowledgments

Sincere thanks to Dr. Myron Horn of Cities Service Oil Company, who gave me permission to use data that were collected while working at the Cities Service Laboratory in Tulsa, Oklahoma. Additional thanks go to Dr. E. W. (Chris) Christensen of Chevron U.S.A., Inc., who provided both representative thin sections and appropriate reference material for the chapter dealing with California.

Thanks are also accorded Mr. Daniel Schafer of ARCO Alaska, Inc., who was kind enough to provide thin sections from cores taken within the vast Prudhoe Bay field. Particular gratitude is due to Dr. John C. Griffiths who encouraged the author and kindly supplied the thin sections of the Third Bradford Sand.

It is also fitting that Eleanor Zindler and Betty Layton receive a significant share of the credit for typing a difficult manuscript.

Other colleagues who deserve special thanks for their organic geochemical work are Bennie Heinz, Dr. Robert Miller, Ted Murray, and Wayne Fallgatter. Also, Mr. William Batten deserves a particular word of appreciation for his work in producing the thousands of thin sections that form the centerpiece of this study.

E.W.B.

Contents

1	**Introduction**	1
	Sampling and Data Gathering	2
	Geographic Areas	3
	Reference	5
2	**The Appalachian Basin**	6
	Bradford Oil Field, Bradford Third Sand, McKean County, Pennsylvania, Geologic Background	7
	Mineralogy	7
	Grainsize	13
	Oil Recovery Projects, Polymer Flooding	13
	Oil Saturation Study	15
	Explanation of the Correlation Index	17
	Summary of Important Features	19
	References	22

3 Kanawha Forest Field, Newburg Gas Sand, Kanawha County, West Virginia 23

Geologic Background 23
Mineral Composition 24
Geologic Interpretation 24
Exploration 31
References 31

4 Walton Field, Big Injun Sandstone, Roane County, West Virginia 32

Geologic Background 32
Capillary Pressure Curves 33
Photomicrographs and Petrographic Descriptions 36
Geologic Interpretations 40
Exploration 40
References 40

5 Thornwood Gas Field, Oriskany Sandstone, Pocahontas County, West Virginia 41

Geologic Background 41
Photomicrographs and Petrologic Detail 43
Geologic Interpretation 45
References 49

6 Rodney Oil Field, Dundee Formation, Elgin County, Ontario, Canada 50

Geologic Background 50
Photomicrographs 51
Geologic Interpretation 53
References 57

7 Silver City Pool, Hardinsburg Sandstone, Butler County, Kentucky 58

General Geologic Setting 58
Photomicrographs 59
Geologic Interpretation 60
Petrographic Description 61
References 66

8 North Wise Field, Rogers City–Dundee Formation, Isabella County, Michigan 67

Geologic Background 67
Geologic Interpretation 73
References 73

Contents

9 Mid-Continent, El Dorado Oil Field, Admire Formation, Butler County, Kansas — **75**
- Geologic Background — 77
- Geologic Interpretation — 79
- Origin of Admire Oil — 82
- Tertiary Oil Recovery Project — 83
- Review of Technology — 85
- References — 86

10 El Dorado Oil Field, Viola Formation, Butler County, Kansas — **88**
- General Geologic Setting — 88
- Capillary Pressure Curves — 90
- Geologic Interpretation — 91
- References — 96

11 El Dorado Oil Field, Simpson Sand Formation, Butler County, Kansas — **97**
- Geologic Background — 97
- Mineral Composition — 98
- Oolitic Zone — 98
- Geologic Interpretation — 98
- Source of the Oil at El Dorado — 105
- Exploration — 106
- Recent Sediment Equivalents — 106
- References — 108

12 Yeage Field, Hunton Formation, Riley County, Kansas — **109**
- Geologic Background — 109
- Exploration — 112

13 Oklahoma City Field, Wilcox Sandstone, Oklahoma County, Oklahoma — **113**
- Geologic Background — 113
- Sample Location — 114
- Source of the Oil — 120
- Recent Equivalents — 120
- Geologic Interpretation — 121
- References — 121

14 Victory Oil Field, Lansing–Kansas City Group, Haskell County, Kansas — **122**
- Geologic Background — 122
- Photomicrographs and Description — 124

	Exploration	124
	Recent Equivalents	127
	References	129
15	**Panhandle Field Reservoirs of Texas**	**130**
	Panhandle Field, Granite Wash Formation, Carson County, Texas—Geologic Background	131
	Mineral Composition—Amorphous Materials	133
	Sampling	133
	Exploration	133
	References	137
16	**The Brown Dolomite Panhandle Field, Hutchinson County, Texas**	**138**
	Geologic Background	138
	Geologic Interpretation	138
	References	141
17	**Panhandle Field, The Red Cave Formation, Carson County, Texas**	**142**
	Geologic Background	142
	Source of Oil for the Panhandle Field	143
	References	148
18	**West Texas and Eastern New Mexico**	**149**
	General Geologic Setting	150
	Central Basin Platform	152
	Delaware Basin	153
	The Ellenburger Formation TXL Field, Ector County, Texas—Geologic Background	154
	Photomicrographs and Description	155
	Comments	155
	References	161
19	**The Spraberry Formation, Spraberry Field, Reagan County, Texas**	**162**
	Photomicrographs and Description	163
	Fractures in the Spraberry	163
	Exploration	166
	Source of Spraberry Oil	166
	References	169

Contents

20 The San Andres Formation, Welch Field, Dawson County, Texas — 170

Organic Matter — 172

Photomicrographs and Description — 173

Reservoir Engineering Problems — 173

Permeability and Porosity — 176

Exploration — 179

References — 179

21 The San Andres Formation, Goldsmith Field, Ector County, Texas — 180

Photomicrographs and Description — 181

Source of Oil at Goldsmith — 181

Exploration — 188

References — 188

22 The Yates Formation, West Teas Field, Lea County, New Mexico — 190

Recent Equivalents — 191

Source of Oil at West Teas Field — 191

Summary — 191

References — 194

23 Gulf Coast — 195

Introduction — 195

General Geologic Setting — 196

The Yegua Formation, Romeo Field, Jim Hogg County, Texas—Geologic Background — 199

Geologic Interpretation — 200

References — 202

24 Frio Formation, The May Field, Kleberg County, Texas — 203

Mineralogy — 206

Mud Balls — 213

Heavy Minerals — 214

Pyrite — 220

Metamorphic Heavy Minerals — 220

Ilmenite and Magnetite — 220

Thin Sections — 222

Grainsize — 222

Source Rocks — 222

Comparison with Samples from Greta–Carancahua Barrier Strand–Plain System at Corpus Christi — 228

	Recent Equivalents	231
	Summary	232
	References	232
25	**The Cadeville Sandstone, Schuler Formation, Calhoun Field, Jackson Parish, Louisiana**	**234**
	Geologic Background	234
	Petrology	236
	Environment of Deposition	236
	Recent Sediment Equivalents	239
	Source Rocks	239
	References	240
26	**The James Limestone, Pearsall Formation, Fairway Field, Anderson and Henderson Counties, Texas**	**241**
	Geologic Background	241
	Crude Oil Characteristics	244
	Geologic Interpretation	244
	Engineering Studies	247
	Summary	249
	References	250
27	**Rocky Mountains and Northern Great Plains**	**251**
	General Geologic Setting	251
	The "Basal Sand" Des Moines Series (Pennsylvanian), Sleepy Hollow Field, Red Willow County, Nebraska—Geologic Background	252
	Basal Sand, Pennsylvania, Blackwood Creek Field, Hayes County, Nebraska—Geologic Background	258
	Geologic Interpretation	258
	References	262
28	**Wyoming**	**263**
	The Geologic Background Frontier Formation	265
	Salt Creek Field	265
	Source Beds	268
	Geologic Environment	270
	Mineralogy	271
	Geologic Interpretation	271
	References	275
29	**Tensleep Formation, Central Wyoming**	**276**
	General Mineralogy	277
	Geologic Interpretation	277

| Contents | xvii |

	Fractures	277
	Porosity and Permeability	278
	Discussion	280
	Source Rocks	281
	References	281

30 Phosphoria Formation, Central Wyoming — 282
Geologic Setting — 282
Photomicrographs — 283
The Phosphoria as a Source Rock — 286
Reference — 288

31 Paddle River Gas Field, Nordegg Member, Alberta, Canada — 289
Mineral Composition and Grainsize — 292
Source Materials — 292
Geologic Interpretation — 292
References — 299

32 Pembina Oil Field, Basal Belly River Formation, Alberta, Canada — 300
General Geologic Setting — 300
Petrologic Detail — 301
Geologic Interpretation — 305
Exploration — 310
References — 310

33 Alaska — 311
Geologic Background — 311
Prudhoe Bay Field, Sadlerochit Formation, North Slope, Alaska — 313
Mineral Composition and Grainsize — 316
Geologic Interpretation — 321
Source Rocks — 322
Recovery Project — 324
Exploration — 324
References — 324

34 Dineh-bi-Keyah Oil Field, Apache County, Arizona — 325
General Geologic Setting — 325
Mineralogy — 327
General Comments — 328
References — 333

35 West Coast — 334

- General Geologic Background — 335
- Lost Hills Field Monterey "Diatomite", Kern County, California — 337
- Geologic Background — 338
- Mineralogy — 339
- Structure — 340
- Fracturing — 340
- Source Rocks — 341
- Reservoir Characteristics — 341
- References — 341

36 McKittrick Field, Diatomite Oil Mine, Kern County, California, and Fractured Monterey Chert — 342

- The Dravo Process — 344
- Fractured Monterey Chert (California) — 345
- Comments — 345
- References — 347

37 West Elk Hills Field, Stevens Sandstone, Kern County, California — 348

- Geologic Background — 348
- Recent Equivalents — 353
- References — 354

38 Guijarral Hills Field, Gatchell–McAdams Sandstone, Fresno County, California — 355

- Geologic Background — 355
- Mineral Composition — 356
- Recent Equivalents — 356
- References — 359

39 Kettleman Hills Field, Kings and Fresno Counties, California — 360

- Geologic Background — 360
- Discussion — 364
- Recent Sediment Equivalents — 364
- Crude Oil Maturation and Source Rocks — 364
- References — 367

40 Thoughts and Conclusions — 369

- References — 377

Appendix A Stratigraphic Charts — 378

Appendix B Conversion Factors	382
Author Index	**387**
Subject Index	**391**

1

Introduction

The working petroleum geologist, geophysicist, and reservoir engineer all need information concerning the mineralogy, pore geometry, and grainsize of petroleum reservoir rocks. The fundamental target of their efforts must be well defined; yet all too often access to core samples from the actual reservoirs is not possible. All of these technical people need to know exactly what the reservoir looks like so that they can draw conclusions concerning both exploration possibilities and the prospects for secondary and tertiary recovery operations. The reservoir engineer may benefit most from such material because it is he or she who must design recovery projects that will show favorable economics. The choice of which technology to use hinges on how the reservoir rock is structured and what it is made of.

Older fields have frequently changed hands several times and much, if not all, the core materials have been lost. Prior to drilling and coring new wells, it is very instructive to study the reservoir rocks from other fields in the same general area to obtain some idea of what the mineral composition is like, how the cementation occurs, and what kind of porosity exists.

This book is designed so that a few of the producing rock types are discussed for each of the major oil and gas productive provinces.

Several reservoirs from Canadian sources are also discussed; in most cases, they form logical extensions of provinces that are productive in the United States.

Sampling and Data Gathering

Cores from petroleum reservoirs are costly to obtain and difficult to store, yet for every newly discovered reservoir, it is desirable to have core materials. Key parameters, as far as the reservoir engineer is concerned, are porosity, permeability, and oil saturations. Once these data have been obtained, additional samples can be obtained on a foot-by-foot basis for thin sections. Oddly enough, the field geologist may not take advantage of the opportunity except at the well site, where routine geologic descriptions of the core are made prior to its being shipped to the laboratory for the determination of the engineering data.

Once the thin sections have been obtained, time must be set aside for the sedimentary petrologist to examine each section and describe and photograph appropriate features. Considering the fact that the cost of obtaining core samples from 1 or 2 miles down in the earth is significant and further that these samples are somewhat rare, it is surprising that more attention is not paid to their long-term study. It is hoped that this work will help stimulate effort in this area.

The initial photomicrographs were obtained with a Zeiss polarizing microscope that had a built-in 35-mm camera. Additional photomicrographs were taken with a Nikon photomicrograph system. The author has attempted to photograph those features that are both typical of the particular reservoir and unusual. In this regard, the mineralogy, the packing of the grains, their size and the size of the pores, and the matrix materials are particularly important.

Examination of cores with the naked eye or a hand lens can reveal much about the fabric of a rock; however, the microscopic examination can be even more rewarding, particularly when anomalies occur in the engineering or geophysical data. At this stage, the petrographic microscope can frequently reveal the cause of the anomaly. Such interesting samples as leached oolites or chert- or anhydrite-cemented sands are clearly obvious in thin section. Whether or not a tight reservoir will respond to acidizing is another case where a brief examination of the thin sections will probably tell the story.

In the pages that follow, an attempt is made to provide some idea of the mineral composition and the grainsize where this is appropriate. Grainsize data from the thin section are measurements of the long axes of 100 randomly selected quartz grains. Although this may seem to be an unusual way of getting an estimate of grainsize, it has the advantage of being relatively quick to do, and it keeps the mineralogy constant for comparisons to other samples. In some cases, actual sieving data is used and this is noted accordingly.

The mineralogy was obtained by point-counting 100 randomly selected locations along random traverses. Greater numbers of points are obviously desirable, but the size and scope of this book does not allow this much detail. The author obtained both the point count and the grainsize data; therefore, these data should be relatively consistent, and whatever biases are introduced in mineral identification or length measurement are maintained throughout this work.

Clay mineral identifications are made on the basis of x-ray diffraction studies. In some cases, the coarseness of the kaolinite books allows the substantiation of the identification by optical techniques. Where possible, references have been made to Recent sedimentary features that resemble the setting of the reservoir. In a number of instances, there are no equivalent Recent sediments that can be referred to for comparison purposes. For instance, the shallow epicontinental seas that provided clean blanket sands are not readily found under today's conditions.

The attempt has been made to avoid jargon where possible and to communicate in such a way that the nonsedimentologist and interested student will be able to use most of the information. In this way, it is hoped that the situation is avoided where, in the words of P. D. Krynine, "a complete triumph of terminology over common sense" occurs.

After examining the reservoir rocks from all over the North American continent, it becomes clear that each major productive area has its own geologic style. In the case of the Appalachian basin, there is a geologic background that imprints itself on the reservoir rocks. In other words, a Bradford field is not likely to be reproduced in the Mid-Continent or west Texas. Similarly, the Mid-Continent differs dramatically from the Gulf Coast in terms of geologic style. Organization of this work on the basis of topics such as porosity and pore morphology could be done so that case histories fit into appropriate classes; however, this traditional approach tends to downgrade the importance of the other geologic considerations such as structure and stratigraphy, both of which have a great deal to do with the formation of oil and gas reservoirs.

This book begins near the birthplace of the petroleum industry, the Bradford field in Pennsylvania, and then moves to other reservoir rock types in the Appalachian basin. The logical extension westward covers production from the Ontario peninsula, the channel sands of the Illinois basin, and the carbonate mud mounds of the Michigan basin. The transition from these interior cratonic basins to the Nemaha ridge type of production in Kansas and Oklahoma is not profound. Nor is it particularly difficult to move from consideration of the basement-faulted Oklahoma City field to the numerous reservoirs that occur along the granite ridge that helps form the vast Texas Panhandle field. Such granite basement uplifts also form the original backbone for the Central basin platform of west Texas.

The Gulf Coast, on the other hand, provides a significant departure from the previously mentioned areas in terms of both sedimentary and structural styles. Again, it is the geologic background that brands the area and leaves a unique imprint on the reservoir rocks.

It is also clear that the Rocky Mountain province is geologically and structurally dissimilar to the other areas and that its character is spread from Arizona into Canada. Alaska and California also differ in the geologic aspect of their oil provinces; hence, the organization has become geographic rather than being based on microscopic rock properties or some other type of classification. This approach has the advantage of allowing the geologist or petroleum engineer who is new in an area to obtain a glimpse of the difficulties he or she is likely to face. As the emphasis of the future moves toward various methods of tertiary recovery, the microscopic petrology is going to become one of the keys to unlocking the remaining oil. It is hoped that the chapters that follow will contribute significantly to the understanding of how our oil and gas resources can be tapped both now and in the future.

Geographic Areas

This section deals with the broad general view of the geology of the seven geographic areas included in this volume. In general, the discussion provides from east to west with detours across the U.S.–Canadian border to oil fields located in or on geologic features that extend into both countries.

The overall distribution of oil and gas resources is best understood by examining the tectonic framework of the North American continent. Briefly stated, the continent is made up of a stable interior surrounded by zones that are more highly deformed. The tectonically stable rocks that form the continental core are those outcrops forming the Canadian Shield (see Figure 1.1). This core is composed of igneous and metamorphic rocks which are not believed to be prospective for hydrocarbons. In the Central Stable Interior south of the shield area, the igneous and metamorphic rocks slope down beneath a relatively thin cover of sedimentary rocks. These latter strata fill a number of relatively shallow basins that are separated from each other by broad arches. In areas like the Colorado Plateau and in portions of the Rocky Mountains, regional deformation has taken place. Those basins located at the edges of the Central Stable Interior appear to be structurally more complex and deeper.

Belts of rock that exhibit considerable folding and thrusting border much of the central platform, specifically, as shown in Figure 1.1, the Cordilleran thrust belt stretching from Alaska to Mexico, the Appalachian thrust belt, the Ouachita thrust belt, and the Marathon thrust belt.

The Appalachian thrust belt is bounded on the east by the crystalline rocks of the Piedmont, which are largely metamorphic. On top of these are deposited Mesozoic and Cenozoic sedimentary rocks that form the Atlantic plain. This wedge of sediments thickens to the east along a large part of the Atlantic coast reaching a maximum thickness of over 40,000 ft (12,192 m).

The Gulf Coastal Plain is formed by Mesozoic and Cenozoic sedimentary rocks that lap onto the Marathon and Ouachita thrust belts. This block of sediments also thickens toward the Gulf of Mexico.

Looking to the West Coast and west of the Cordilleran thrust belt, the main features are the Basin and Range province portions of the Rocky Mountains and the Pacific Margin System. This part of the continent appears to have been tectonically active through much of the geologic record.

Examples of oil- and gas-producing reservoir rocks discussed in this volume come from the major productive areas of the United States, as illustrated in Figure 1.2. With the exception of regions 6 and 2, which include most of the Gulf Coast other than Florida and California, production is obtained from rocks of Paleozoic age.

The thick sedimentary sequence of the Gulf Coast (region 6) has been highly productive and contains rocks of Mesozoic and Cenozoic age. Continuous

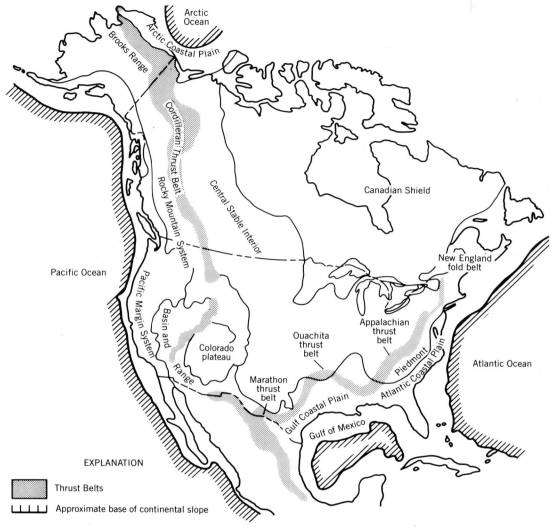

Figure 1.1. Tectonic map of North America showing major structural features. (After Dolton et al. 1981, p. 10.)

deposition in this region provides an excellent area for comparisons between the oil-productive sediments and those being deposited today.

In the regions extending from Montana to New Mexico, there are a number of oil- and gas-productive basins that contain sedimentary rocks ranging from the Paleozoic to the Cenozoic. These regions (3 and 4) tend to have fewer broad areas where production is widespread.

The far western United States, with the exception of the prolific Cenozoic sediments of California, contains little oil and gas production to date.

The time comes in the study of a petroleum reservoir rock when it is necessary to look at the actual grains, matrix, and cementing materials of which it is composed. It is not enough to measure the electric log response or to model the reservoir according to numbers developed by someone else. A close look at these interesting rocks must be made by study of the thin sections with the polarizing microscope. Such an effort is not done to name or classify the sediment, although this helps; it is rather to understand it. In

Reference

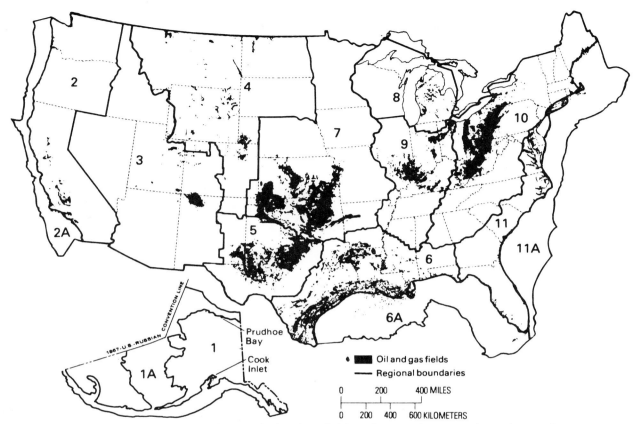

Figure 1.2. Map showing the areal distribution of oil and gas fields in the United States. (After Dolton et al. 1981, p. 11.)

order to do this, it is necessary to know the minerals it is composed of and how they are formed together.

Where possible, the following study will attempt to expand our understanding of the probable sources of the sediments, the environment in which it was deposited, the nature of the rock fabric and pore system, the cementation, and the alterations of the mineral grains that change the porosity and permeability of the rock.

Reference

Dolton, G. L., K. H. Carlson, R. R. Charpentier, A. B. Coury, R. A. Crovelli, S. E. Frezon, K. S. Khan, J. H. Lister, R. H. McMullin, R. S. Pike, R. B. Powers, E. W. Scott, and K. L. Varnes, 1981, *Estimates of Undiscovered Recoverable Conventional Resources of Oil and Gas in the United States,* Geological Survey Circular 860, 87 pp.

2

The Appalachian Basin

It is fitting that the first reservoirs to be discussed are from the Appalachian basin, birthplace of the petroleum industry. The Appalachian basin cuts across from New York State southwest to an area west of the Great Smoky Mountains in east Tennessee (Figure 2.1). On the western rim of the basin is the Cincinnati arch, which splits into two segments at the northern end providing structural highs on both the eastern and western sides of the Michigan basin. West of the Cincinnati arch is the Eastern Interior coal basin. East of the Appalachian basin lies the Appalachian Highlands and the onlapping Atlantic Coastal Plain plus the adjoining continental shelf. Unfortunately, no examples of reservoir rocks from the Atlantic offshore are available; however, this province provides a challenge for the future.

Commercial oil and gas production from the Appalachian basin has been obtained from Paleozoic rocks ranging in age from Cambrian to Pennsylvanian. The reservoir rocks to be discussed are from Pennsylvania and West Virginia and in general reflect the geologic style of the area.

This initial group of reservoirs also includes examples from the Michigan basin, the Illinois basin, and the extension of the Cincinnati arch into southwestern Ontario. The writer recognizes that these few reservoir examples from this vast area cannot be claimed to be truly representative; nevertheless, it is hoped that the reader will appreciate the overall geologic character and will be able to contrast and compare the reservoirs with those from the other provinces.

Mineralogy

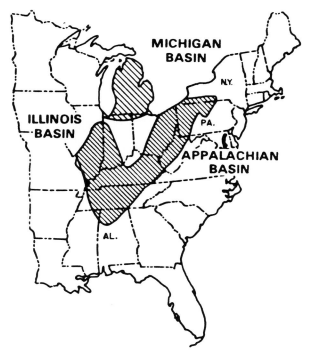

Figure 2.1. Map showing sedimentary basins presented by reservoir rocks covered in the first major division of this book. (After Tetra Tech 1981, p. 3.)

Bradford Oil Field, Bradford Third Sand, McKean County, Pennsylvania

Geologic Background

The first "giant" oil field in North America was discovered near Bradford, Pennsylvania, in 1871. By 1881, it had expanded to contain more than 11,000 wells covering an area of 84,000 acres (Figure 2.2) and reached a peak annual primary production of almost 23 million barrels. In 1907, water flooding began contributing to the overall production, which reached its peak in 1937. More recent efforts aimed at recovering additional oil by tertiary methods have focused on oil displacement with micellar solutions. This latter phase will be discussed after the section dealing with petrology.

Probably the most exhaustive study of the whole Bradford field was that carried out by Fettke (1938). The book by Fettke contains some petrologic information; however, the most detailed report concerning all aspects of the mineralogy and its interpretation was authored by Krynine (1940), and this latter report supplies some of the material used in this discussion.

Halbouty et al. (1970), in their paper "Factors Affecting Formation of Giant Oil and Gas Fields," point out that the Bradford Sandstone is a relatively poor quality reservoir rock. In this instance, however, the Bradford oil field still rates among the giants because of the great areal extent of the productive area.

The Bradford field is located at the convergence of two anticlines that trend northeastward and merge to form a wide dome. This structure is part of a system of gentle foreland folds occurring within the basinal structure of the Appalachian plateau. Petroleum production in the Bradford field as a whole is associated with the Bradford dome; however, the commercial success of any specific well is controlled by local conditions such as lensing and sand thickness (Krynine 1940).

Practically all of the production comes from the Bradford Third Sand; therefore, the detailed petrology of this sand is particularly germane. As illustrated in Figure 2.2, it covers a large area and varies from 25 to 130 (7.6 to 39.6 m) ft in thickness with 30–45 ft (9.1–13.7 m) being the most common range. Its position in the stratigraphic column in McKean County, Pennsylvania, is shown in Figure 2.3.

Mineralogy

The major minerals that compose the Bradford Third Sand are quartz, rock fragments, micas, clay minerals, feldspar, and carbonates. Heavy minerals also occur in considerable variety.

A large part of the sediment can be recognized as having been derived from older Paleozoic sediments. The rock fragments represented are from a wide variety of geologic terrains and include low-rank metamorphic rocks (major), phyllites, slates, schists, and metamorphic quartzites. Unfortunately, the study of these rock fragment types is rather difficult because of the fine grainsize and the degree of weathering.

A brief glance at the photomicrographs reveals that the quartz grains are very angular (Figures 2.4–2.9). Krynine (1940) suggests that this may represent the original shape of the grains, in which case it is metamorphic quartz; or it may have been produced by violent abrasion and the breaking up of rounded grains. According to Krynine (1940, p. 55), the Brad-

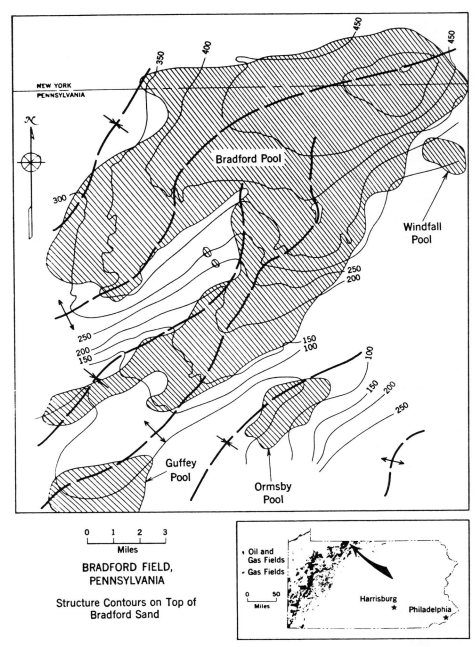

Figure 2.2. Productive area of the Bradford field (diagonal lines) superimposed on the 50-ft structure contours. (After Landes 1970, p. 17, and Fettke 1934, p. 193.)

Mineralogy

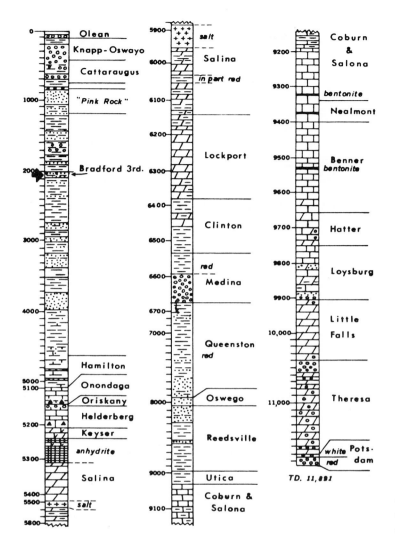

Figure 2.3. Stratigraphy of the Minard Run Oil Co. no. 1 well, McKean Co., Pa. (From Lytle and others (1966), courtesy of Pennsylvania Geological Survey.)

ford Third Sand has the following average mineral composition:

Rock fragments	30%
Large biotite and muscorite flakes	2–3%
Clayey matter	5–12%
Feldspar	1½%
Quartz	60–65%

Later work by Smith (1964) related mineral composition and texture to bulk density for the Bradford Third Sand in the Bradford field where bulk density was used as a reflector of the oil-bearing sands. The objective of this study was to make a statistical comparison of the barren Lewis Run Sand and the oil-bearing Bradford Third Sand; both sands being in the same stratigraphic setting and close in distance. Smith found that there were significant correlations between bulk density of the oil-bearing sand and carbonate cement, silica cement, and feldspar. Quartz grainsize was also correlated with bulk density.

The multiple regression analysis of the compositional data showed that bulk density is not independent of the textural and compositional variables. Furthermore, 72% of the variation in bulk density is explained by an "economical set" composed of quartz, micaceaous material, quartzose rock fragments, feldspar, and the sorting of quartz in a-phi units. These relationships must be kept in mind when attempting to design innovative tertiary recovery projects.

Figure 2.4. Bradford Third Sand, 1560 ft (475.6 m), Well No. F-10, Bradford oil field, McKean County, Pennsylvania (0.5 in. in the photo equals 0.08 mm). General view of the typical sand with a large muscovite mica flake slightly bent at the center of the photomicrograph. Clay minerals and dark organic matter fill most of the spaces between grains. Some quartz cementation shows in the cluster of grains on the right side of the photomicrograph. Most of the quartz grains are angular and give the impression of being dumped into place. Permeability, 1.5 millidarcies.

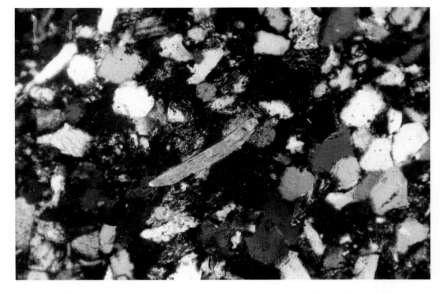

Figure 2.5. Bradford Third Sand, 1539 ft (469.2 m), Well No. Exp. 4, Bradford oil field, McKean County, Pennsylvania (0.44 in. in the photo equals 0.08). The overall grain-size is coarser than in Figure 2.4. Again organic matter and clay minerals clog the spaces between the quartz grains. Some of the quartz grains are cemented with quartz cement (upper right corner). The dust particle outlines, which reveal the quartz overgrowths, suggest that some grains were rounded prior to cementation. Permeability, 16.6 millidarcies.

Figure 2.6. Bradford Third Sand, 1835.5 ft (559.6 m), Well No. U11L, Bradford oil field, McKean County, Pennsylvania (0.44 in. in the photo equals 0.08 mm). This photomicrograph shows a diagonal orientation of the quartz-cemented grains. Most of the porosity has been filled with quartz cement. Some weathered feldspar grains are included (grayish brown grain on the right). Again the overall impression is one of a highly angular fine-grained sand. Permeability, 0.84 millidarcies.

11

Figure 2.7. Bradford Third Sand, 1560 ft (475.6 m), Well No. F-10, Bradford oil field, McKean County, Pennsylvania (0.44 in. in the photo equals 0.08 mm). Fine-grained quartz sand is cemented with both quartz and some chert cement. Both clay minerals and biotite mica are involved in the pore filling. Permeability, 1.5 millidarcies.

Figure 2.8. Bradford Third Sand, 1539 ft (469.2 m), Well No. Exp. 4, Bradford oil field, McKean County, Pennsylvania (0.44 in. in the photo equals 0.004 mm). Higher magnification reveals how the feldspar grain in the middle (dark gray with straight cleavages) has been surrounded and cemented into place by quartz overgrowths. Phyllite grains (straw colored) are in various stages of being broken up. Permeability, 16.6 millidarcies.

Figure 2.9. Bradford Third Sand, 1539 ft (469.2 m), Well No. Exp. 4, Bradford oil field, McKean County, Pennsylvania (0.44 in. in the photo equals 0.04 mm). Mica flake (right center) has a splayed out end indicating the breakup that was in process at the time of deposition. Quartz grains (near the bottom of the photomicrograph) are quartz cemented and surround feldspar. Plagioclase feldspar (striped), at middle to lower left, appears to be angular and reasonably fresh. Permeability, 1.16 millidarcies.

12

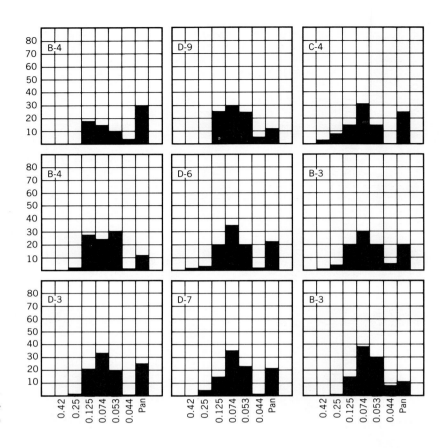

Figure 2.10. Bradford field, Pennsylvania. Histograms of fine and very fine sandstones. (After Krynine 1940, p. 42.)

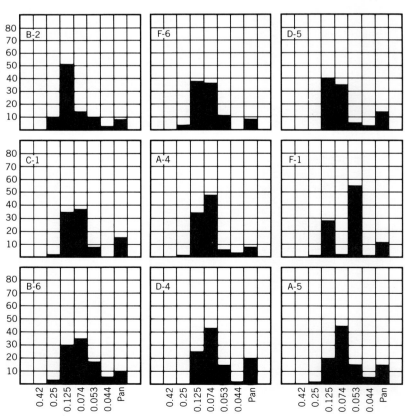

Figure 2.11. Bradford field, Pennsylvania. Histograms of coarse and medium-grained sandstones of the Third Bradford Sand. (After Krynine 1940, p. 41.)

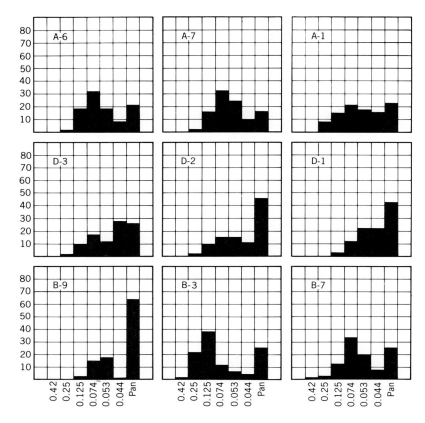

Figure 2.12. Bradford Field, Pennsylvania. Histograms of very fine grained sandstones, siltstones, shales, and calcareous sandstones. (After Krynine 1940, p. 43.)

Grainsize

Defining and illustrating a "typical" grainsize distribution is a problem. Krynine (1940, p. 63) discussed the distribution of rock types exhibited by the Bradford reservoir and indicated that

graywackes compose 70 percent, siltstones and microconglomerates for 5 to 10 percent, and shales 25 to 50 percent. Single specimens contain a complex mixture of grainsizes, compositions, angularities and cementation and matrixes.

The grainsize data provided in Krynine's report (1940, pp. 41–43) are the results of mechanical sieving and therefore differ significantly from the thin-section measurements of the long axes of quartz grains presented in most of the remainder of this work. They are reproduced in Figures 2.10, 2.11, and 2.12, because they provide an excellent overview of the variability and range that occurs in the Bradford Third Sand.

It is this high degree of variability, whether it is on the scale of the individual sand lens and shale streaks or that of the thin section, that makes tertiary oil recovery a frustrating task.

Oil Recovery Projects, Polymer Flooding

The Bradford Third Sand in the Bradford field has probably received more attention in terms of secondary and tertiary oil recovery projects than any other U.S. oil field. Some idea of how important artificial water flooding has been to the total amount of oil recovered from the Bradford field can be obtained from Figure 2.13, which shows the crude oil production history through 1965. All sorts of tertiary recovery concepts have been attempted in the Bradford field with limited success. The more recent work has involved the use of micellar solutions to dislodge the oil followed by a mobility buffer to increase sweep

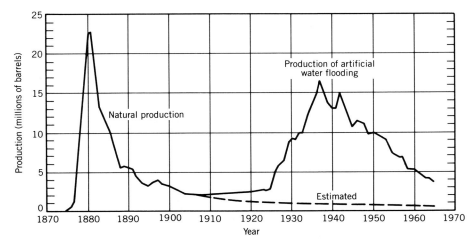

Figure 2.13. Crude oil production plot for the Bradford District, Pennsylvania and New York, but excluding the Music Mountain field. (After Lytle et al. 1966, p. 12.)

efficiency and water to drive both slugs through the sand to the producing wells. The fact that this method of tertiary recovery has been more successful than the approaches used previously in the Bradford field suggests that a brief summary of the technology is in order.

Danielson et al. (1976) describe the micellar solution as containing primarily hydrocarbon, surfactant (sulfonates), and water. The solutions are designed to displace a high percentage of the residual oil in the reservoir. To prevent early breakthrough of the oil to the producing well and the resultant loss of sweep

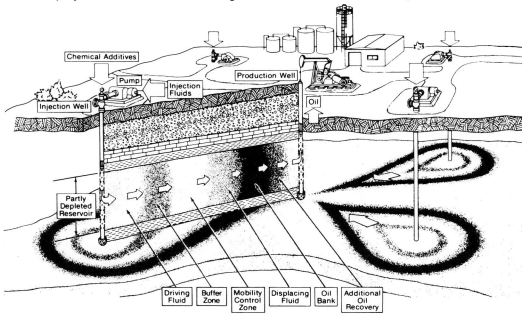

Figure 2.14. Generalized diagram of a micellar–polymer flood system such as that used in the Bradford field, Pennsylvania. (After Linville et al. 1979b, p. 21.)

efficiency, a mobility buffer is injected. The mobility buffer is a bank of a specially designed polymer (polyacrylamides) solution that increases the viscosity and lowers the mobility of the micellar slug. The extra viscosity of the buffer allows time for more of the reservoir sand to be contacted by the displacing fluid. The last part of the technique involves the injection of the water bank that drives the two slugs ahead of it to the producing well (see Fig. 2.14).

Oil Saturation Study

Danielson et al. (1976) initially sought to establish a firm value for the oil saturation remaining in the Bradford Third Sand after water flooding (see Tables 2.1 and 2.2). What becomes clear when comparing the oil saturations from the pilot test with those of the expanded test is that there are significant areal differences across the flooded area. Danielson et al. (1976, p. 130) concluded that the data obtained from the pressure transient tests was closest to the true reservoir conditions and therefore used a post-water-flood oil saturation in the range 36–43%.

Table 2.3 provides information concerning the overall size and scope of the expanded Bingham test, and Figure 2.15 provides a production forecast for the Farm A part of the test. Some of the important conclusions from these tests were as follows:

1. Additional oil can be successfully recovered from previously water flooded Bradford Third Sand using a micellar–polymer flood.
2. The ultimate oil recovery could be approximately 50% of the oil in place at the time of the initiation of the project.
3. Injectivity is the primary limiting factor in the application of the mobility buffer used in the Bradford reservoir due to polymer retention in the lower permeability zones.
4. The main zone effectively flooded by the micellar–polymer technique is a high-permeability section in the middle of the reservoir.

In 1977, another micellar–polymer flood test was initiated in the Bradford field. According to Linville et al. (1979a,b), the objective of this test was "to demonstrate the efficiency and economics of recovering tertiary oil from a low-permeability depleted water flood using a micellar fluid followed by a mobility-controlled buffer zone (polymer solution) and dis-

TABLE 2.1 Oil Saturation Data Obtained for the Bingham Pilot Test of the Micellar Flooding Method (Danielson et al. 1976, p. 130)

	Range of Oil Saturations	Midpoint of Range
Routine commercial core analysis	0.27–0.36	0.32
Water–oil ratio and laboratory relative permeabilities	0.28–0.43	0.35
Transient test and laboratory relative permeabilities	0.28–0.35	0.31
Well log calculations	0.22–0.50	0.36

placed through the reservoir by a driving fluid (water)." Figures 2.16 and 2.17 show the history for the injection fluids and the history of the oil produced from the test area.

The Penn Grade Micellar Displacement Project, as this later operation was called, was initiated in May of 1977 and contained 25 production wells plus 16 injection wells. The whole design consisted of 16 inverted five-spots. All wells were cored, completed open hole, and shot with nitroglycerin (Ondrusek and Paynter 1982).

The initial injection of 10% of the pore volume involved a "reservoir conditioning" brine preflush containing 50,000 ppm NaCl. All the individual five-spot patterns had received at least 10% of a pore volume brine injection by the time the switch was made to the micellar slug injection. Plans called for

TABLE 2.2 Oil Saturation Data Obtained for the Expanded Bingham Test of the Micellar Flooding Method (Danielson et al. 1976, p. 131)

	Range of Oil Saturations	Midpoint of Range
Routine commercial core analyses	0.24–0.53	0.39
Water–oil ratio and laboratory relative permeabilities	0.38–0.41	0.40
Transient test and laboratory relative permeabilities	0.36–0.43	0.40

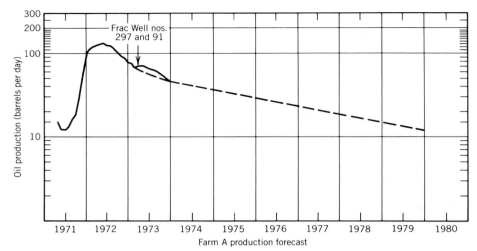

Figure 2.15. From micellar–polymer flooding of the expanded Bingham test in the Bradford field, Pennsylvania (After Danielson et al. 1974, p. 223.)

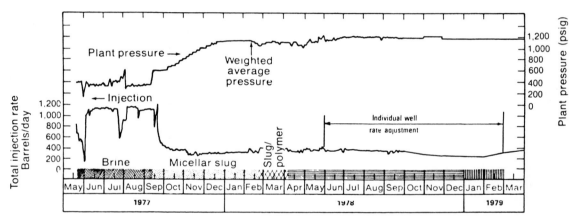

Figure 2.16. Bradford field, Pennsylvania micellar–polymer flood, Lawry project total fluid injected from May 1977 to February 1979. (After Linville et al. 1979b, p. 43.)

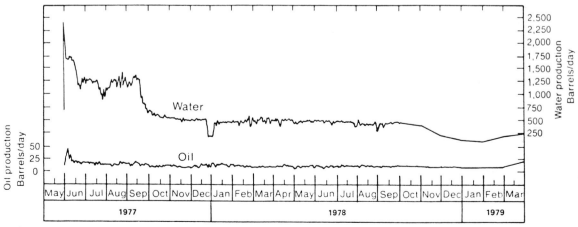

Figure 2.17. Bradford field, Pennsylvania micellar–polymer flood, Lawry project total fluids produced from June 1977 to March 1979. (After Linville et al. 1979b, p. 44.)

Explanation of the Correlation Index

TABLE 2.3 Bingham Expansion Test (Danielson et al. 1976, p. 134)

General information	
Reservoir	Bradford Third Sand
Depth	1866 ft
Pore volume	1.5×10 bbl
Pattern	2.9 acre five-spot
Wells	16 injectors and 25 producers
Total size	46.5 acres
Injection	
Slug injection	
Started	March 23, 1971
Injected	79,990 bbl
Size	5% PV
Mobility buffer injection	
Started	June 3, 1971

Polymer Schedule

Injection (% PV)	Concentration (ppm)	Average Viscosity (cP at 6 rpm and 70 °F)
0–2	1500	50.0
2–20	1150	35.0
20–26	980	25.0
26–31	900	21.5
31–37	840	19.0
37–43	760	16.5
43–48	700	14.2
48–65	650	12.5
65–71	590	10.7
71–77	540	9.4
77–82	480	8.2

The total thickened water injection was 1,257,000 barrels or 82% PV for the week ending January 1, 1974.

Planned Schedule

82–85	480	8.2
85–90	440	7.2
90–95	380	6.2
95–100	350	5.4
100–105	320	4.7
105–110	280	4.1
110–115	255	3.6
115–120	225	3.2
120–125	195	2.8
125	150	2.2

the injection of a 9% pore volume slug. The period for micellar slug injection lasted until fill-up, which was essentially completed in April 1978.

The next phase involved the injection of a biopolymer for mobility control, and plans called for the injection of a three-step tapered viscosity sequence consisting of a 20% volume "spike," a 52% pore volume "body," and a 30% pore volume "tail." The cumulative spike injection for the project actually amounted to 15.3% of the pore volume when the change over to the body phase took place in February 1979. The body injection phase continued through the end of the project operations in November 1980.

According to Ondrusek and Paynter (1982, p. 3), a number of wells showed diminishing injection rates that, although treated with hydrochloric acid and sodium hypochlorite, showed little or no improvement. The inability to maintain injection rates over the life of the project without pressure parting limited the economic potential of the project.

In addition, an evaluation core taken in July 1980 showed that the viscous micellar slug and polymer had displaced oil from only about 5 ft of the highest permeability sand, leaving the remaining tighter sand untouched.

Since mobility control was obviously not maintained and could not be achieved by increasing the viscosity because of decreasing injection rates, the decision was made to terminate the project in November 1980. The cumulative oil recovery was 5.29% of the estimated pre-project oil in place; therefore, the project must be considered a failure.

It is clear that the more permeable zones, which are not highly cemented with carbonate or silica cements, are the best targets for micellar–polymer recovery. Unfortunately, these are also the zones that respond best to secondary recovery methods; hence, much of the oil remains in the finer grained, tighter sands.

Smaller-scale projects where many different sand lenses are not involved appear to have a better chance of succeeding.

This veteran oil field still has considerable oil that is worth recovering, and it is likely that in the future many new recovery ideas will continue to be tested in the Bradford Third Sand.

Explanation of the Correlation Index

Where possible, throughout this book an attempt is made to provide some idea of the chemical character of the produced crude oil. In the case of the Bradford crude, the type of crude oil is particularly important because it has been used as the classic example of a highly paraffinic oil against which other oils are compared. Some years ago the Bureau of Mines developed a technique that provided an easily understandable view of the chemical composition of crude oils. What perhaps is more important, they carried out this analysis for many of the important crude oils from all over the United States. This work allows the comparison of crude oils on a statistical basis.

The correlation index, as the analysis is called, provides a uniform measurement that is available for a number of the crude oils from fields and reservoirs discussed in this volume, and it is appropriate at this point to discuss what the correlation index is and how it is derived.

Studies that focus on problems associated with the sources for various crude oils invariably run up against the question concerning what measurement should be used to compare crude oils. This problem was recognized at an early date by personnel at the U.S. Bureau of Mines and the correlation index approach was developed whereby the analyses of crude oils could be visually compared with relative ease. The correlation index is a measurement that has been recorded for most of the crude oils analyzed by the Bureau of Mines since 1940. Smith (1940) describes the significance of the correlation index as follows:

The correlation index is a number whose magnitude indicates certain characteristics of a crude-oil distillation fraction. If a fraction were composed exclusively of normal paraffin hydrocarbons (n-alkanes), the value of the index number would be zero. If the fraction be from a paraffin-base crude oil of the usual type, its index will not be zero but will be small, while fractions from intermediate- and naphthene-based crude oils will have increasingly greater values for the indexes.

The range of correlation index numbers normally obtained is from close to zero up to 75 for very naphthenic distillates. The index system is based on average boiling points and the specific gravities of the Hampel fractions. It has been so arranged that pure benzene has an index value of 100.

For comparisons of crude oils, it is relatively easy to plot the correlation index numbers against the fraction numbers or the actual temperature of the distillation cut. Throughout this book the correlation index number for each distillation cut will be plotted against temperature, and the volume percentage of each distillation fraction will be plotted on the same illustration as shown in Figure 2.18 for the Bradford crude.

It might well be asked why this rather simple approach is used when modern gas chromatograph mass spectrometer data are so much more detailed. The answer is that the Bureau of Mines correlation index data are available for a reasonable number of the fields studied in this book, and comparisons of these data are useful. The second reason for the use of the correlation index is that the meaning is relatively clear and the plots of the data are simple enough to be readily understood by those who are unfamiliar with the organic geochemistry of crude oils.

Figure 2.18 contains two sets of measurements that represent two independent samples of the same Bradford Sand crude oil taken and analyzed roughly 9 years apart. The two sets of curves indicate that the

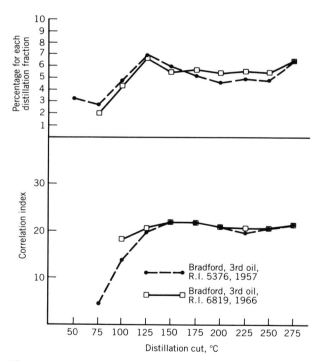

Figure 2.18. Correlation index curves and distillation fractions for crude oil from the Bradford Third Sand, Bradford oil field, McKean County, Pennsylvania. (Data from McKinney and Garton 1957, p. 170, R. I. 5376, and McKinney, Ferrero, and Wenger 1966, p. 204, R. I. 6819.)

analytical procedure remained relatively uniform over this period of time. Thus, the comparisons between the crude oils from various fields on the basis of correlation indexes appear to be valid over time.

Summary of Important Features

The significant characteristics of the Bradford Third Sand as outlined by Krynine (1940, p. 81) and others are

1. poor mineralogic sorting,
2. exceptional angularity,
3. lack of mineral alteration,
4. cross-bedding, and
5. lensing and channeling on both a microscopic and a large scale.

Other important features include the fairly uniform gray color and the presence of marine fossils and carbonized plants.

In Krynine's view, the sediment experienced a "tumultuous but fairly violent erosion." The source area involved steep, but not necessarily high, relief. Numerous streams contributed to broad rivers that carried the fine-grained sediment into a shallow marine sea. The resultant deltas or series of deltas are preserved as the main sand bodies in the Bradford oil field.

Figure 2.19 shows the cross section of the Catskill delta from the Hudson Valley, New York, to Erie, Pennsylvania. Having walked over and pondered the meaning of the outcrops of the Upper Devonian from the red beds in subaerially deposited parts of the delta through the Genesee Formation and the Rhinestreet and Chemung facies (Figure 2.20), the writer is impressed with the horizontal extent and relatively flat lying nature of the whole section. The recent discussion of the sedimentary processes believed to have been at work in the Catskill Sea by Woodrow and Isley (1983) is particularly interesting in view of what is observed in the field. These authors point out that finding a present-day equivalent to the Catskill delta and the associated sediments may not be realistic. The slopes normally associated with delta fronts, distributaries, and so on, are not clearly defined. In the words of Woodrow and Isley (1983, p. 464), "Regional patterns are clear, however, and they seem to require a seafloor surface of low relief and minimal gradient." The models proposed by these authors are illustrated in Figure 2.21. If the Bradford Third Sand is to be found in this model, it would most likely occur near the "distal shelf sand accumulations" of the progradational model (A). Krynine (1940, p. 83) thought of the whole complex as a "series of coalescing deltas formed by a series of fairly small or at best medium-sized streams. Such deltas, individually small, but very extensive when added together, are a feature of subsiding coastal plains in regions of heavy rainfall such as found at the present time in parts of Central America."

The cross section of the Bradford field drawn by Wilson (1950) (Figure 2.22) suggests that the multiple-source hypothesis may be right. It is likewise conceivable that the shifting lobes of a major delta might also produce the observed distribution. Additional work is desirable to clear up this point.

Exploration for shallow Devonian sands in the Appalachian basin will probably continue for many years; however, the discovery of another field in this area with the dimensions of the Bradford field seems unlikely.

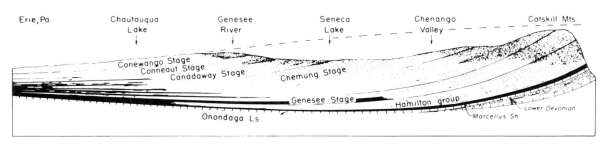

Figure 2.19. Catskill delta cross section from the Catskill Mountains to Erie, Pennsylvania. (After Dunbar and Waage, 1969, p. 234.)

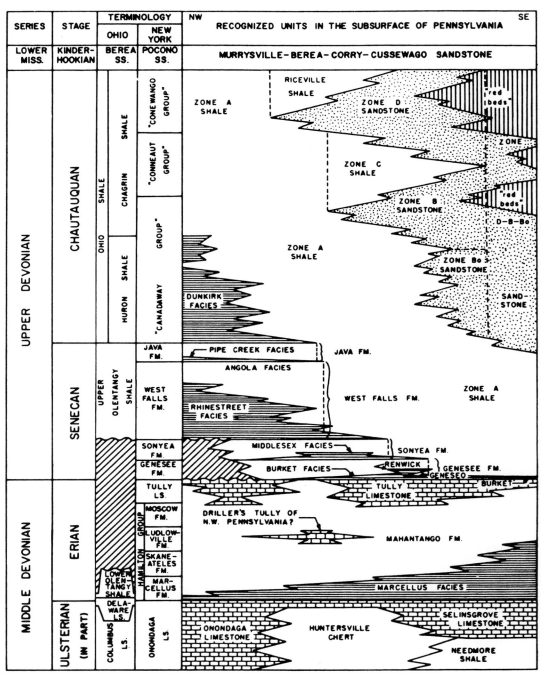

Figure 2.20. Schematic diagram showing Middle and Upper Devonian stratigraphic units as correlated in the subsurface through northwestern Pennsylvania with the New York and Ohio sections. (After Tetra Tech 1981, p. 8.)

Summary of Important Features

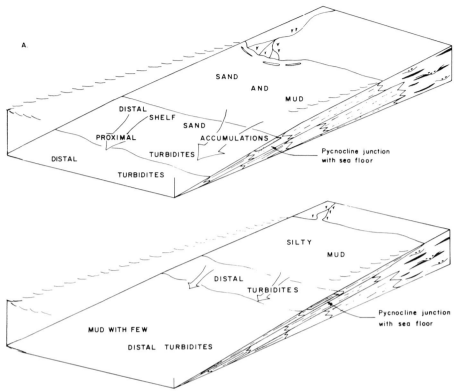

Figure 2.21. Suggested models of deposition for sediments of the Upper Devonian in New York and Pennsylvania. (a) Progradation is well advanced. (b) Transgression where silt and sand deposition is greatly reduced. (After GSA Bull., D. L. Woodrow and Ann E. Isley, 1983.

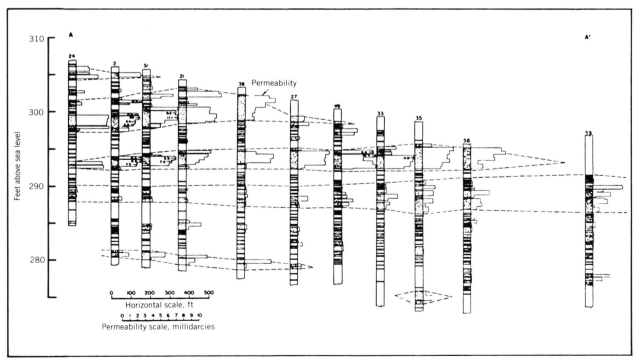

Figure 2.22. Cross section showing the complexity and discreteness of individual sand reservoirs in Bradford field, Pennsylvania. (After Wilson 1250, Courtesy API)

References

Danielson, H. H., W. T. Paynter, and H. W. Milton, Jr., 1974, *Tertiary Recovery by the Maraflood Process,* preprint, Society of Petroleum Engineers of AIME, SPE 4753, pp. 209–223.

Danielson, H. H., W. T. Paynter, and H. W. Milton, Jr., 1976, Tertiary Recovery by the Maraflood Process in the Bradford Field, *Journal of Petroleum Technology,* February, pp. 129–138.

Dunbar, C. O. and K. M. Waage, 1969, *Historical Geology,* 3rd ed., John Wiley & Sons, New York, p. 234.

Fettke, C. R., 1934, Physical Characteristics of Bradford Sand, Bradford Field, Pennsylvania and Relation to the Production of Oil, *American Association of Petroleum Geologists Bulletin,* Vol. 18, No. 2, p. 193.

Fettke, C. R., 1938, *Bradford Oil Field Pennsylvania and New York,* Bulletin M21, Pennsylvania Geological Survey Fourth Series, 453 pp.

Halbouty, M. T., R. E. King, H. D. Klemme, R. H. Dott, Sr., and H. A. Meyerhoff, 1970, Factors Affecting Formation of Giant Oil and Gas Fields and Basin Classification, in *Geology of Giant Petroleum Fields,* M. T. Halbouty, Ed., American Association of Petroleum Geologists, Memoir 14, pp. 505–555.

Krynine, P. D., 1940, *Petrology and Genesis of the Third Bradford Sand,* Mineral Industries Experiment Station Bulletin 29, The Pennsylvania State College, State College, PA, 134 pp.

Landes, K. K., 1970, *Petroleum Geology of the United States,* Wiley-Interscience, New York, 571 pp.

Linville, B., J. Lindley, and J. Whaling, 1979a, *Enhanced Oil and Gas Recovery and Improved Drilling Technology,* Vol. 18, Progress Review BETC-79/2, U.S. Department of Energy, pp. 43–44.

Linville, B., J. Lindley, and J. Whaling, 1979b, *Enhanced Oil Recovery and Improved Drilling Technology,* Vol. 19, Progress Review BETC-79/3, U.S. Department of Energy, p. 21.

Lytle, W. S., J. H. Goth, Jr., D. R. Kelley, W. G. McGlade, and W. R. Wagner, 1966, *Oil and Gas Developments in Pennsylvania in 1965,* Progress Report 172, Bureau of Topographic and Geologic Survey, Commonwealth of Pennsylvania Department of Internal Affairs, 66 pp.

McKinney, C. M. and E. L. Garton, 1957, *Analyses of Crude Oils from 470 Important Oilfields in the United States,* U.S. Bureau of Mines Report of Investigations 5376, 276 pp.

McKinney, C. M., E. P. Ferrero, and W. J. Wenger, 1966, *Analyses of Crude Oils From 546 Important Oilfields in the United States,* U.S. Bureau of Mines Report of Investigations 6819, 345 pp.

Ondrusek, P. S. and W. T. Paynter, 1982, *Penn Grade Micellar Displacement Project, Final Report,* U.S. Department of Energy Report DOE/ET/08002-26, 246 pp.

Smith, C. M., Jr., 1964, Quantitative Petrographic Comparison of the Bradford Third and Lewis Run Sands, unpublished Ph.D. thesis, The Pennsylvania State University, 111 pp.

Smith, H. M., 1940, Correlation Index to Aid in Interpreting Crude-Oil Analyses, Tech. Paper 610, U.S. Bur. Mines, 34 pp.

Tetra Tech, Inc., 1981, *Evaluation of Devonian Shale Potential in Pennsylvania,* U.S. Department of Energy Report DOE/METC-119, 78 pp.

Wilson, W. W., 1950, Supplement to Gettke, C. R., 1950, Influence of Geological Factors on Secondary Recovery of Oil, in *Secondary Recovery of Oil in the U.S.,* New York American Petroleum Institute, New York, p. 211.

Woodrow, D. L. and A. M. Isley, 1983, Facies, Topography and Sedimentary Processes in the Catskill Sea (Devonian), New York and Pennsylvania, *Geological Society of America Bulletin,* Vol. 94, pp. 459–470.

3

Kanawha Forest Field, Newburg Gas Sand, Kanawha County, West Virginia

According to Fenstermaker (1966, p. 158), "the first commercial Newburg sand production in West Virginia was in October 1964, when Cities Service Oil Company completed the 14 W. C. Hardy for a flow gauge of 7,838 Mcfd from perforations [at] 5,260–68 ft in the Newburg sand section. Subsequently, 17 wells have been drilled and completed, ranging in potentials from 1.2 MM cfd to over 52 MM cfd."

The core materials from the W. C. Hardy well are the focus of this discussion.

Geologic Background

The geographic location of the Kanawha Forest field is shown in Figure 3.1. The generalized geologic column for northern West Virginia showing the relative position of the Newburg Sandstone is shown in Figure 3.2. The Newburg Sandstone occurs in the middle of the Silurian section below the Wills Creek Formation and the Tonoloway Limestone. Below the Newburg is the McKenzie Limestone.

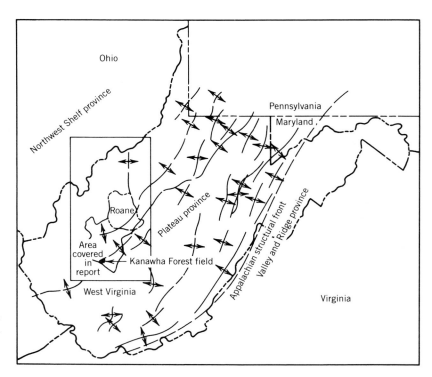

Figure 3.1. Location of the Kanawha Forest field, Kanawha County, West Virginia. (From Fenstermaker 1966, p. 158.)

The gas accumulation in the Kanawha Forest field is a stratigraphic trap where the Newburg Sand pinches out updip over the plunging nose of the Warfield anticline. Figure 3.3 provides a general view of the structure for the Kanawha Forest field contoured on top of the Newburg Sandstone.

In the Kanawha field, it is usual for only a show of gas to be recorded prior to acidizing and fracturing; therefore, it is instructive to examine the petrology of this sand and to see the appropriate characteristics so that they can be recognized in other exploratory wells.

Mineral Composition

The mineralogic point count data for thin sections from six different footages are recorded in Table 3.1. The samples were chosen to show the rather striking changes in mineral composition and the resultant effects on the reservoir properties. The two key measures of the latter are permeability and porosity, which are provided for each footage along with the mineralogic point count data. Figures 3.4 through 3.9 provide a visual summary of the mineralogical differences that occur in the Newburg along with the changes in fluid flow properties.

Recorded in Figure 3.10 are the means and standard deviations for the long axes of 100 randomly selected quartz grains. Examination of these data in comparison with the porosity and permeability data (see Table 3.1 bottom) suggests that sorting, as represented by the standard deviation, can be quite important. In the present cases, the samples with the highest values for permeability and porosity are those with the lowest standard deviation for quartz grainsize.

The other factors that appear to be critical to the permeability and porosity are the amount of clay minerals and the amount of dolomite. It is not surprising that these variables are highly significant; however, the degree to which they appear to control these important properties is noteworthy.

Geologic Interpretation

That the Newburg Sand is a thin but extensive sandstone that is present in the subsurface throughout much of the state of West Virginia places it in the blanket-sand category of Krynine (1948). In the Kanawha Forest field, it has the fine-grained, well-sorted character typical of the blanket sands (Figure 3.11). Numerous well-rounded grains are mixed in

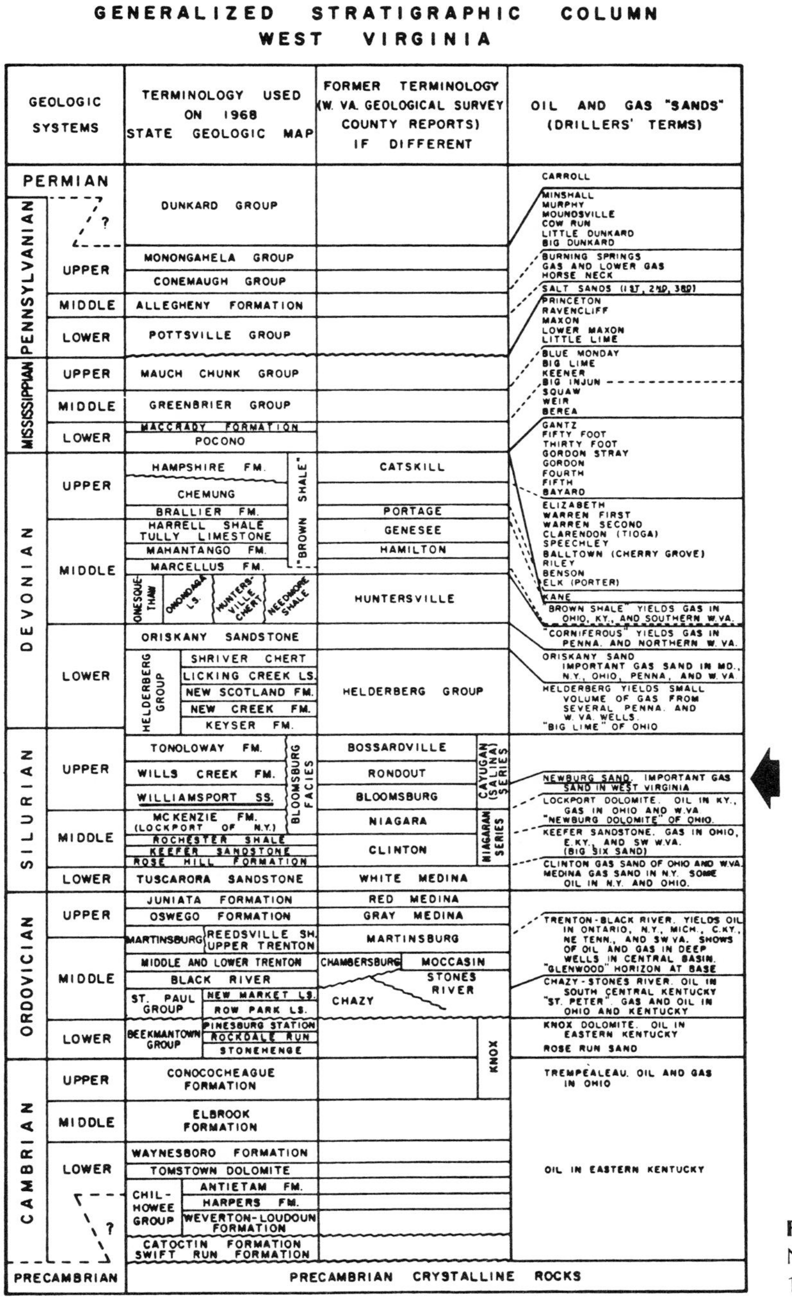

Geologic Interpretation

Figure 3.2. Stratigraphic positions of the Newburg Sandstone. (From Patchen et al. 1977, p. 5.)

with quartz grains that vary from subangular to subrounded (Figure 3.12).

The lithofacies distribution of the Newburg shows that shale extends from central Pennsylvania to northwest Virginia and grades into sandstone. Ultimately, the Newburg Sand changes into carbonate in the western counties (Figure 3.11). The shale facies is composed mostly of red beds and, according to Chen (1977, p. 94), was deposited rapidly in brackish water as part of a coalescing delta system. The sandstone facies is believed to have been deposited in a "shallow sea on broad wave-cut benches" (Chen 1977, p.

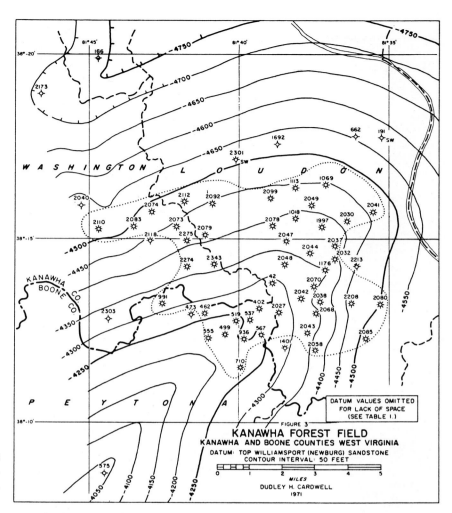

Figure 3.3. Structure contour map of the top of the Newburg Sandstone, Kanawha Forest field, West Virginia. (From Cardwell 1971, p. 20.)

TABLE 3.1 Petrologic Point Count of the Mineral Composition, Kanawha Forest Gas Field, Newburg Sand, C.S.O. No. 14 W. C. Hardy, Jr., Kanawa County, West Virginia

Minerals	5272 ft (1607 m)	5274 ft (1608 m)	5278 ft (1609 m)	5282 ft (1610 m)	5286 ft (1611 m)	5288 ft (1612 m)
Quartz grains	80	62	66	62	80	74
Quartz cement	20	6	21	21	19	10
Clay minerals	—	—	10	—	—	16
Chert cement	—	—	1	—	—	—
Pyrite	—	—	1	—	1	—
Tourmaline	—	—	1	—	—	—
Dolomite	—	32	—	17	—	—
Total	100	100	100	100	100	100
Permeability	183.00	0.28	0.48	0.00	25.30	0.00
Porosity (%)	15.1	6.4	7.9	3.9	12.2	1.5
Long-axis quartz, Mean in mm	0.141 mm	0.137 mm	0.182 mm	0.191	0.173	0.183
Standard deviation of long-axis measurements	0.039	0.058	0.059	0.057	0.049	0.070

Figure 3.4. Newburg Sandstone, 5272 ft (1607 m), Cities Service No. 14, W. C. Hardy, Jr., Kanawha Forest field, Kanawha County, West Virginia (0.44 in. in the photo equals 0.30 mm) The Newburg Sand in this photomicrograph is composed almost totally of rounded, well-sorted, quartz grains and quartz cement. The cleanness of the sands helps to provide the best type of reservoir (permeability, 183 millidarcies; porosity, 15.2%) that is illustrated in this well. In Krynine's (1948) sandstone classification, this would be called an orthoquartzite.

Figure 3.5. Newburg Sandstone, 5274 ft (1608 m) Cities Service No. 14, W. C. Hardy, Jr., Kanawha Forest field, Kanawha County, West Virginia (0.44 in. in the photo equals 0.14 mm). The major difference between this sand and the one occuring at 5272 ft is the dolomite content. The dolomite rhombs have filled in much of the porosity and permeability (porosity, 6.4%; and permeability, 0.28 millidarcies). The grainsize is almost the same as in the sample from 5272 ft (1607 m), although it is skewed toward the coarser grains.

The light gray quartz grain near the left center of the photomicrograph shows the high degree of rounding and a clear overgrowth.

Figure 3.6. Newburg Sandstone, 5278 ft (1609 m), Cities Service No. 14, W. C. Hardy, Jr., Kanawha Forest field, Kanawha County, West Virginia (0.44 in. in the photo equals 0.30 mm). As in the previous samples, the sand is highly cemented with quartz; however, clay minerals (dark brown) are beginning to fill in some of the pore space. (Porosity, 7.9%; permeability, 0.48 millidarcies) Although not clear in the photomicrograph, there are also patches of pyrite that grew into the original pore space.

Figure 3.7. Newburg Sandstone, 5282 ft (1610 m), Cities Service No. 14, W. C. Hardy, Jr., Kanawha Forest field, Kanawha County, West Virginia (0.44 in. in the photo equals 0.14 mm). Dolomite again has filled in much of the pore space. In several cases, the dolomite rhombs are suspended in the quartz cement. The high degree of rounding of some of the quartz grains suggests that they are derived from older sands of probable Cambrian age (see Figure 3.2). The carbonate has filled in the available pore space to the point where there was no measured permeability and only 3.9% porosity.

Figure 3.8. Newburg Sandstone, 5286 ft (1611 m), Cities Service No. 14, W. C. Hardy, Jr., Kanawha Forest field, Kanawha County, West Virginia (0.44 in. in the photo equals 0.30 mm). The sand is almost 100% quartz grains and quartz cement. The only other material present is pyrite, which occurs in small patches. As in the previous thin sections of the Newburg, there are well-rounded quartz grains with quartz overgrowths. Both the porosity (12.2%) and the permeability (25.30 millidarcies) are improved.

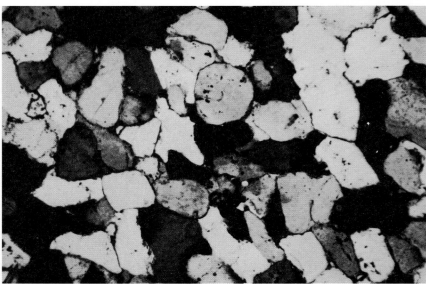

Figure 3.9. Newburg Sandstone, 5288 ft (1612 m), Cities Service No. 14, W. C. Hardy, Jr., Kanawha Forest field, Kanawha County, West Virginia (0.44 in. in the photo equals 0.14 mm). Clay minerals (light brown) have formed coatings on many of the quartz grains and have filled in the available pore space. Patches of pyrite are present. The clay filling is so complete that there is no measured permeability and very low porosity (2.1%).

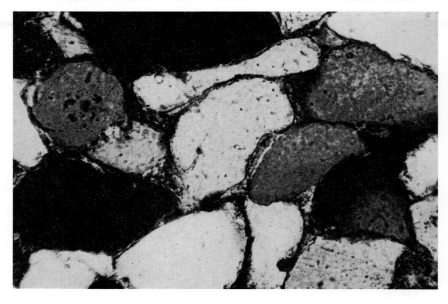

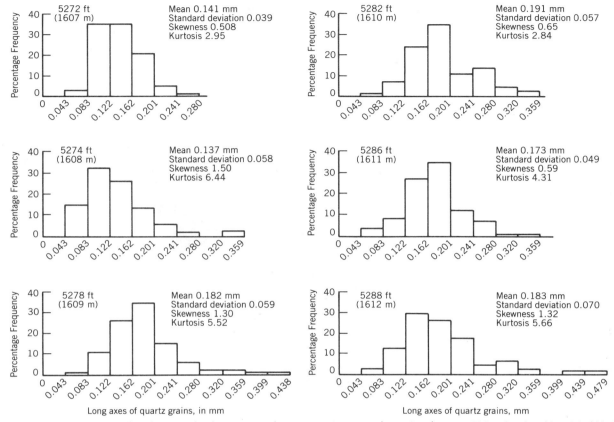

Figure 3.10. Grainsize distributions for long axes of quartz grains, Newburg Sandstone, Cities Service No. 14, W. C. Hardy, Jr., Kanawha Forest Field, Kanawha County, West Virginia.

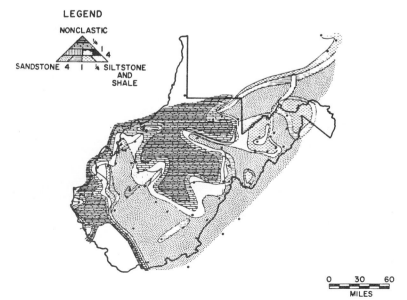

Figure 3.11. Lithofacies map of the Newburg Sandstone showing major areas of deposition within West Virginia. (After Patchen 1973, p. 261.)

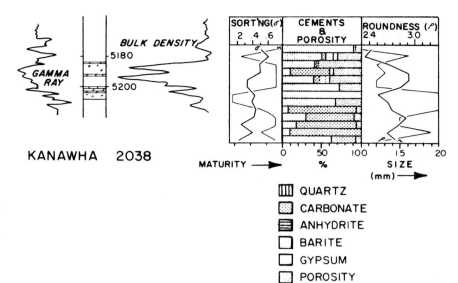

Figure 3.12. Typical gamma ray and bulk density logs along with textural and mineralogic properties for the Newburg Sandstone, Kanawha County, West Virginia. (After Patchen 1973, p. 257.)

94). The major source of the Newburg Sand facies was from the southeast (Chen 1977, p. 95), as illustrated by sand thicknesses of over 100 ft.

Locating a modern equivalent where a clean, almost monomineralic quartz sand is being deposited is not easy. Perhaps the closest example of a modern environment where such a sand is being deposited is along the beaches and offshore shelf sediments of the Florida panhandle (see Figure 3.13).

Patchen (1973, pp. 262–264) suggests three possible options for the origin of the Newburg. The first is that the sand was swept to the west by marine currents and deposited as a marine spit. The second model is similar to the Gulf Coast west of the Mississippi delta, where a chenier plain and a barrier island system occur. The third model, which is favored by Patchen, resembles the Nile delta, which includes a beach–barrier system in the southeast. The isopachous map of the Newburg shown in Figure 3.14 supports the conclusion that a significant influx of sediment probably produced a delta. The shallow, stable shelf appears to have permitted the deposition of the thin clean sand over a considerable distance to the west.

Figure 3.13. Clean, white quartz beach sand near Destin on the Florida panhandle.

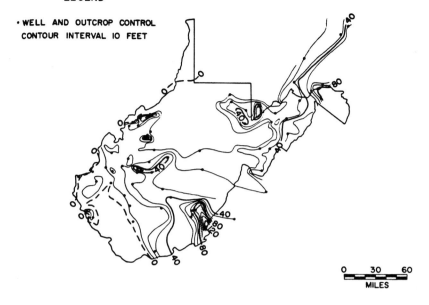

Figure 3.14. Isopach map of the Newburg Sandstone. (From Patchen 1973, p. 261.)

Exploration

Since the discovery well discussed in this section was drilled, considerable drilling for Newburg gas has taken place and at least 10 fields have been discovered (Cardwell 1971). In 1982, only five exploratory tests of the Newburg were conducted (Patchen et al. 1983, p. 1577). The fact that the Newburg is sandwiched between the Oriskany (Lower Devonian) and the Tuscarora (Lower Silurian) means that more testing of the Newburg is likely in the future.

References

Cardwell, D. H., 1971, *The Newburg of West Virginia*, Bulletin 35, West Virginia Geological and Economic Survey.

Chen, P., 1977, *Lower Palezoic Stratigraphy Tectonics, Paleogeography and Oil/Gas Possibilities in the Central Appalachians, West Virginia and Adjacent States*, Part I, *Stratigraphic Maps*, Report of Investigation No. RI 26-1, West Virginia Geological and Economic Survey.

Fenstermaker, C. D., 1966, Large Newburg Area Awaits New Play in West Virginia, *The Oil and Gas Journal*, March 14, pp. 158–162.

Kornfeld, J. A., 1970, Record Gas Strike Spurs West Virginia Drilling, *World Oil*, Vol. 170, No. 3, pp. 122–126.

Krynine, P. D., 1948, The Megascopic Study and Field Classification of Sedimentary Rocks, *Journal of Geology*, Vol. 56, pp. 130–165.

Patchen, D. D., K. A. Schwarz, T. A. Debrosse, E. P. Bendler, M. McCormac, J. A. Harper, W. W. Kelly, Jr., and K. L. Avary, 1983, Oil and Gas Developments in Mid-Eastern States in 1982, *American Association of Petroleum Geologists Bulletin*, Vol. 67, No. 10, pp. 1570–1592.

Patchen, D. G., 1973, Stratigraphy and Petrography of the Upper Silurian Williamsburg Sandstone, West Virginia, *Proceedings of the West Virginia Academy of Science*, Vol. 45, No. 3, pp. 250–265.

Patchen, D. G., P. C. Kline, M. A. Behling, 1977, *Catalog of Subsurface Information for West Virginia*, Mineral Resources Series No. 6, West Virginia Geological and Economic Survey.

4

Walton Field, Big Injun Sandstone, Roane County, West Virginia

Geologic Background

The Big Injun Sandstone (Lower Mississippian) of eastern Ohio and western West Virginia is described in the literature as a highly productive, massive, coarse, conglomeratic, and cross-bedded sandstone (Levorsen 1956, p. 58). The Big Injun as it occurs in the Walton field in Roane County is finer grained and less massive. The general location of the Walton field is shown in Figure 4.1. The stratigraphic position of the Big Injun Sand is provided in Figure 4.2.

Roane County is positioned near the deepest part of the Appalachian basin. The structures are relatively gentle and generally trend northeast–southwest. The exception to this rule is the major structure in the county, the Burning Springs anticline, which trends north–south.

The Pocono interval, which includes the Big Injun Sand, is the lowermost group in the Mississippian System. In the area of the Walton field, the Pocono is overlain by the Middle Mississippian Greenbriar Group and underlain by the Upper Devonian Catskill. The Big Injun Sandstone is the top unit of the Pocono Group and can be confused with what is called the Greenbrier Big Injun, which is a sandy dolomite or limestone (Moore 1976, p. 12). In the case of the Walton field, we are concerned with the Pocono Big Injun, which can be classified as a subgraywacke.

The Big Injun interval is responsible for more than 50% of the larger oil and gas fields in West Virginia; however, only 20% of the original oil in place is characteristically produced from this reservoir. The reservoir rock clearly deserves investigation.

Capillary Pressure Curves

33

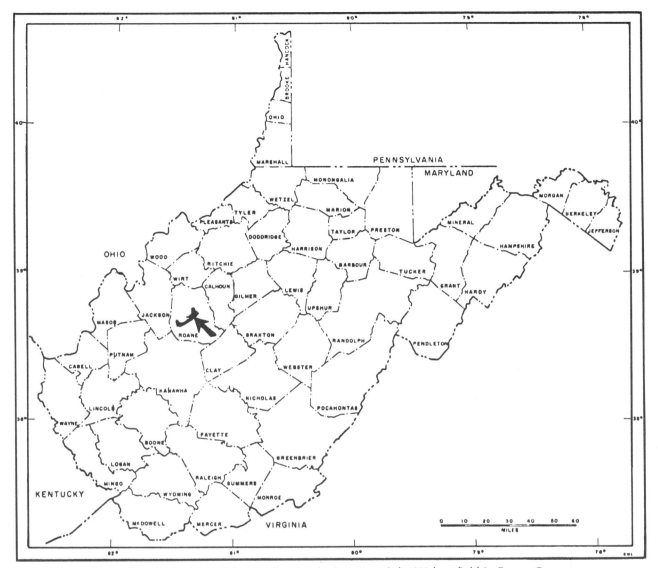

Figure 4.1. Map of West Virginia showing the location of the Walton field in Roane County.

Capillary Pressure Curves

The nature of capillary pressure curves and how to interpret them requires a brief discussion prior to examining the next section and other sections later in this book. In order to produce the curves shown in Figures 4.3–4.7, a sample of water-wet core from the reservoir is used. Oil is forced into the sample under pressure, and this is recorded as pounds per square inch along the vertical axis of the graph. The ease with which the water is displaced is what we observe by studying the shape of the curve. For instance, if the pressure needed to force the oil into the sample is relatively low, the initial displacement pressure is low and the curve begins relatively close to the horizontal axis at the far right. (See Figure 4.3, point B.) The distance up the vertical axis from point A or zero pressure to point B is the measure of the displacement pressure. If we are working with a clean, well-sorted quartz sand with a permeability between 100 and 200 millidarcies, the curve will look like that shown in Figure 4.3, which progresses from B to C. When the curve turns and parallels the vertical axis, it has

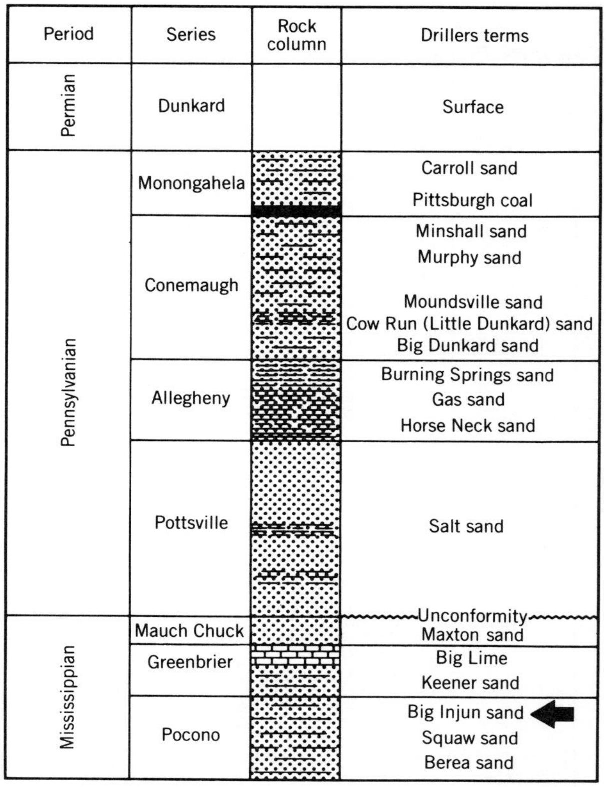

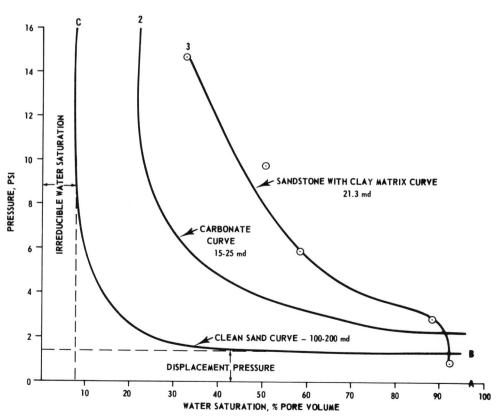

Figure 4.2. Generalized stratigraphic section in the vicinity of the Walton field, Roane County, West Virginia. (After Wasson et al. 1964, p. 6.)

Figure 4.3. Examples of capillary pressure curves for various reservoir rock types.

Capillary Pressure Curves

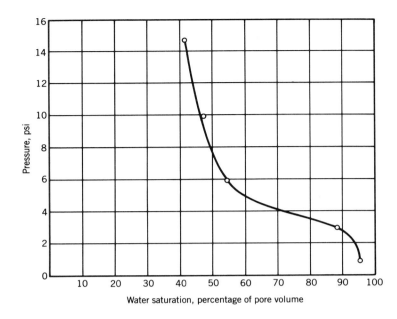

Figure 4.4. Taylor No. 8W, Roane County, West Virginia, Big Injun Sand. Capillary pressure curve. Depth, 2059 ft; air permeability, 43.2 millidarcies; porosity, 20.4%.

reached what is called the irreducible water saturation. In the case of the clean quartz sand, this may be down, in the 8% range. The water that remains is in the pendular areas where individual grains come in contact.

The second curve in Figure 4.3 is typical of a limestone or dolomite reservoir rock with a permeability between 15 and 25 millidarcies. The lower permeability slows down the progress of the oil, and greater pressures are required in order to obtain a less efficient displacement of water.

The third curve is an actual curve taken from the Cities Service No. 8-W Taylor core at 2065 ft.

It will be observed that even though we are dealing with a quartz sand that is reasonably well sorted, the presence of clay matrix produces a curve that indicates that there is a considerable gradation in pore size (see Figure 4.8). In other words, a steady increase in pressure is required in order to force oil into the sandstone. Another result is that the residual water content is high because more water is held in the smaller pores.

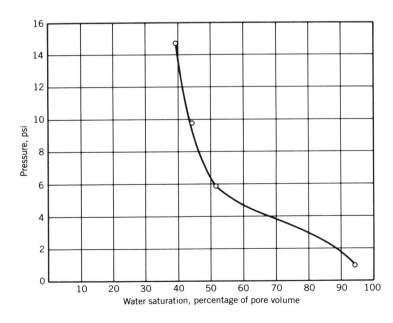

Figure 4.5. Taylor No. 8W, Roane County, West Virginia, Big Injun Sand. Capillary pressure curve. Depth, 2049 ft; air permeability, 66 millidarcies; porosity, 22.4%.

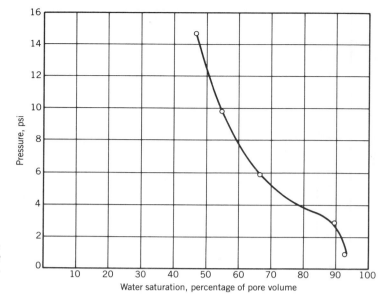

Figure 4.6. Taylor No. 8W, Roane County, West Virginia, Big Injun Sand. Capillary pressure curve. Depth, 2063 ft; air permeability, 10.5 millidarcies; porosity, 18.7%.

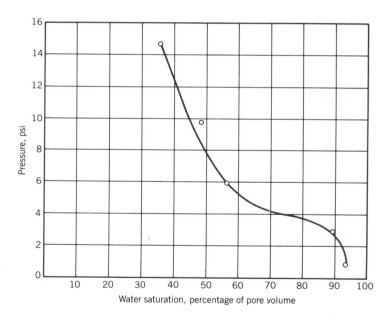

Figure 4.7. Taylor No. 8W, Roane County, West Virginia, Big Injun Sand. Capillary pressure curve. Depth, 2071 ft; air permeability, 25.8 millidarcies; porosity, 19.8%.

Photomicrographs and Petrographic Descriptions

The photomicrographs in this chapter (Figures 4.9–4.14) were selected at those intervals for which capillary pressure curves are available. By extrapolating these curves, it becomes clear that the irreducible water saturation for the Big Injun is between 30 and 40% of the pore volume. This also indicates that the pore size distribution is far from uniform. Table 4.1 provides mineral composition data on the six samples for which photomicrographs are presented (Figures 4.9–4.14).

From the reservoir properties point of view, the clay matrix and carbonate contents are the most important mineralogic factors.

Photomicrographs and Petrographic Descriptions

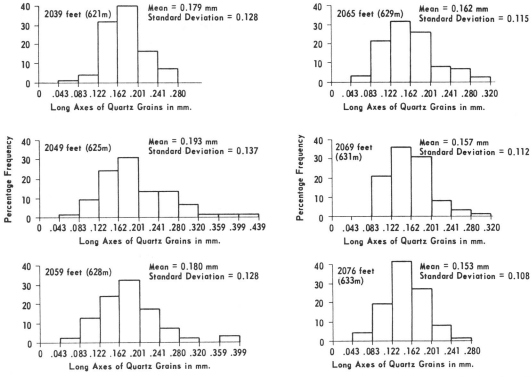

Figure 4.8. Quartz grainsize distributions, Big Injun Sandstone, Cities Service No. 8W Taylor, Walton field.

TABLE 4.1 Mineral Composition Data (%) for the Big Injun Sand, Cities Service, No. 8-W Taylor, Walton Field, Roane County, West Virginia

Minerals	2039 ft (621 m)	2049 ft (625 m)	2059 ft (628 m)	2065 ft (629 m)	2069 ft (631 m)	2076 ft (633 m)
Carbonate	7	1	4	4	39	1
Quartz	72	70	66	53	37	54
Clay matrix (includes chlorite, illite and kaolinite)	8	14	16	15	5	12
Chert	5	1	2	4	6	1
Quartz cement	2	5	5	3	1	11
Feldspar	1	4	2	4	1	5
Micaceous rock fragments	4	—	1	5	6	6
Muscovite	1	—	—	3	1	2
Metaquartzite rock fragments	—	5	4	9	4	9
Total	100	100	100	100	100	100
Air permeability, millidarcies	NA	66.0	43.2	21.3	NA	NA
Porosity, %	NA	22.4	20.4	20.5	NA	NA

Figure 4.9. Big Injun Sandstone, 2039 ft (621 m), Cities Service No. 8W Taylor, Walton field, Roane County, West Virginia (0.44 in. in the photo equals 0.26 mm). The Big Injun Sand is composed mainly of the fine-grained well-sorted, angular quartz with minor amounts of clay minerals and carbonate cement. Muscovite mica, chert, and micaceous rock fragments are also common. Quartz cement is the main cementing material. Although not visible in this photomicrograph, chlorite is frequently observed as a coating on the quartz grains (Moore 1976, p. 40).

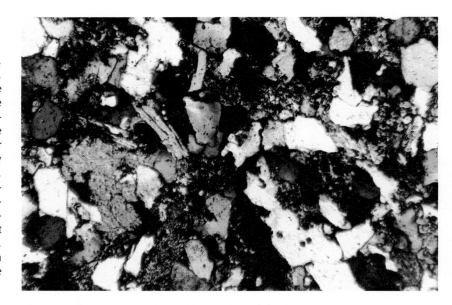

Figure 4.10. Big Injun Sandstone, 2049 ft (625 m), Cities Service No. 8W Taylor, Walton field, Roane County, West Virginia (0.44 in. in the photo equals 0.36 mm). The quartz grains are highly angular and not as well sorted as the previous sample. Carbonate cement patches are few in number and quartz cement is the major cementing agent. Some metaquartzite rock fragments are present.

Figure 4.11. Big Injun Sandstone, 2059 ft (628 m), Cities Service No. 8W Taylor, Walton field, Roane County, West Virginia (0.44 in. in the photo equals 0.36 mm). The photomicrograph shows fine-grained angular quartz and small patches of carbonate cement. Quartz is still the most important cementing material. Not shown in the photomicrograph but present in thin section are shale-like fragments that are included within the clay matrix classification in the mineral composition table.

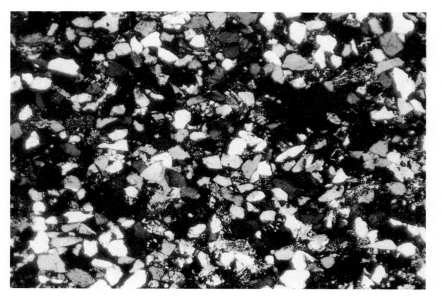

Figure 4.12. Big Injun Sandstone 2065 ft (629 m), Cities Service No. 8W Taylor, Walton field, Roane County, West Virginia (0.44 in. in the photo equals 0.36 mm). Fine-grained, angular quartz is still dominant mineral type; however, more micaceous rock fragments and metaquartzite rock fragments are observed.

Figure 4.13. Big Injun Sandstone, 2069 ft (631 m), Cities Service No. 8W Taylor, Walton field, Roane County, West Virginia (0.44 in. in the photo equals 0.26 mm). Carbonate cement becomes a dominant mineral. Fine-grained quartz remains angular and reasonably well sorted. Chert and rock fragments of both the micaceous and metaquartzite varieties are becoming significant.

Figure 4.14. Big Injun Sandstone, 2076 ft (633 m), Cities Service No. 8W Taylor, Walton field, Roane County, West Virginia (0.44 in. in the photo equals 0.26 mm). Quartz cement returns as the dominant cementing material. Clay matrix fills in much of the available pore space and carbonate cement occurs in small patches. Quartz grains remain well sorted and highly angular.

Geologic Interpretations

In looking at the photomicrographs, one is reminded of the Third Bradford Sand in Pennsylvania; however, in this case, the overall shape of the reservoir suggests a different origin. The reservoir sand in the Walton field is elongated in a northeast–southeast direction, which suggests some form of channel deposit. Moore (1976, p. 12) states that the system is fluvial in origin with the source to the north. Williamson (1974) further postulated that the Big Injun is part of a delta complex that spread out from the eastern side of the Appalachian basin.

Although not observed in this work, Moore (1976) reports that small zones containing glauconite have been observed. Since glauconite is an indicator of a marine environment, the deltaic interpretation fits the observation.

Exploration

Because of the scattered nature of Big Injun reservoirs, it seems likely that exploration efforts will continue to pay off. The rather complex clay mineralogy suggests that the reservoir rocks will be a challenge to tertiary recovery projects for years to come.

References

Levorsen, A. I., 1956, *Geology of Petroleum,* W. H. Freeman and Company, San Francisco, 701 pp.

Moore, S. B., 1976, A Petrographic Study of the Big Injun Sandstone in Roane County, West Virginia, unpublished Masters Thesis, West Virginia University, Morgantown, West Virginia, 109 pp.

Williamson, N. L., 1974, Depositional Environments of the Pocono Formation in Southern West Virginia, unpublished Ph.D. dissertation, West Virginia University, Morgantown, West Virginia, 127 pages.

5

Thornwood Gas Field, Oriskany Sandstone, Pocahontas County, West Virginia

Geologic Background

The Oriskany Sandstone (Lower Devonian) is one of the classic blanket sand deposits of North America. It is also one of the major gas producers of the Appalachian basin. The fact that it has been productive in New York, Maryland, Pennsylvania, Ohio, and West Virginia makes it of particular interest as deeper targets are probed in overthurst areas. In the eastern fields of West Virginia, of which the Thornwood field is one, the Oriskany is overlain by the dark gray to black Needmore Shale. Proceeding to the west, the Oriskany is overlain by the Huntersville Chert and farther west it is overlain by the Onandaga Limestone. Throughout West Virginia the Helderberg Limestone lies beneath the Oriskany Sandstone. Figure 5.1 shows the position of the Oriskany in the overall stratigraphic column.

In the literature, the Oriskany is usually described as a clean, pure, calcareous orthoquartzite. In such areas as central Pennsylvania, the Oriskany is composed of 94% quartz sand with quartz cement adding 4.5% to the total mineral composition (Mapleton Quarry, Huntingdon County, Pennsylvania; Wood 1960, p. 39.) This is not the case for the Oriskany Sand as it occurs in the Thornwood field. The general location of the Thornwood field in relation to other eastern West Virginia Oriskany gas fields is illustrated in Figure 5.2.

SYSTEM	SERIES		STAGE	ROCK UNIT
DEVONIAN	MIDDLE	Erian	Cazenovia	Marcellus Shale
				Tioga Metabentonite
		ULSTERIAN	Ones-quethaw	Onondaga Limestone / Huntersville Chert / Needmore Shale
	LOWER		Deer Park	Oriskany Sandstone ◀
			Helder-berg	Includes Shriver Chert, Licking Creek Ls., New Scotland Cherty Ls., New Creek Formation, Keyser Ls., etc.
SILUR-IAN	Cayugan			Tonoloway Ls.

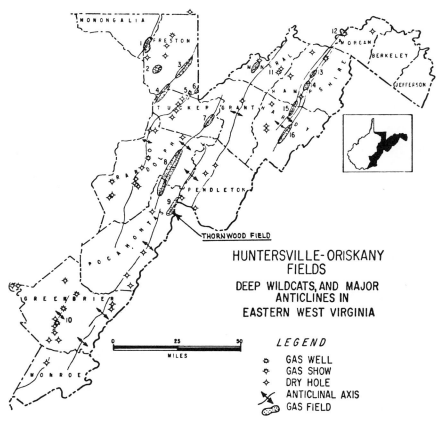

Figure 5.1. Stratigraphic nomenclature for the Middle and Lower Devonian Ulsterian Series, which contains the Oriskany Sandstone. (After Cardwell 1982, p. 7.)

Figure 5.2. Location of Thornwood gas field. (Modified after Patchen 1968, p. 3.)

Photomicrographs and Petrologic Detail

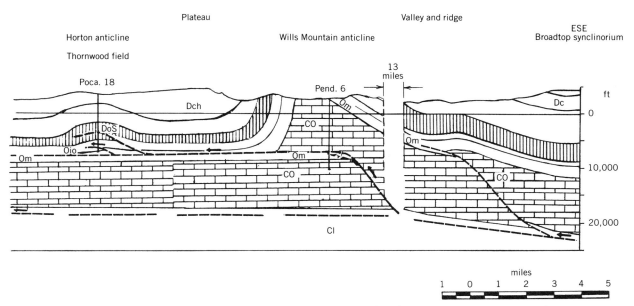

Figure 5.3. Cross section of the plateau and northwestern Valley and Ridge province in Virginia and West Virginia, showing thrust faults and repeated Oriskany section (dos). (Modified from Gwinn 1964, plate 2.)

Geologically, the Thornwood field is located on the surface axis of the Horton anticline. The Oriskany at this location is quite variable both in mineral composition and in thickness. On average, it is reported to be 170 ft thick (Patchen 1968, p. 20). Gwinn (1964, p. 884) indicated that the discovery well in the Thornwood field penetrated thrust faults branching off an Ordovician sole fault that caused a repeat of the Oriskany section as shown in Figure 5.3. Obviously, such structure could explain the variability in thickness.

Photomicrographs and Petrologic Detail

The mineralogic point count data for six core samples from the Cities Service GW-1468, USA R-1 well in the Thornwood field are provided in Table 5.1. What is immediately clear is that the classical clean quartz sand typical of the Oriskany in other areas is not present. It is only the top part of the core that contains appreciable amounts of cemented quartz sand. Farther down the core, large amounts of carbonate cement and detrital carbonate debris become dominant features.

The quartz grainsize information in this particular well differs from that described in the literature and therefore requires some discussion. When the initial work was done, a severe deadline had to be met. In order to provide some concept of how the quartz grainsize was changing, an extreme value approach was used. This involved measuring the long axes of the 10 largest quartz grains found in each thin section. It was reasoned that the largest grains reflected the overall competence of the depositing currents; therefore, these were the most sensitive measures of depositional changes. Figure 5.4 shows a profile plot of the maximum, mean, and minimum values for the 10 largest quartz grains in each thin section. The conclusion that emerges is that there were wide changes in the strength of the paleocurrents. The summary of the Devonian paleocurrents as obtained by Kepferle et al. (1977) is shown in Figure 5.5. Whether examined in New York, Pennsylvania, or West Virginia, the resulting direction of current flow is from east to west. Looking at the stratigraphic relationships and

TABLE 5.1 Mineralogic Point Count Data (%), Oriskany Sand, CSO-GW-1468 USA R-1

Minerals	1508 m (4949 ft)	1510 m (4953 ft)	1512 m (4960 ft)	1516 m (4974 ft)	1517 m (4977 ft)	1518 m (4980 ft)
Quartz	68	69	50	47	16	44
Quartz overgrowths	2	6	10	4	3	3
Carbonate cement	21	16	25	35	39	39
Detrital shell carbonate	2	7	13	13	41	10
Dark organic matter	7	1	—	1	1	4
Microcline	—	1	1	—	—	—
Pyrite	—	—	1	—	—	—
Total	100	100	100	100	100	100
Air permeability, millidarcies	0.00	0.00	0.00	0.00	0.00	0.00
Porosity, %	1.2	0.3	1.2	0.8	1.0	2.0

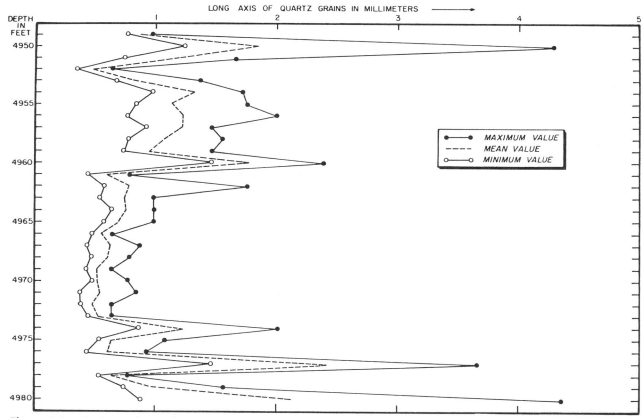

Figure 5.4. Grainsize for 10 largest quartz grains in thin section, Cities Service GW-1468, U.S.A. R-1, Pocahontas County, West Virginia.

Geologic Interpretation

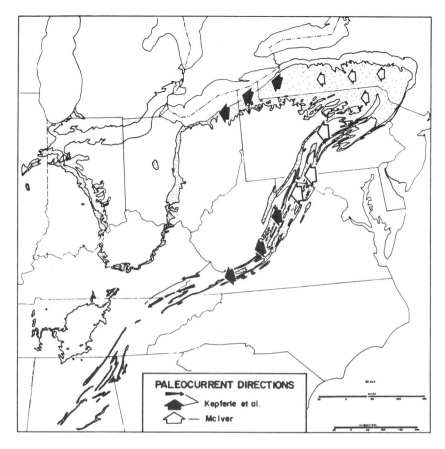

Figure 5.5. Map of the Appalachian basin showing a summary of the Devonian paleocurrents as they occur in the Upper (stippled) and Middle Devonian outcrop zone. (After Kepferle et al. 1977, p. 441.)

the thickness of the Oriskany Sandstone in particular, it becomes clear from Figure 5.6 that the Oriskany thickens markedly toward its source area to the east. With this overall picture in mind, it is instructive to examine the photomicrographs of the thin sections (see Figures 5.7–5.12).

Geologic Interpretation

The photomicrographs in Figures 5.7–5.12 provide some idea of the highly cemented nature of this reservoir rock; however, it is the permeability and porosity measurements that make the picture clear. Of the 208 core plugs (both vertical and horizontal) taken over the 155 ft of core, not one provided any measurable permeability. Furthermore, the highest porosity value obtained was 3.9%. The gas production in this case is obviously coming from the fracture system associated with the structure. This agrees with the concept that the valley and ridge part of West Virginia gas accumulation is dependent on fracture porosity and anticlinal entrapment (Diecchio 1985).

From both the geologic and engineering points of view, this type of reservoir is not an attractive target for future exploration.

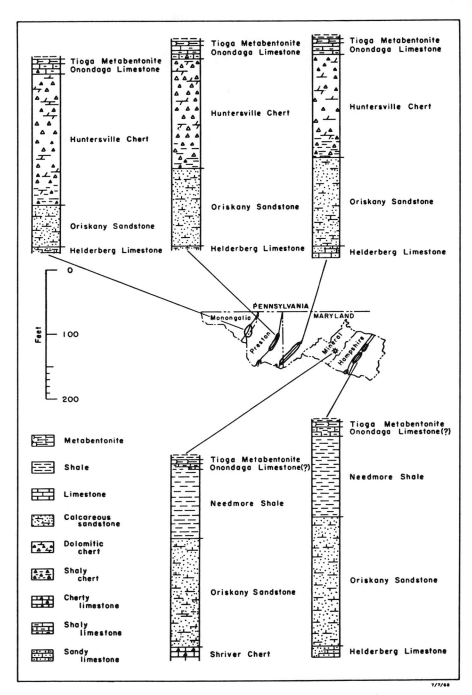

Figure 5.6. Regional stratigraphic relationships and facies changes in the Onesquethaw. (After Patchen 1968, p. 8.)

Figure 5.7. Oriskany Sandstone, 4949 ft (1508 m), Cities Service GW-1468, U.S.A. R-1, Thornwood field, Pocahontas County, West Virginia (0.44 in. in the photo equals 0.42 mm). The quartz grains are poorly sorted and mixed in with detrital carbonate (shell) debris. Evidence of previous fracturing and healing includes bubble trains that cross from one quartz grain to the next. A number of the grains have strain shadows and Boehm lamellae similar to those illustrated by Schoelle (1979, p. 165). Many of the larger quartz grains are well rounded.

Figure 5.8. Oriskany Sandstone, 4953 ft (1510 m), Cities Service GW-1468, U.S.A. R-1, Thornwood field, Pocahontas County, West Virginia (0.44 in. in the photo equals 0.42 mm). In contrast to Figure 5.7, the Oriskany quartz sand at this level is well rounded and well sorted. Except for the relatively large amount of carbonate cement and detrital carbonite, it resembles the classical description of the Oriskany Sand as it occurs in gas reservoirs throughout a large part of the Appalachian basin. Clear evidence of fracturing is not present.

Figure 5.9. Oriskany Sandstone, 4960 ft (1512 m), Cities Service GW-1469, U.S.A. R-1, Thornwood field, Pocahontas County, West Virginia (0.44 in. in the photo equals 0.42 mm). The quartz grains are poorly sorted and highly cemented with carbonate quartz. Swarms of oriented bubble trains cross from grain to grain indicating that there has been considerable strain and fracturing associated with this zone. In the upper right corner is a highly rounded quartz grain with an overgrowth. Quartz grains near the center show areas with quartz cement filling the pore space.

Figure 5.10. Oriskany Sandstone, 4974 ft (1516 m), Cities Service GW-1468, U.S.A. R-1, Thornwood field, Pocahontas County, West Virginia (0.44 in. in the photo equals 0.42 mm). Relatively large quartz grains appear to "float" in a medium composed of much smaller quartz grains, carbonate shell fragments, and carbonate cement. This appears to be a good example of a bimodal size distribution for quartz grains. The rather even size distribution for the smaller grains suggests that they may have been transported by wind rather than by water.

Figure 5.11. Oriskany Sandstone, 4977 ft (1517 m), Cities Service GW-1468, U.S.A. R-1, Thornwood field, Pocahontas County, West Virginia (0.44 in. in the photo equals 0.42 mm). The quartz grains are poorly sorted and dispersed among large carbonate shell fragments. Carbonate cement dominates and fills in the pore space. Several quartz grains show strain shadows, however, the sets of bubble trains observed in the thin section from 4960 ft are missing.

Figure 5.12. Oriskany Sandstone, 4980 ft (1518 m), Cities Service GW-1468, U.S.A. R-1, Thornwood field, Pocahontas County, West Virginia (0.44 in. in the photo equals 0.42 mm). A sharp break in quartz grain-size seems to separate the upper and lower halves of the photomicrograph. The relatively coarse grained dolomite-cemented bottom half appears to be separated from the finer grained carbonate-cemented top half by a dark brown heavy organic zone. Note the presence of a large chert grain (speckled black and white) on the extreme right.

References

Cardwell, D. H., 1982, *Oriskany and Huntersville Gas Fields of West Virginia,* West Virginia Geological and Economic Survey, Mineral Resources Series MRS-5A, 180 pp.

Diecchio, R. J., 1985, Regional Controls of Gas Accumulation in Oriskany Sandstone, Central Appalachian Basin, *American Association of Petroleum Geologists Bulletin,* Vol. 69, N. 5, pp. 722–732.

Gwinn, V. E., 1964, Thin-Skinned Tectonics in the Plateau and Northwestern Valley and Ridge Provinces of Central Appalachians, *Geological Society of America Bulletin,* Vol. 75, pp. 863–900.

Kepferle, R. C., P. Lundegard, J. B. Maynard, P. E. Potter, W. A. Pryor, N. Samuels, and F. J. Scharef, 1977, *Paleocurrent Systems in Shaly Basins: Preliminary Results for Appalachian Basin (Upper Devonian),* U.S. Department of Energy, First Eastern Gas Shales Symposium, Morgantown, West Virginia, October 1977, Proceedings, MERC/SP-77/5, p. 441.

Patchen, D. G., 1968, Oriskany Sandstone–Huntersville Chert Gas Production in the Eastern Half of West Virginia, *West Virginia Geological and Economic Survey Circular,* Vol. 9, p. 38.

Scholle, P. A., 1979, *A Color Illustrated Guide to Constituents, Textures, Cements, and Porosities of Sandstones and Associated Rocks,* Memoir 28, American Association of Petroleum Geologists, Tulsa, OK, p. 20.

Wood, G. V., 1960, A Comparison of Three Quartzites, Unpublished Ph.D. thesis, The Pennsylvania State University, 159 pp.

6

Rodney Oil Field, Dundee Formation, Elgin County, Ontario, Canada

The Rodney oil field is interesting because it is a combination of reservoir types. More specifically, some zones are limestone, some are a mixture of limestone, dolomite, and quartz sand, and others are mostly quartz sand. From the engineering standpoint, this leads to large permeability and porosity, variations that are difficult to deal with when employing enhanced recovery techniques.

The Rodney cores also provide good examples of dolomitization and thereby open up an opportunity to achieve a better understanding of the mechanisms involved.

Geological Background

Southwestern Ontario is situated on the broad structural feature that is the northeastward extension of the Cincinnati arch, as shown in Figure 6.1. The region is underlain by a relatively thin layer of sediments ranging in age from Cambrian to Late Devonian. The areal distribution of these sedimentary rocks and the location of the Rodney field is provided in Figure 6.2. As the regional cross section indicates, the regional dip is around 20–30 ft/mile toward the south and southwest.

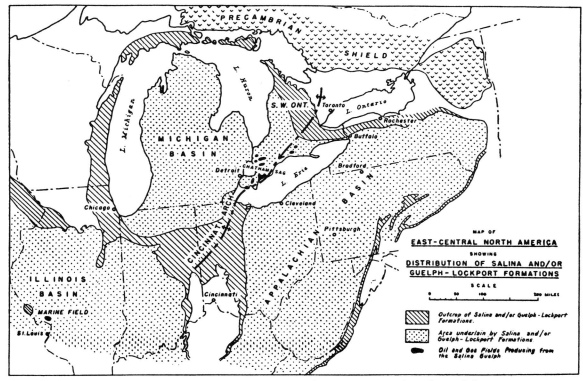

Figure 6.1. Relationship of southwestern Ontario to the Cincinnati arch. (After Roliff 1949, p. 156.)

A summary of the sedimentary rocks represented in the stratigraphic column is as follows: limestones and dolomites make up 55.5%, shales 40%, evaporites 2.5%, and sandstones 2% (Roliff 1949, p. 155).

Photomicrographs

The Rodney field is located in Elgin County and produces from the Dundee horizon of Middle Devonian age (or the Dundee equivalent) as shown in Figure 6.3. It should be noted that the production is from a very shallow reservoir only 300–400 ft deep. The particular thin sections pictured in the photomicrographs in Figures 6.4–6.9 are chosen to show the average characteristics of the producing zone and the various porous streaks. In addition, they are chosen to provide some idea of why and how the highly porous and permeable zones were formed. The mineral composition data presented in Table 6.1 provide some of the answers. In the sample from 336 ft, cal-

TABLE 6.1 Point Count Mineral Composition of the Rodney Field, Elgin County, Ontario, Canada, Dundee Formation, Cities Service Rodney Unit 24W-TR5

Minerals	336 ft (102.4 m)	339 ft (103.3 m)	342 ft (104.3 m)	346 ft (105.5 m)	350 ft (106.7 m)	352 ft (107.3 m)
Calcite	84	73	17	33	17	18
Dolomite	14	25	66	52	39	41
Quartz grains	2	2	17	15	45	41
Porosity, %	9.6	13.2	18.5	11.3	20.0	18.5
Permeability, millidarcies	0.20	51.00	210.00	4.60	420.00	1000.0

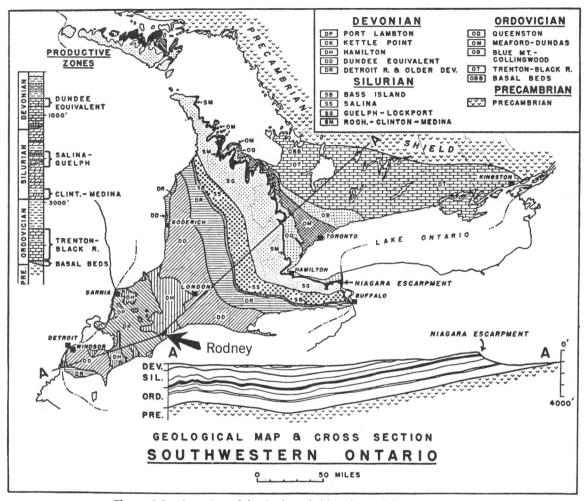

Figure 6.2. Location of the Rodney field. (After Roliff 1949, p. 157.)

cite is the dominant mineral and the amount of pore space is relatively low (9.6%). Permeability and porosity are improved in the sample from 339 ft, (permeability, 51.00 millidarcies; porosity, 13.2%) and the major change in mineralogy is the increase in the percentage of dolomite.

The sample from 342 ft illustrates the case where dolomite becomes dominant and increases in permeability (210.00 millidarcies) and porosity (18.5%) are again observed. The return of significant amounts of calcite (33%) tend to reduce the permeability and porosity. Finally, in the remaining samples, from 350 and 352 ft, there is a significant influx of rounded quartz grains along with the dolomite. These two minerals in both samples account for over 80% of the mineral content. The important point is that the permeability has increased markedly (even up to 1 darcy) with the increased sand content.

The grainsize distributions for the long axes of the quartz grains are presented in Figure 6.10. In examining the difference between the sample from 350 ft (106.7 m) and the sample from 352 ft (107.3 m), it becomes clear that there is a significant increase in quartz grainsize in the sample from 352 ft (107.3 m). The relatively coarse grainsize of the quartz grains in this sample plus their increase in percentage helps to explain the very good (1-darcy) permeability. In other words, the zone is composed of close to half coarser sand and the other half is heavily dolomitized. The photomicrographs in Figures 6.4–6.9 illustrate the robustness of the dolomitization that has taken place but also show the role played by the quartz sand.

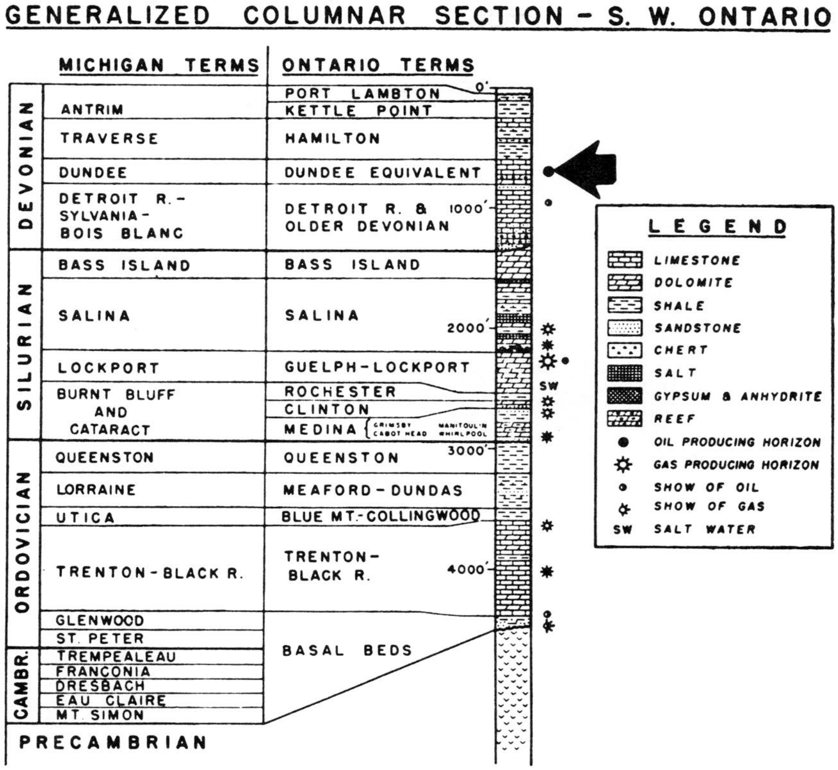

Geologic Interpretation

GENERALIZED COLUMNAR SECTION – S. W. ONTARIO

	MICHIGAN TERMS	ONTARIO TERMS
DEVONIAN	ANTRIM	PORT LAMBTON — 0'
	TRAVERSE	KETTLE POINT
		HAMILTON
	DUNDEE	DUNDEE EQUIVALENT
	DETROIT R.-SYLVANIA-BOIS BLANC	DETROIT R. & OLDER DEVONIAN — 1000'
SILURIAN	BASS ISLAND	BASS ISLAND
	SALINA	SALINA — 2000'
	LOCKPORT	GUELPH-LOCKPORT
	BURNT BLUFF AND CATARACT	ROCHESTER
		CLINTON
		MEDINA (CRIMSBY, CABOT HEAD, MANITOULIN, WHIRLPOOL)
ORDOVICIAN	QUEENSTON	QUEENSTON — 3000'
	LORRAINE	MEAFORD-DUNDAS
	UTICA	BLUE MT.-COLLINGWOOD
	TRENTON-BLACK R.	TRENTON-BLACK R. — 4000'
	GLENWOOD	
	ST. PETER	BASAL BEDS
CAMBR.	TREMPEALEAU	
	FRANCONIA	
	DRESBACH	
	EAU CLAIRE	
	MT SIMON	
PRECAMBRIAN		

LEGEND
- LIMESTONE
- DOLOMITE
- SHALE
- SANDSTONE
- CHERT
- SALT
- GYPSUM & ANHYDRITE
- REEF
- ● OIL PRODUCING HORIZON
- ✷ GAS PRODUCING HORIZON
- ○ SHOW OF OIL
- ✧ SHOW OF GAS
- SW SALT WATER

Figure 6.3. Stratigraphic position of the Dundee Limestone and Dolomite. (After Roliff 1949, p. 158.)

The photomicrographs are chosen to illustrate the various types of reservoir rock present. The samples cover the productive zone and are not representative of the less porous zones above and below the reservoir.

In order to obtain a more accurate point count of the mineral composition, a staining technique that stains calcite pink was used.

Geologic Interpretation

The portion of the Rodney sedimentary sequence that is commercially productive occurs at the base of what has been called a biostrome, a bioherm, or a reef. The fact that quartz sand occurs in a well-defined layer strongly suggests that at that particular stage there was no reef or even a bioherm. The recognizable fossil remains (at least in thin section) are mainly crinoid columnals; therefore, it appears that this most permeable zone began as a local accumulation of quartz sand and crinoid fragments. Examination of the sand from the sample at 352 ft (107.3 m) indicates that it is well rounded. The source of such well-rounded quartz is likely to be basal sands associated with the base of the Ordovician, which today outcrops 200 miles to the northeast.

It should also be observed that the grainsize of the sand grains found in the upper portions of the producing zone is fine grained and quite uniform in its size distribution (Figure 6.10).

One is left with the impression that at least part, if not all, of the fine-grained sand in the upper zone may be wind blown.

Further examination of the highly porous zones indicates that they may intersect two or more wells; however, these zones cannot be traced uniformly across the field. In some cases, there is only one well with a highly porous section. The core data for the producing zone show very few impermeable barriers. This suggests that communication is likely to exist between the highly permeable zones in adjacent wells so that a water flood model may be described as a stratified reservoir with cross-flow.

It is also instructive to examine the description of the 160 ft of core materials above the productive zone. A large part of the core is "fossiliferous" with

Figure 6.4. Dundee Formation, 336 ft (102.4 m). Cities Service Rodney Unit 24W-TR 5, Rodney field, Elgin County, Ontario, Canada (0.44 in. in the photo equals 0.08 mm). Numerous dolomite rhombs (light brown) appear to be floating in calcite (stained pink). Individual crinoid columnals (not shown in photomicrograph) are all calcite. All fossil fragments tend to be calcite. Zones with greater porosity appear to have more dolomite around them. Dolomite rhombs appear to be zoned with numerous inclusions in the center and clearer rims.

Figure 6.5. Dundee Formation, 339 ft (109.3 m), Cities Service Rodney Unit 24W-TR 5, Rodney field, Elgin County, Ontario, Canada (0.44 in. in the photo equals 0.08 mm). A few quartz grains are present (gray and white grains, upper left). Zoned dolomite crystals again appear to surround the larger pores. Crinoid fragments, not shown in photomicrograph, are all calcite.

Figure 6.6. Dundee Formation, 342 ft (104.3 m), Cities Service Rodney Unit 24W-TR 5, Rodney field, Elgin County, Ontario, Canada (0.44 in. in the photo equals 0.08 mm). The whole sample is heavily dolomitized with quartz grains becoming more numerous. Calcite (pink) is reduced.

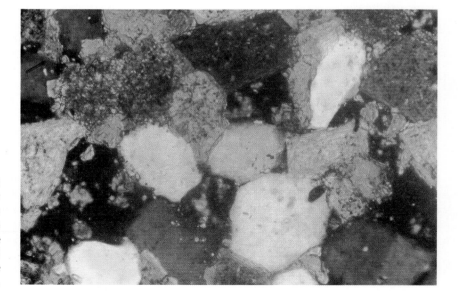

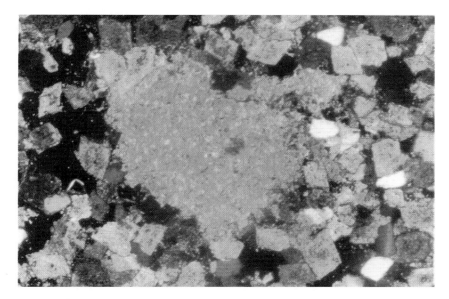

Figure 6.7. Dundee Formation, 346 ft (105.5 m), Cities Service Rodney Unit 24W-TR 5, Rodney field, Elgin County, Ontario, Canada (0.44 in. in the photo equals 0.17 mm). The whole sample is moderately dolomitized with fossil fragments such as that above remaining calcite (pink). Fine-grained sand is clearly present (white and gray grains).

Figure 6.8. Dundee Formation, 350 ft (106.7 m), Cities Service Rodney Unit 24W-TR 5, Rodney field, Elgin County, Ontario, Canada (0.44 in. in the photo equals 0.17 mm). Sand becomes much more important. Most of the carbonate has been dolomitized. Large rounded quartz grains (white) occur in what looks like two populations of quartz.

Figure 6.9. Dundee Formation, 352 ft (107.3 m), Cities Service Rodney Unit 24W-TR 5, Rodney field, Elgin County, Ontario, Canada (0.44 in. in the photo equals 0.17 mm). Large rounded quartz grains become important members, along with dolomite and some calcite cements, in supporting the very open framework that provides good porosity and excellent permeability.

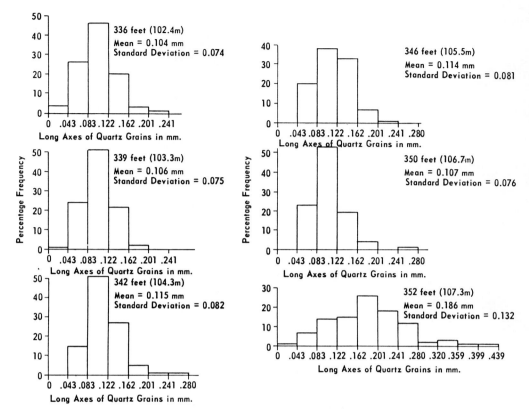

Figure 6.10. Grainsize distributions for long axes of quartz grains Dundee Formation, Cities Service Rodney Unit 24W-TR 5, Rodney field, Elgin County, Ontario, Canada.

shows of oil; however, the porosity and permeability are low.

The presence of corals is also mentioned (335 ft). Therefore, the upper part of what began as a crinoidal bank appears to have evolved into a reef structure.

Recent studies by Legall et al. (1981, p. 526) indicate that the depths of burial for most of the Paleozoic sequence inferred from paleotemperature estimates based on conodonts and palynomorphs was minimal. Furthermore, the Dundee Formation is believed to have been in a zone that remained at temperatures of less than 60°C (Legall et al. 1981, p. 525). All of the above suggests that reservoirs of the Rodney type were subjected to relatively low temperatures and pressures. This also suggests that the oil migrated into the porous zone. The beautifully zoned dolomite crystals that are common features clearly were not short-term high-temperature phenomena.

Finding a present-day equivalent of the Rodney field is not an easy task.

Few studies of Recent sediments have been carried out that discuss the mixture of environments such as those encountered in the Rodney field. One recent work by Fryberger et al. (1983) comes close. In this modern example, the authors discuss an "offshore prograding sand sea . . . which exists along portions of the Arabian coastline near Dhahran, Saudi Arabia" (p. 280). More specifically, eolian sand is blown into the marine carbonate environment so that sand-rich layers occur in the sequence of marine carbonate muds. In the case of the Rodney field, the eolian sand is blown into the carbonate bank composed of crinoidal fragments; however, the concept is basically the same.

The area has been heavily explored and today is populated to such a degree that the drilling of onshore prospects becomes a legal challenge. On the other hand, the shallow depths are so attractive that additional drilling will continue and new oil will be found.

Perhaps the most important possibilities are in the area of research, where both the origins of the oil and the secondary dolomite are attractive targets.

References

Fryberger, S. G., A. M. Al-Sari, and T. J. Clisham, 1983, Eolian Dune, Interdune, Sand Sheet, and Siliciclastic Sabkha Sediments of an Offshoe Prograding Sand Sea, Dhahran Area, Saudi Arabia, *American Association of Petroleum Geologists Bull.*, Vol. 67, No. 2, pp. 280–312.

Legall, F. D., C. R. Barnes, and R. W. MacQueen, 1981, Thermal Maturation, Burial History and Hotspot Development, Paleozoic Strata of Southern Ontario–Quebec, from Conodont and Acritarch Colour Alteration Studies, *Bulletin of Canadian Petroleum Geology,* Vol. 29, No. 4, pp. 492–539.

Roliff, W. A., 1949, Salina–Guelph Fields of Southwestern Ontario, *American Association of Petroleum Geologists Bulletin,* Vol. 33, No. 2, pp. 153–188.

7

Silver City Pool, Hardinsburg Sandstone, Butler County, Kentucky

Frequently, small fields such as the Silver City pool are not covered at all in the literature. The well records and production data are kept by the companies involved, and ultimately when the field is abandoned, very little remains to tell about the type of field it was or anything about the mineralogy of the reservoir. In this instance, a minimum amount of information is available; however, the reservoir appears to be typical of the numerous small fields that are found at shallow depths in the Eastern Interior area.

General Geologic Setting

Throughout most of the Paleozoic, the Upper Mississippi Valley region was characterized by a gently subsiding platform that forms the southcentral flank of the Canadian Shield. On this base, a series of strata were deposited that are typical of cratonic sedimentation. The Illinois basin is the largest of several intracratonic basins, and during the later Paleozoic, it became progressively more important. Its overall outline and extension into Kentucky is shown in Figure 7.1.

Photomicrographs

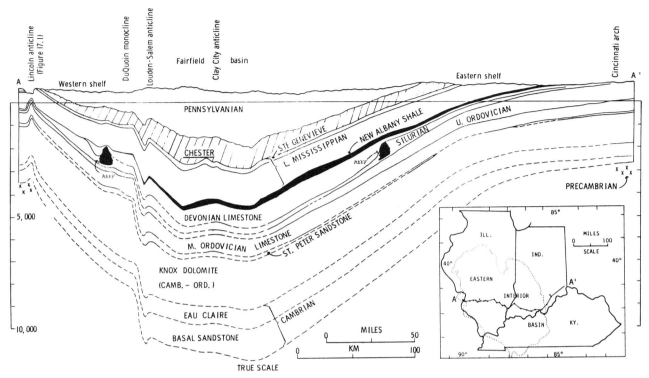

Figure 7.1. Outline and cross section of the Illinois basin showing the relative stratigraphic position of the Chester Series. (After Landes 1970, p. 61.)

The Hardinsburg Sandstone is one of a series of sandstones that form part of the Upper Mississippian Chester Series. An electric log illustrating the relative position of the Hardinsburg Sandstone as it occurs within the Chester is shown in Figure 7.2. The location of Butler County relative to some of the major geologic features in Kentucky is provided in Figure 7.3.

The dominant lithologic type for the Chester Series is shale, which forms more than 50% of the whole series. The next most abundant rock type is limestone, and this forms almost 25% of the Chester section.

Sandstones also compose almost 25% of the Chester, and they vary considerably in distribution around the Illinois basins, exhibiting a much greater lateral variability than the shales and limestones. According to Potter et al. (1958, p. 1016), nearly all Chester sandstones have sharp erosional channels at their base and rarely exceed 50 ft in depth. Plant casts are common locally in most of the sandstones (Potter et al. 1958, p. 1016).

We now look at the photomicrographs (Figures 7.4–7.9).

Photomicrographs

In terms of texture, the sands are usually fine to very fine grained and are usually well sorted. In the case of the Silver City pool, sand-size variations occur on the sedimentary unit level (see Figure 7.10).

The mineralogy of the Chester-type sandstones and of the Hardinsburg in particular is almost all quartz; therefore, point counting is not particularly informative. As Potter et al. (1958, p. 1017) point out, there is usually less than 1% feldspar with zircon and tourmaline composing the dominant heavy minerals. These sandstones, therefore, have the typical othoquartzite composition but not the blanket-type distribution common to some of the Appalachian basin sands, such as the Oriskany and Tuscarora sandstones of Pennsylvania. Potter et al. (1958, p. 1017) conclude that the Chester sandstones were almost entirely derived from pre-existing sediments. Furthermore, according to these researchers, it seems quite likely that most of the Chester sandstones were deposited in "environments ranging from shallow-marine shelf, through littoral to fluvial." Of these three,

Weller and Sutton (1940, p. 847) indicate that the fluvial environment is dominant.

In the case of the Silver City pool, the oil production appears to have relatively little to do with structure. As illustrated in Figure 7.11, the structural highs and other features do not seem to be controlling factors.

Geologic Interpretation

From other published information (Potter et al. 1958), it is clear that the Hardinsburg frequently exhibits cross-bedding and occurs in erosion channels. The clean sandstone, furthermore, along with the down-channel cross-bedding makes the fluvial deposition model the most attractive hypothesis. The presence of kaolinite in the pore space and the absence of marine indicators such as glauconite also tend to support the freshwater origin.

From the reservoir engineering standpoint, this field posed a problem. Fractures in the reservoir allowed the fluids from water input wells in some cases to reach the producing wells in as little as 2 days time, as indicated by tracer studies. Obviously, much of the oil in the less porous sand was bypassed. It has been noted in the case of very shallow fields such as this one and the Rodney field of Ontario, Canada, that water flooding commonly runs into fracture problems where (1) the communication between input and producing wells occurs in too short a time or (2) large volumes of pumped-in water are being lost from the flooded zone and are not causing a response at the producing well in terms of either oil or water.

From the exploration point of view, the shallowness of these reservoirs can be attractive in terms of reduced drilling costs and numbers of wells that can be drilled in a brief time. In attempting to predict where these channels are trending, it is instructive to remember the major directions of sand transport shown in Figure 7.12.

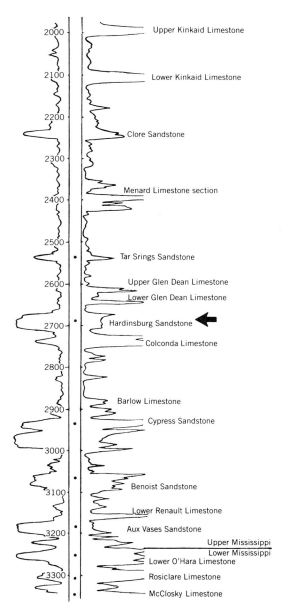

Figure 7.2. Generalized electric log showing the relative position of the Hardinsburg Sandstone. (After Landes 1970, p. 67.)

Petrographic Description

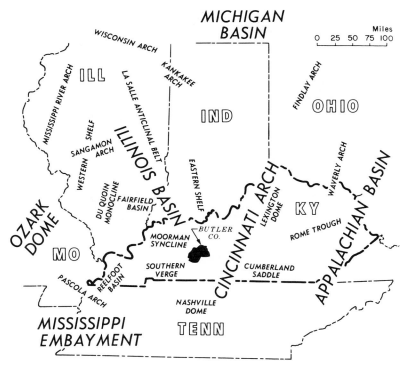

Figure 7.3. Map showing the location of Butler County in relation to the major regional tectonic features. (Reprinted from Schwalb, H. R., 1975, Oil and Gas in Butler County, Kentucky: Kentucky Geological Survey, Report of Investigations 16, Ser. 10, 65 p.)

Petrographic Description

The samples available from the Hardinsburg Sandstone in the Silver City pool were not labeled for specific depth. However, it is still useful to examine the photomicrographs (Figures 7.4–7.9) because this reservoir is only 25 ft thick. As mentioned previously, the mineralogy is almost totally quartz; therefore, no mineral point counts are included.

A number of the available samples have been impregnated with red plastic to emphasize the pore distribution. From the engineering standpoint, it is the pores and minerals that sometimes fill these pores that are important. From the explorationist's point of view, almost every scrap of information may be useful in deciphering the history of an oil reservoir.

62

Figure 7.4. Hardinsburg Sandstone, 500+ ft (151.4 m) Cities Service Hawes North No. 14, Silver City pool, Butler County, Kentucky (0.44 in. in the photo equals 0.30 mm). The Hardinsburg Sand is composed mainly of quartz. As can be observed, it is also highly cemented with quartz. The overall sorting on the scale of the thin section is good. Some grains show quartz overgrowths that were present on the grains prior to deposition, indicating that the source of the sand is probably on older sediment that has been eroded and redeposited.

Figure 7.5. Hardinsburg Sandstone, 500+ ft (151.4 m), Cities Service Hawes North No. 14, Silver City pool, Butler County, Kentucky (0.44 in. in the photo equals 0.42 mm). The connected pores have been impregnated with red plastic to illustrate the distribution. It is clear that the available pore space is quite patchy and that many parts are highly cemented.

Figure 7.6. Hardinsburg Sandstone, 500+ ft (152.4 m), Cities Service Hawes North No. 14, Silver City pool, Butler County, Kentucky (0.44 in. in the photo equals 0.30 mm). Large pore in the center of the photomicrograph contains kaolinite "books" or vermicular kaolinite. Under the scanning electron microscope, these appear as stacks of pseudohexagonal plates. Scholle (1979, p. 138) suggests that the intercrystalline permeability associated with these stacks is very low.

Figure 7.7. Hardinsburg Sandstone, 500+ ft (152.4 m), Cities Service Hawes North No. 14, Silver City pool, Butler County, Kentucky (0.44 in. in the photo equals 0.42 mm). The lower magnification shows clusters of highly quartz cemented grains (lower right and lower left). This again emphasizes the overall patchiness of the permeable zones.

Figure 7.8. Hardinsburg Sandstone, 500+ ft (151.4 m), Cities Service Hawes North No. 14, Silver City pool, Butler County, Kentucky (0.44 in. in the photo equals 0.42 mm). In a few places, calcite fossil fragments and very minor amounts of calcite cement are observed (light brown). These fragments help define the bedding place as illustrated by the diagonal lineation.

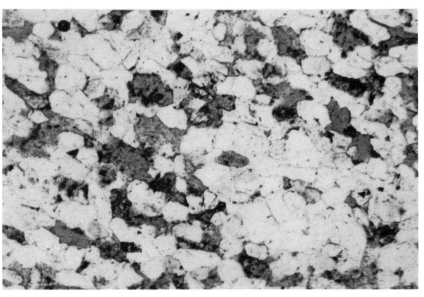

Figure 7.9. Hardinsburg Sandstone, 500+ ft (152.4 m), Cities Service Hawes North No. 14, Silver City pool, Butler County, Kentucky (0.44 in. in the photo equals 0.42 mm). Some lineation is observable, and the connected pores impregnated with red plastic tend to follow these zones. Again, patches of grains that are heavily cemented with quartz shows no porosity.

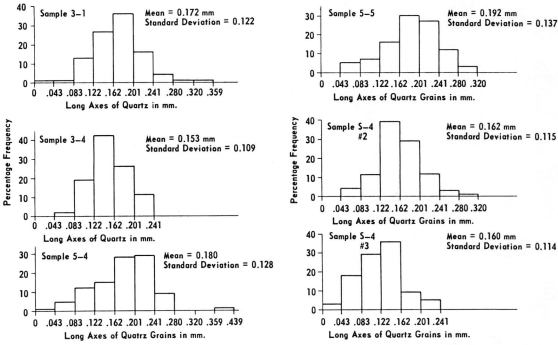

Figure 7.10. Grainsize distributions of long axes of quartz grains, Hardinsburg Formation, Cities Service Hawes North No. 14, Silver City pool, Butler County, Kentucky.

Petrographic Description

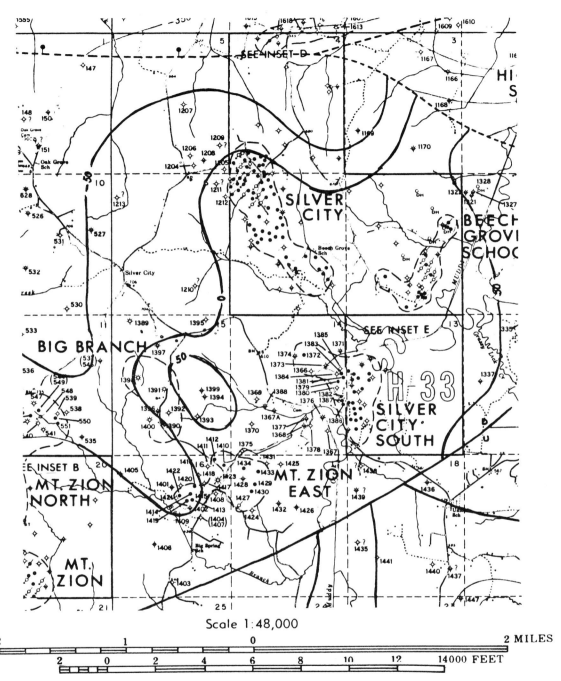

Figure 7.11. Oil, gas, and structure map showing Silver City pool area, structure contour on the base of the Beech Creek Limestone contour interval, 50 ft, 15.2 m. (After Schwalb 1975.)

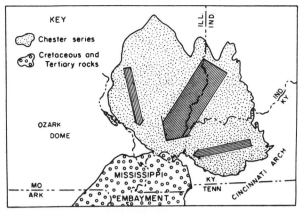

Figure 7.12. Interpretation of major and minor directions of sand transport. (After Potter et al., 1958, p. 1039.)

References

Landes, K. K., 1970, Petroleum Geology of the United States, Wiley-Interscience, New York, 571 pp.

Potter, P. E., E. Nosow, N. M. Smith, D. H. Swann, and F. H. Walker, 1958, Chester Cross-Bedding and Sandstone Trends in Illinois Basin, *American Association of Petroleum Geologists Bulletin*, Vol. 42, No. 5, pp. 1013–1046.

Schwalb, H. R., 1975, *Oil and Gas in Butler County, Kentucky*, Kentucky Geological Survey Series X, Report of Investigations 16.

Weller, J. M. and Sutton, A. H., 1940, *Mississippian Border of Eastern Illinois Basin, Bulletin of the American Association of Petroleum Geologists*, Vol. 24, No. 5, pp. 765–858.

8

North Wise Field, Rogers City–Dundee Formation, Isabella County, Michigan

The North Wise field is situated at roughly the same stratigraphic horizon as the Rodney field in southwestern Ontario, Canada (see Figure 8.1); however, the petrology of the producing zone is quite different. For instance, the porosity seems to have a different origin. There are no quartz-rich zones providing a pathway for dolomitizing fluids and the carbonate is for the most part a rather fine grained mud. The field, therefore, provides a good example of a contrasting environment of deposition.

Geologic Background

The North Wise field is located in the northeast corner of Isabella County, Michigan. On the basis of regional structure, contoured on the top of the Rogers City–Dundee Formation, the field lies near the center of the Michigan basin (Figure 8.2). According to Landes (1970, p. 77), the Rogers City–Dundee producing belt is from three to five counties wide across central Michigan. He furthermore states:

These limestones, where productive, have porosity due either to secondary leaching or to local dolomitization.

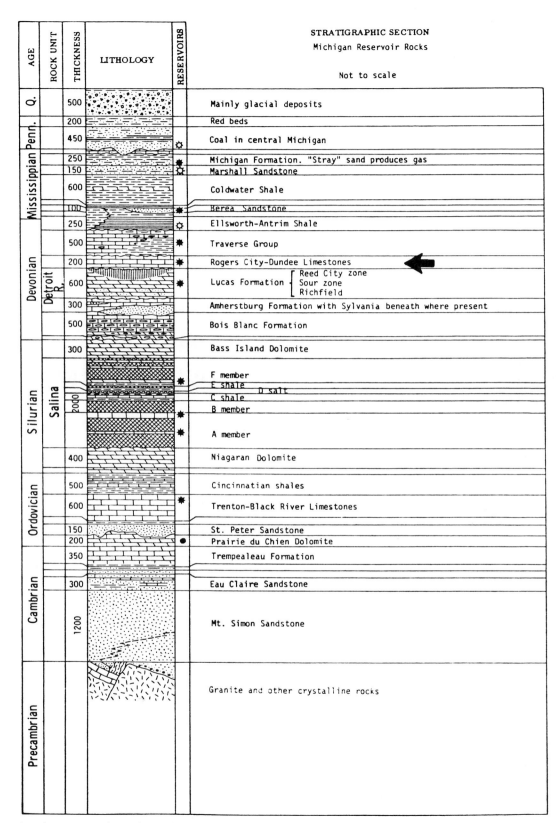

Figure 8.1. Showing the stratigraphic relationship of the Rogers City–Dundee Limestone in the Michigan basin. (After Landes 1970, p. 79.)

Geologic Background

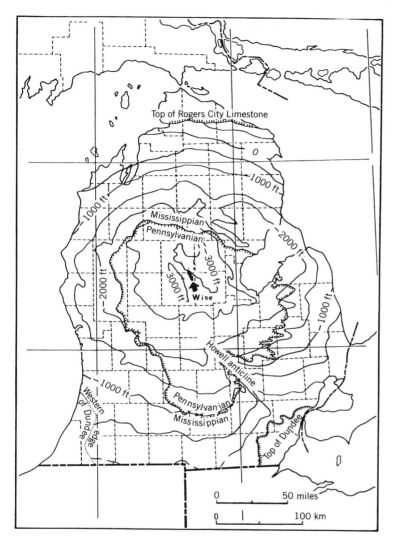

Figure 8.2. Location of the Wise field in relation to the structure of the Michigan basin contoured on top of the Rogers City–Dundee Limestone, interval, 500 ft. (Modified from Landes 1970, p. 78.)

If the above is correct, the North Wise field appears to be an exception that deserves further study.

Another feature of this field that differs from the Dundee observed in the Rodney field (Ontario) is the large number of persistent stylolites. The significance of these features as interpreted by recent research will be discussed later in this chapter.

The thin sections have not been stained to help differentiate between calcite and dolomite; however, where dolomite occurs, it exhibits the characteristic rhombic shape. These dolomite rhombs do not form a very appreciable part of the sediment; therefore, point counts of the mineral composition were not carried out. Similarly, most of the specimens involve fine-grained lime mud, and grainsize measurements provide relatively little useful data.

It might be assumed from these limitations that the reservoir is a featureless rock with very little character, but this is not the case.

Once again, the author is woefully short of supporting data such as electric logs and permeability plus porosity data; nevertheless, the reservoir is of interest because it differs significantly from the other carbonates, which will be covered later on in this book.

In recent years, the carbonate mud-mound-type deposit has received increased attention (Pratt 1982, Petta 1980). The North Wise reservoir has many of the features that characterize mud mounds, and several additional photomicrographs have been added to illustrate the various textural types as well as typical pore configurations (Figures 8.3–8.11).

Figure 8.3. Rogers City–Dundee Formation, 3716 ft (1133 m), Cities Service North Wise Unit No. 4, North Wise field, Isabella County, Michigan (0.44 in. in the photo equals 0.17 mm). Bioclasts are shown floating in lime mud. The large circular crinoid columnal in the lower left does not appear to be abraided. Fragments of conodonts, sponge spicules, and brachiopods are also present. This sample appears to represent the disturbed sediment texture produced by a large storm and could be classified as a biomicrite (Folk).

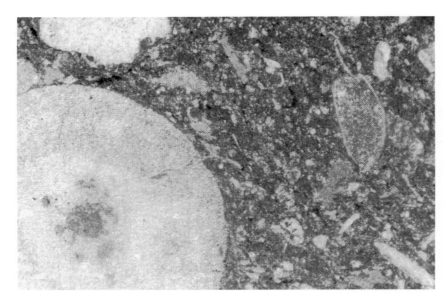

Figure 8.4. Rogers City–Dundee Formation, 3722 ft (1135 m), Cities Service North Wise Unit No. 4, North Wise field, Isabella County, Michigan (0.44 in. in the photo equals 0.17 mm). Micrite with spar-filled bioclastic fragments and walled calcispheres (of possible green algal origin). Moldic porosity occurs that is frequently not interconnected.

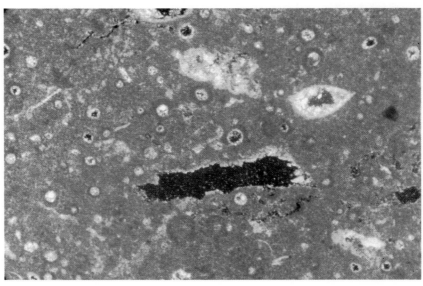

Figure 8.5. Rogers City–Dundee Formation, 3725 ft (1136 m), Cities Service North Wise Unit No. 4, North Wise field, Isabella County, Michigan (0.44 in. in the photo equals 0.17 mm). The original intergranular porosity formed by pellets and walled calcispheres provides excellent porosity. Note the small dolomite rhomb growing into the pore space (right center). Some cementation took place at an early stage so that the open framework was allowed to persist through time.

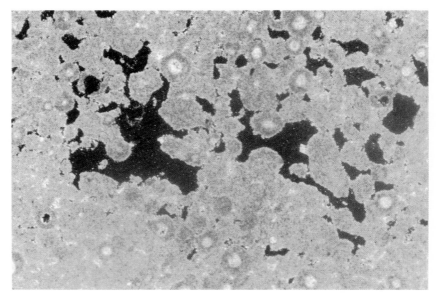

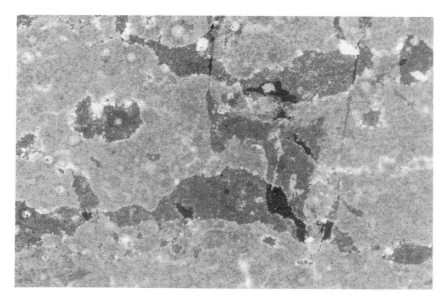

Figure 8.6. Rogers City–Dundee Formation, 3727 ft (1136.3 m), Cities Service North Wise Unit No. 4, North Wise field, Isabella County, Michigan (0.44 in. in the photo equals 0.17 mm). The lime mudstone contains some vermiform microstructure along with what appear to be parts of anastomosing thrombolites (dark colored). Some walled calcispheres filled with sparry calcite are also observable.

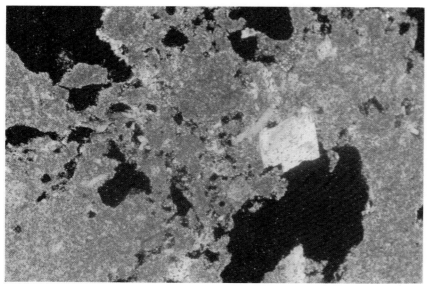

Figure 8.7. Rogers City–Dundee Formation, 3732 ft (1138 m), Cities Service North Wise Unit No. 4, North Wise field, Isabella County, Michigan (0.44 in. in the photo equals 0.17 mm). Again excellent porosity appears to have been produced by moving fluids that dissolved some of the micrite matrix as well as the large dolomite crystal (center right).

Figure 8.8. Rogers City–Dundee Formation, 3734 ft (1138.4 m), Cities Service North Wise Unit No. 4, North Wise Field, Isabella County, Michigan (0.44 in. in the photo equals 0.17 mm). Stylolites such as those pictured above are very common in the Rogers City–Dundee Formation as it occurs in the Wise field and they deserve special attention. As Nelson (1983, p. 315) states: "Stylolitization is a major rock deformation mechanism initiated in response to both overburden and tectonic stress." In this case, the major stresses are associated with the adjustments that occur within the mud mound. Typically the dark organic matter is concentrated along the stylolitic boundary.

Figure 8.9. Rogers City–Dundee Formation, 3738 ft (1140 m), Cities Service North Wise Unit No. 4, North Wise field, Isabella County, Michigan (0.44 in. in the photo equals 0.17 mm). Mudstone shows evidence of having been deposited with good intergranular porosity produced in part by pellets and walled calcispheres. The mud also appears to have been penetrated by numbers of worm burrows that have been filled with sparry cement. All of the pores show evidence of some sparry cement in filling.

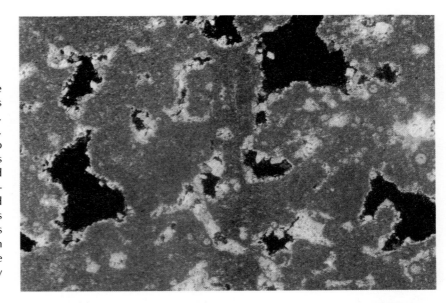

Figure 8.10. Rogers City–Dundee Formation, 3748 ft (1143 m), Cities Service North Wise Unit No. 4, North Wise field, Isabella County, Michigan (0.44 in. in the photo equals 0.17 mm). A dense pelmicrite shows evidence of circulating fluids that have almost completely filled some channels with sparry carbonate and left others open.

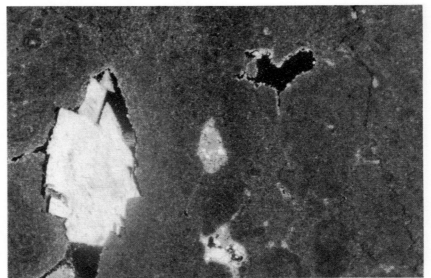

Figure 8.11. Rogers City–Dundee Formation, 3753 ft (1144 m), Cities Service North Wise Unit No. 4, North Wise field, Isabella County, Michigan (0.44 in. in the photo equals 0.17 mm). Large sparry crystals, some with dolomite rhomb outlines, have completely filled a large channel. The micrite framework shows both calcispheres and pellets along with a few open pores.

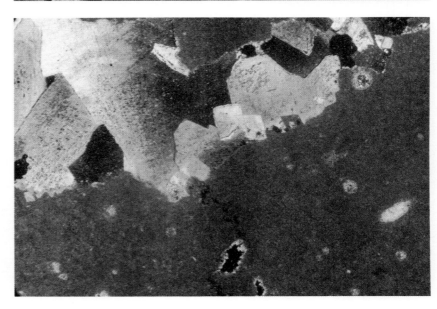

Pratt proposes "that the bioclastic lime mudstone of mud-mounds comprises a reef framework of unlaminated stromatolites (thrombolites) that arose from organic (probably blue-green algal) binding and submarine cementation of locally generated sediment deposited from suspension. Framework and cavity morphology was controlled by algal-mat distribution, sedimentation rate, and subsequent winnowing of unbound sediment" (1982, p. 1203).

The environment where mud mounds were formed appears to be in waters that are relatively deep compared to the shallow-water reef that occurs within the fair-weather wave base. This environmental setting fits the mid-basin location of the North Wise field. As pointed out by Pratt (1982, p. 1204), mud mounds can form in water depths that sometimes exceed 100 m. He also indicates that giant storms can stir up the fine-grained bottom sediment and introduce bioclastic debris. In the photomicrographs in Figures 8.3–8.11, an attempt will be made to cover the major sediment varieties encountered at the North Wise field.

Geologic Interpretation

The history of the North Wise field reservoir appears to involve the formation of a mud mound structure that was cemented early enough to preserve some of the intergranular porosity from the effects of compaction. At a later date, spar was deposited in some of the major channels completely filling them; however, other porous areas sustained only minor crystal growth along the surfaces of the pores.

Stylolitization appears to have occurred sometime after the spar was deposited in that spar crystals are found at the stylolite interface (Figure 8.8). Gillett (1983, p. 217) suggests that a minimum overburden thickness is required for stylolitization to take place. In the cases mentioned by Gillett, a thickness of 750 m is suggested as the point at which compaction is complete. Whether or not this holds true in the Wise field is not clear; however, the present overburden is sufficient for stylolites to have formed. If stylolites are concentrated in areas of stress as postulated by Nelson (1983), then the compaction produced over the mud mound would appear to be a favorable condition.

The generalized picture of a hypothetical reconstructed mud mound as proposed by Pratt (1982, p. 1223) is shown in Figure 8.12. From the explora-

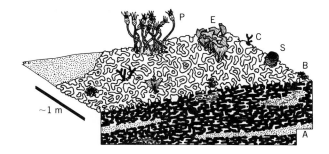

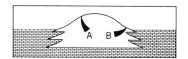

Figure 8.12. Reconstruction of hypothetical mud-mound showing living surfaces and internal framework. Shaded area indicates living algal mats; black refers to algal-bound sediment; dots stand for bioclastic grainstone and blank areas are unbound sediment and cavities. Diagram A shows reticulate framework from the mud-mound core along with a patchy algal mat, discontinuous grainstone layers, unbound sediment, cavities and fossil "nests." Diagram B shows the laminar framework on a mud-mound flank with sheetlike unbound sediment, cavities and grainstone layers. The algal mat is almost a continuous cover. (After Pratt, 1982, p. 1223; Reprinted with permission of Journal of Sedimentary Petrology, v. 52, N.4, p. 1203–1227.)

tionists point of view, there may be little distinction between a mud mound and a reef in overall outline; however, the fact that porous and productive mud mounds can occur in relatively deep water toward the center of a basin such as the Michigan basin should be noted. Other basins noted for production along the margins may not have been thoroughly tested for mud mound reservoirs.

References

Gillett, S. L., 1983, Major Through-Going Stylolites in the Lower Ordovician Goodwin Limestone, Southern Nevada: Petrography with Dating from Paleomagnetism, *Journal of Sedimentary Petrology,* Vol. 53, No. 1, pp. 209–219.

Landes, K. K., 1970, *Petroleum Geology of the United States,* Wiley-Interscience, New York, 571 pp.

Nelson, R. A., 1983, Localization of Aggregate Stylolites by Rock Properties, *American Association of Petroleum Geologists Bulletin,* Vol. 67, No. 2, pp. 313–319.

Petta, T. J., 1980, Silurian Pinnacle Reef Diagenesis—Northern Michigan: Effects of Evaporites on Pore Space Distribution: "Notes" for SEPM Core Workshop No. 1, *Carbonate Reservoir Rocks,* R. B. Halley and R. G. Loucks, Eds., American Association of Petroleum Geologists, Tulsa, Oklahoma, pp. 32–42.

Pratt, B. R., 1982, Stromatolitic Framework of Carbonate Mud-Mounds, *Journal of Sedimentary Petrology,* Vol. 52, No. 4, pp. 1203–1227.

9

Mid-Continent, El Dorado Oil Field, Admire Formation, Butler County, Kansas

The mid-continent part of the United States has been a classic area for oil production since the beginnings of the oil industry. It is in this area that geology became an extremely important part of oil exploration, and it is in this same area that many of the classic examples of oil field structures and sedimentary traps were found.

The province consists of a stable platform with basins located between granite ridges or other structural highs. If the west Texas area is included, as Perrodon (1983) has done (Figure 9.1), the large downwarps that characterize the Delaware and Midland basins are part of the mid-continent platform. The basins of the province are filled with Paleozoic sediments with the major emphasis on Carboniferous and Permian sequences. These sediments reveal a number of unconformities that aid in oil accumulation both in terms of source rocks and trap formation. The regional transgression of the Pennsylvania Cherokee sediments provides one of the best examples of this type of sedimentary change.

In terms of petroleum exploration, the structural highs associated with the basement uplifts have been particularly productive. The fields that will be discussed in this section in large measure are associated with such features, and at least in one case, a sample of the fractured basement granite is provided.

For the purposes of this study, the basins in Kansas, Oklahoma, and the

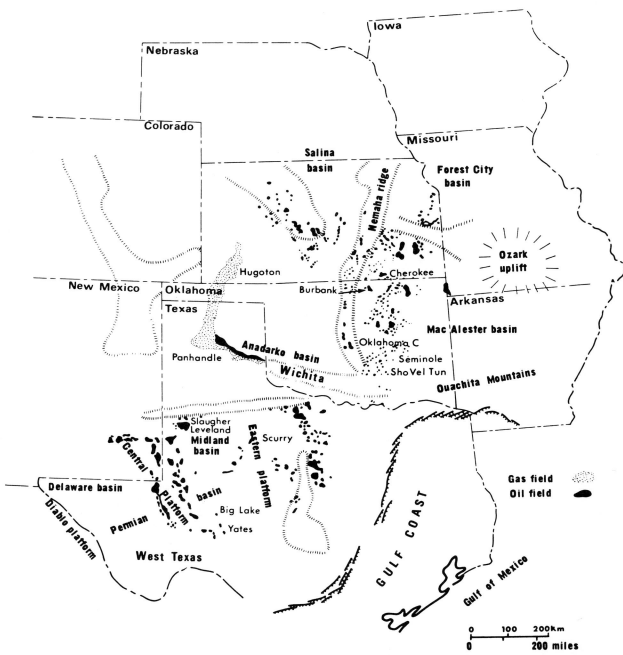

Figure 9.1. Map of the major mid-continent petroleum and natural gas provinces. (After Perrodon, 1983, p. 281.)

Texas panhandle are treated separately from the Delaware and Midland basins of west Texas. It should be kept in mind, however, that they are all part of the interior platform and grade into one another.

The El Dorado field is one of the classic cases of a field containing several producing zones of different ages stacked one on top of another. The first reservoir to be discussed is the relatively shallow zone formed by the Admire Sand of Permian age. Other reservoirs from the same field, which will also be described, include the Viola Limestone and the Simpson Sand of Ordovician age.

Geologic Background

The size and shape of the El Dorado production is summarized briefly by Landes (1970, p. 113) as follows:

The El Dorado field (Fath 1921, Reeves 1929) of Butler County, which was discovered in 1914 and flooded the domestic market in 1916, has been an important source of oil in Kansas ever since, standing first in cumulative production (over 270 million barrels) and returning briefly in 1962 to first place in annual production through a resurgence in output brought about by the secondary recovery.

Within the last few years the El Dorado "shallow" or the Admire Sand has been the target for a tertiary recovery experiment. This latter sand will be the first reservoir discussed; however, in order to become properly oriented to the mid-continent, it is necessary to examine the geologic setting.

Geologic Background

The key feature responsible for the oil accumulation at El Dorado is the Nemaha uplift (or anticline), which extends from above Nemaha County in northeastern Kansas to the Oklahoma City field in Central Oklahoma (Figure 9.2). The Nemaha ridge has a granite core and includes a narrow zone 10–25 miles wide (15–40 km) and at least 170 miles (270 km) long (see Figure 9.2). In addition to El Dorado and Oklahoma City, other noteworthy fields attributable to this extended structure include Garber, Augusta, Tonkawa, and West Edmond. Roughly half of the oil is produced from Pennsylvanian and younger reservoirs and half has been produced from Ordovician horizons including the Arbuckle Dolomite, St. Peter Sandstone, and Viola Limestone. The reservoirs from El Dorado described in this study are the Admire Sandstone (Permian), the St. Peter or Simpson Sandstone, and the Viola Limestone. The relative stratigraphic positions of these producing zones are shown in Figure 9.3. The Admire sand is at the base of the Permian and occurs in the El Dorado field at depths ranging from 600 to 700 ft. From the point of view of the petroleum engineer, this zone has been a frustrating problem. The oil saturation data for the sand zones suggests that secondary and tertiary recovery projects might be attractive. Indeed, in recent years a special exploratory project involving improved oil recovery by micellar–polymer flooding has been in progress. More will be said about this project at the end of this section.

A brief look at the photographs of the 3-in. cores taken from the Admire interval at El Dorado reveals the reasons this reservoir is a challenge (see Figures 9.4 and 9.5). The rapid alternation of sand and shale is such that even on the scale of the thin section, several sand and shale units can be observed. Even where the sand is somewhat thicker (Figure 9.4, top of the core labeled 615, 187.5 m), the bedding is highly disturbed, suggesting that fluid flow would be impeded.

Although containing more sand, the core from the Hegberg lease (Figure 9.5) contains highly disturbed

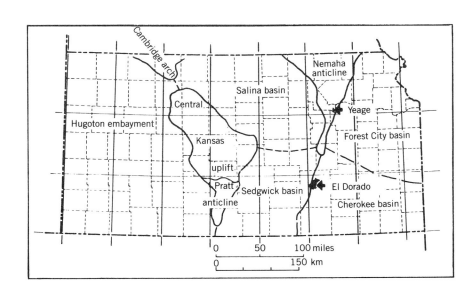

Figure 9.2. Tectonic map of Kansas showing the locations of the El Dorado and Yeage fields in relation to the Nemaha anticline. (Modified after Landes 1970, p. 109, and Hilpman, 1958.)

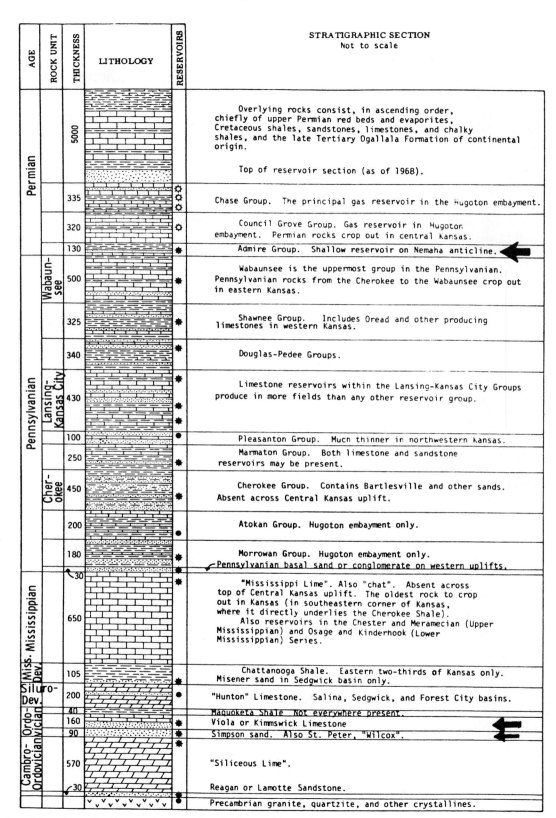

Figure 9.3. Stratigraphic column for Kansas showing the relative positions of the Admire, Viola, and Simpson formations. (After Landes 1970, p. 110. Moore et al. 1951, and Zeller 1968.)

Geologic Interpretation

Figure 9.4. Core of Admire Formation from El Dorado field, core diameter, 3 in.

Figure 9.5. Core of Admire Formation from El Dorado field, core diameter, 3 in.

zones (641 ft, 195.4 m) and thin coal streaks (642.3 ft, 195.8 m). Again, the rapid alternating layers of silt, sand, and shale are present (656.5 ft, 200.2 m), making the problem of fluid communication between wells a difficult one. Even obtaining representative core plugs becomes a troublesome sampling problem.

Thin-section analyses tend to emphasize the fact that this is a highly variable environment that maintains this variability down to the finer details in each layer (Figures 9.6–9.11).

Geologic Interpretation

The major clues to the environment of deposition of the Admire sands can be obtained directly from the photographs of the cores (Figures 9.4 and 9.5). Figure 9.4, in particular, shows the rapidly alternating layers of sand and silty shale that gives portions of the core a zebralike appearance. These characteristics have been described for tidal flats from the area north of Wilhelmshaven and Jade Bay, Germany. Hantzschel (1939) and, more recently, Weimer et al. (1982, p. 198) pointed out that this type of sediment texture (Figure 9.12) occurs in tidal flats where gullies migrate across them. The sand-filled gullies are represented in part of the core pictured in Figure 9.5. Some disturbed salt marsh texture and a thin coal are preserved at the 641–642-ft levels.

The fact that the Nemaha anticline has been active since Ordovician time suggests that the highest areas such as El Dorado probably continued to be elevated somewhat more than the surrounding areas in Late Pennsylvanian and Early Permian time. The El Dorado structure therefore appears to have been a low-relief island or peninsula surrounded by tidal flats that were crossed by sand-filled gullies.

Figure 9.6. Admire Formation, 677 ft (206 m), El Dorado field, Butler County, Kansas (0.44 in. in the photo equals 0.17 mm). Thin, highly organic layers interspersed with silt and fine-grained sand as viewed with uncrossed polarizers in transmitted light. The brownish colored mineral grain in the lower center is a tourmaline grain. Black streaks are organic matter.

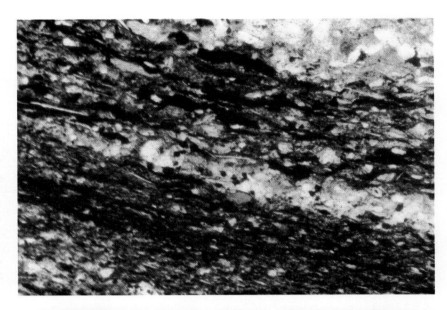

Figure 9.7. Admire Formation, 677 ft (206 m), El Dorado field, Butler County, Kansas (0.44 in. in the photo equals 0.08 mm). Thin clay layer at higher magnification with some organic matter sandwiched between two layers of fine-grained quartz sand. Sand in the lower half shows flakes of both muscovite (lower right) and biotite (orange-brown flake, lower left).

Figure 9.8. Admire Formation, 731 ft (223 m), El Dorado field, Butler County, Kansas (0.44 in. in the photo equals 0.08 mm). High organic layer (black center) is squeezed between clay-filled fine-grained sand layers. Mica flake (lower center) lies along the bedding plane.

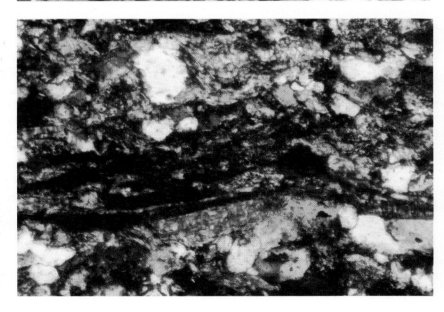

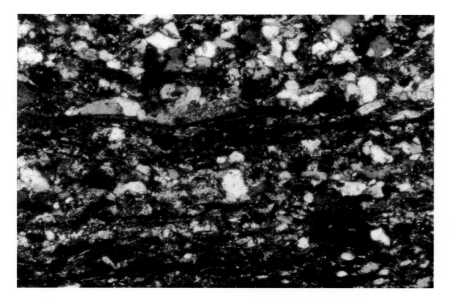

Figure 9.9. Admire Formation, 731 ft (223 m), El Dorado field, Butler County, Kansas (0.44 in. in the photo equals 0.17 mm). General view of angular quartz sand layers alternating with organic-rich microlayers. Mica flakes are mixed in with the clay mineral matter. In several places where the quartz grains become packed together, they are cemented with quartz cement.

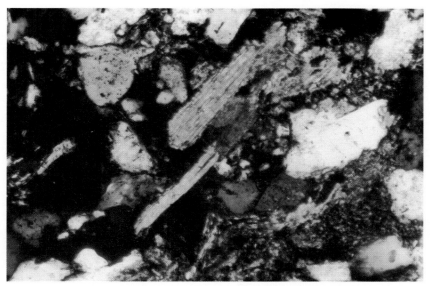

Figure 9.10. Admire Formation, 731 ft (223 m), El Dorado field, Butler County, Kansas (0.44 in. in the photo equals 0.08 mm). Higher magnification view showing chert, mica flakes, clay mineral matter, and cemented quartz grains.

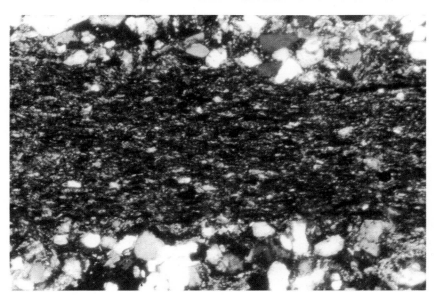

Figure 9.11. Admire Formation, 731 ft (223 m), El Dorado field, Butler County, Kansas (0.44 in. in the photo equals 0.17 mm). Lower magnification view of a thin shale layer between two fine-grained sand layers. This is a typical fabric of much of the Admire Formation as it occurs at El Dorado.

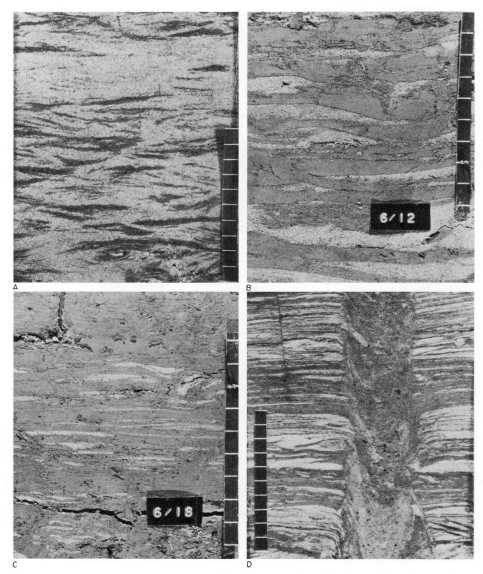

Figure 9.12. Sedimentary structures from mixed tidal flats, Jade Bay, Germany. Light layers are sand, dark layers are mud. (a) Flaser bedding, mud lenses formed in ripple troughs. (b) Wavy bedding. (c) Lenticular bedding. (d) Interbedded sand and mud similar to Admire Sand Cores. (After Weimer et al. 1981, p. 198.)

Origin of Admire Oil

When attempting to find a probable source for oils found in multipay structures such as that at El Dorado, it is tempting to think that there has been vertical communication between reservoirs. In the case of the Admire Sand, the reasons this may not seem to be the case are the following:

1. Five gas sands are reported to have been present between the Admire sands and the next oil-producing formation down the hole (Fath 1921). The timing of the gas accumulation may make this reason invalid.

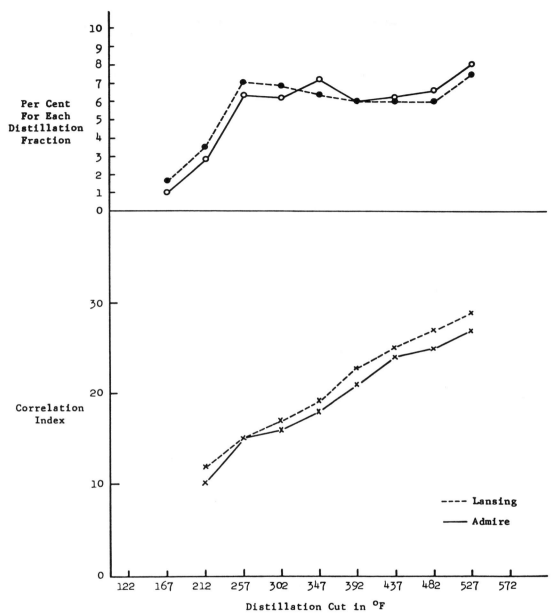

Figure 9.13. Plot of distillation data for Admire versus Lansing crude oils. (From Lane and Garton 1943.)

2. There appear to be some small differences in the distillation characteristics of the Admire oil and the next closest oil from the Lansing (Figure 9.13). As we shall see in a later chapter, these differences are not significant in terms of oils that occur in the same formation in other fields in the same county.

Tertiary Oil Recovery Project

The Admire Sand reservoir at the El Dorado field was selected as a demonstration project by the Department of Energy and the Cities Service Company in order "to determine the economic feasibility of im-

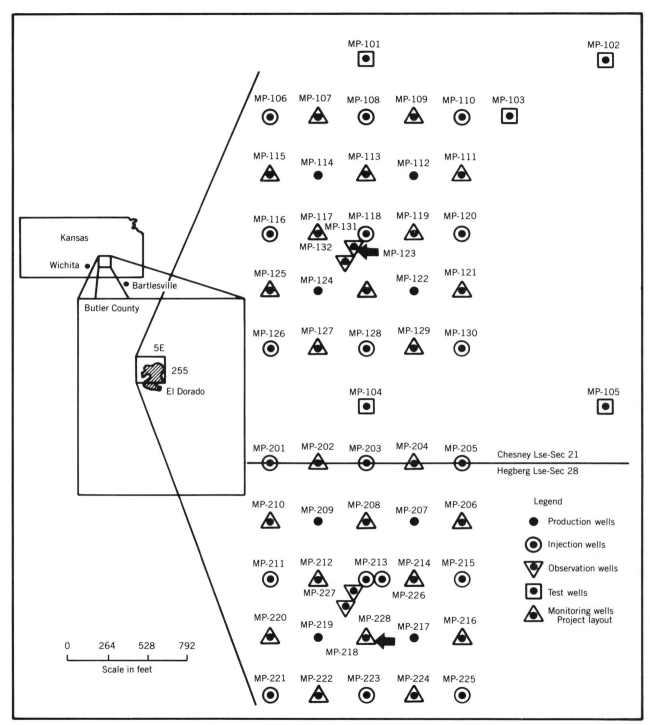

Figure 9.14. Locations of observation wells MP-131 (north lease) and MP-228 (south lease). (After Van Horn 1981, p. 308.)

proved oil recovery using two micellar–polymer processes and to determine the associated benefits and problems of each process" (Van Horn 1981, p. xiv). In the summary of the latest progress report, it is stated that "the El Dorado Demonstration Project is designed to allow a side-by-side comparison of two distinct micellar–polymer processes in the same field so that the reservoir conditions for the two floods are as similar as possible" (Van Horn 1981, p. xiv).

The core materials shown in Figures 9.4 and 9.5 suggest that the objectives are likely to be difficult to obtain; however, the information provided is of interest.

Review of Technology

Enhanced oil recovery using biological agents has been a subject of research for some time; however, no commercial applications have been successful. Problems with this approach include the high cost of the specialized surfactants and polymers and the difficulties of controlling the growth of organisms down in the formation. This latter problem results in the plugging up of the available porosity.

Present practice involves first flooding the oil reservoir sand with a brine solution of petroleum sulfonate surfactant. Next, the surfactant is followed by a slug of water thickened with xanthan gum or polyacrylamide. The primary purpose of the surfactant is to lower the interfacial tension of the oil so that it can be flooded out of the pores without being emulsified.

The project for the so-called north pattern or the Chesney lease (see Figure 9.14) involved the use of xanthan gum as the thickening agent, whereas the Hegberg lease (south) employed the polyacrylamide. In the words of Van Horn (1981), "Severe biodegradation of the north pattern" occurred; however, it was "controlled" by the addition of alcohol to the injection fluids. Another approach for the northern pattern that is being implemented at this time is the use of polyacrylamide following the xanthan gum. According to Van Horn (1981, p. xiv), this design is supposed to provide better mobility control.

Figure 9.15 provides a chart of the oil cut volume

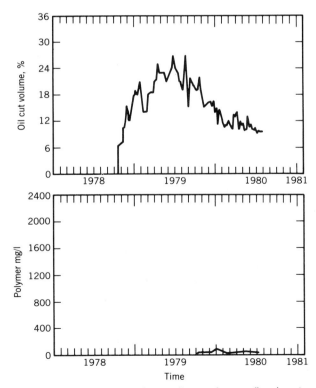

Figure 9.15. Results of micellar–polymer flood using Xanthan gum, Admire Sand, El Dorado field, Butler County, Kansas. (After Van Horn 1981, p. 195.)

Figure 9.16. Results of micellar–polymer flood using polyacrylamide, Admire Sand, El Dorado field, Butler County, Kansas. (After Van Horn 1981, p. 215.)

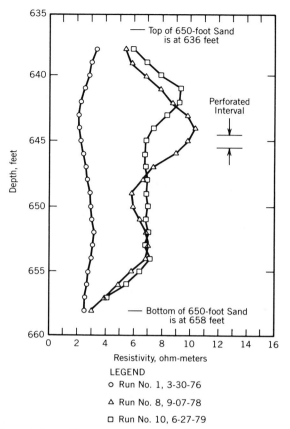

Figure 9.17. Changes in resistivity of Admire Sand showing progress of the flood with time at observation well MP-131 (north lease), El Dorado field, Butler County, Kansas. (After Van Horn 1981, p. 319.)

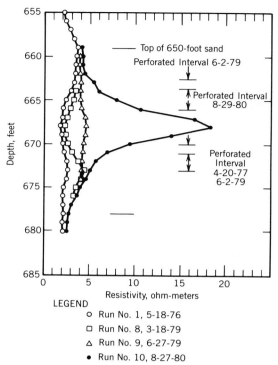

Figure 9.18. Changes in resistivity of Admire Sand showing progress of the flood with time at observation well MP-228 (south lease), El Dorado field, Butler County, Kansas. (After Van Horn 1981, p. 322.)

in percent versus time for producing well MP-131 in the northern pattern (see Figure 9.14 for location). Figure 9.16 provides a similar oil cut chart for producing well MP-228 in the southern (Hegberg) lease. A review of changes in electric log resistivity values for wells MP-131 (Figure 9.17) and MP-228 (Figure 9.18) shows the gradual increase in oil saturation (increased resistivity) with time. At the time of the 1981 report, the project was still continuing and the overall summary and evaluation had not been made.

Although the wells chosen for the study contained more sand than many of the wells at El Dorado, it is clear that the target was probably not the easiest one that could have been selected for a demonstration of tertiary recovery methods.

References

Fath, A. E., 1921, Geology of the El Dorado Oil and Gas Field, Butler County, Kansas, *Kansas State Geological Survey,* Bulletin 7, 187 pp.

Hantzschel, W., 1939, Tidal Flat Deposits (Wattenschlick), in *Recent Marine Sediments,* a Symposium, American Association of Petroleum Geologists, pp. 195–206.

Hilpman, P. L., 1958, Producing Zones of Kansas Oil and Gas Fields, *Kansas Geological Survey,* Oil and Gas Investigations No. 16.

Landes, K. K., 1970, *Petroleum Geology of the United States,* Wiley-Interscience, New York, 571 pp.

Lane, E. C. and E. L. Garton, 1943, *Analyses of Crude Oil from Some Fields in Kansas,* Report of Investigations 3688, U.S. Bureau of Mines, 95 pp.

References

Moore, R. C., J. C. Frye, J. M. Jewett, L. Wallace, and H. G. O'Connor, 1951, The Kansas Rock Column, *Kansas Geological Survey,* Bulletin 89, 132 pp.

Perrodon, A., 1983, *Dynamics of Oil and Gas Accumulations,* Elf Aquitaine, Memoir 5, Pau, France, 368 pp.

Reeves, J. R., 1929, El Dorado Oil Field, Butler County, Kansas, in *Structure of Typical American Oil Fields,* Vol. II, American Association of Petroleum Geologists, pp. 160–167.

Van Horn, L. E., 1981, *El Dorado Micellar–Polymer Demonstration Project, Sixth Annual Report, September 1979–August 1980,* U.S. Department of Energy, Report No. DOE/ET/13070-63, 375 pp.

Weimer, R. J., J. D. Howard, and D. R. Lindsay, 1982, Tidal Flats and Associated Tidal Channels In Sandstone Depositional Environments, *American Association of Petroleum Geologists,* Memoir 31, pp. 191–245.

Zeller, D. E., 1968, The Stratigraphic Succession in Kansas, *Kansas Geological Survey,* Bulletin 189, 81 pp.

10

El Dorado Oil Field, Viola Formation, Butler County, Kansas

The Viola Formation forms part of the major producing zone at El Dorado, which is composed of the eroded edges of the Viola, Simpson, and Arbuckle formations. These three porous zones of Cambro-Ordovician age are overlain by the Pennsylvanian Cherokee shales, as illustrated in Figures 10.1 and 10.2. This latter formation has been thought by many to be the source rock for the main productive zone. The Viola reservoir is interesting because of the variation in rock characteristics, which in turn cause variations in the permeability and porosity. Another interesting feature is the structure and the effects of structural movements in both enhancing and degrading the reservoir properties. The structure of the El Dorado anticline on top of the Ordovician is shown in Figure 10.3.

General Geologic Setting

The Viola at El Dorado ranges in thickness from 15 to 50 ft on the north flank of the structure and has been removed by erosion over a large part of the field. Where the section is complete, the topmost 2 ft contain chert with floating dolomite rhombs. Beneath this chert layer, there is about 4 ft of medium-grained dolomite exhibiting relatively little residual structure. From this point

General Geologic Setting

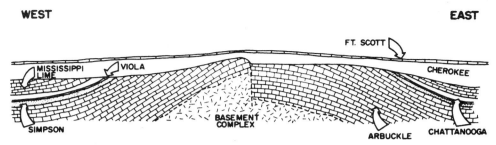

Figure 10.1. Ideal west–east cross section of El Dorado anticline from Sec. 7.T.26S, R.4E, to Sec. 12.T.26.R.5E. Length, 12 miles; height, 1200 ft. (Modified after Reeves 1929, from Biederman 1966, p. 43.)

down to the top of the Simpson Sandstone, there are a number of textures that deserve discussion.

Vuggy dolomite with good permeability occurs for about 10 ft and shows evidence of the leaching and subsequent redeposition of sparry calcite. In some instances, the sparry calcite itself has been dissolved. Approximately 12 ft below the top of the Viola, rounded quartz grains occur that appear to be floating in the dolomite. These reappear in several zones down to the top of the Simpson Formation.

The lower half of the Viola was originally composed of fossil debris. In most instances, dolomitization has destroyed the structure; however, some of the fossils were silicified and remain identifiable. Phosphate pellets are also present that contain fossil remains.

Detailed study of the larger pores and vugs reveals that the character of the fluids in the pores changed several times in response to changes in both sea level and elevation (Biederman 1966, p. 43). The Nemaha anticline has been an active positive element since the Ordovician period (Merriam 1963); therefore, the numerous changes in the pore fluid chemistry and the natural fracturing present are not unexpected features. (See cores in Figures 10.4 and 10.5.) The degree of fracturing revealed in these 3-in. cores shows the reason both water and oil were allowed to flow and accumulate so readily and also why there is communication between the Simpson and Viola reservoirs. This latter situation was revealed during water-flooding operations.

The distributions for porosity and permeability values for one of the major producing leases at El Dorado is shown in Figure 10.6. From these it becomes clear that the probability of dealing with porosity values of less than 3% is less than 5 in 100 and the odds of porosity being greater than 27% are less than 5 in 100.

The permeability distribution presents an example of a Poisson distribution where the most frequent values range from 0 to 10 millidarcies and the higher values are spread over a wide range. The fact that there are numerous fractures in the reservoir (Figures 10.4 and 10.5) suggests that obtaining an accurate picture from routine core analysis is difficult at best.

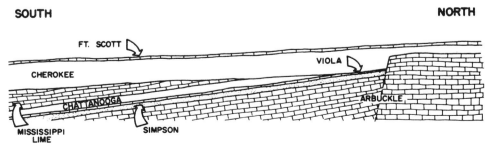

Figure 10.2. Ideal south–north section of southern part of El Dorado anticline from Sec. 31 to Sec. 8.T.26S.R.5E. Length, 5 miles; height, 500 ft. (Modified after Reeves 1929, after Biederman 1966, p. 43.)

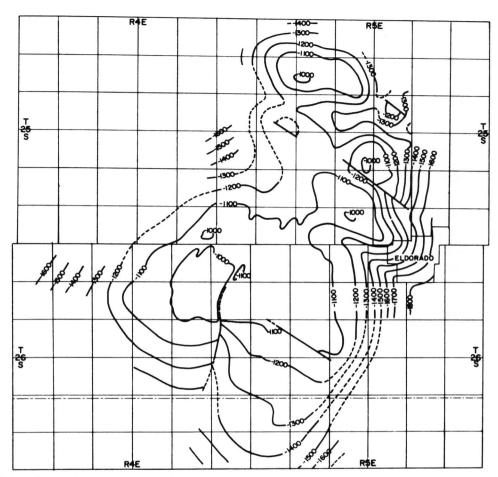

Figure 10.3. Structure of El Dorado anticline. Datum, top of Ordovician; contour interval, 50 ft; width of mapped area, 12 miles. (After Biederman 1966, p. 44.)

Capillary Pressure Curves

Information obtainable from capillary pressure curves reflects the uniformity of pore size and the types of pores present. Examination of Figures 10.7–10.10 shows that the Viola is capable of a wide range of responses. The curve from Figure 10.7 for 2523 ft suggests that the pore size is distributed over a wide range. At 6 lb/in.^{2} the water saturation as a percentage of pore volume is 40%. The fact that this reduces to 15% at 14.5 lb seems to reflect the response from the smaller capillaries. Such a curve could be interpreted as a favorable indicator for successful water flooding.

The curves from Figures 10.8 and 10.9, on the other hand, show an irreducible water saturation of around 40%, which is definitely unfavorable.

Figure 10.10 provides an almost straight-line relationship, which suggests either that a few large pores are involved or a fracture is present because the response is almost directly related to pressure.

From these curves taken from samples that were within 3.5 ft of each other, it becomes clear that a large number of samples are required in order to obtain a reasonable picture of this type of reservoir.

The photomicrographs in Figures 10.11–10.15 provide some idea of why the response is so variable.

Geologic Interpretation

Figure 10.4. Core of Viola Formation, El Dorado field, Cities Service Oil Co. No. 39 Knox, top half.

Figure 10.5. Core of Viola Formation, El Dorado field, C. S. Oil Co. No. 39 Knox, bottom half.

Geologic Interpretation

The southeast Nebraska arch extended into northeastern Kansas prior to Simpson deposition and appears to have been the precursor of the Nemaha anticline. It is probable that small pulsations along the Nemaha trend produced local highs during the Ordovician period. The fact that streaks of sand strongly resembling sand from the Simpson Formation occur in the Viola at El Dorado suggests that nearby shoals and low islands were eroded and the reworked sand redeposited in the Viola. Very fragmentary fossil evidence within the Viola indicates that crinoids and shells with coarse costae were deposited in a shallow agitated environment.

The postdepositional history of the Viola appears to be complex. The fact that the Chattanooga Shale (Late Devonian and/or Early Mississippian) was deposited on top of the Viola indicates that considerable erosion took place between the deposition of the two formations.

Merriam (1963) indicates that the Viola attains a maximum thickness of 310 ft in northeastern Kansas (Figure 10.16). The fact that a maximum thickness of 50 ft occurs at El Dorado along with a more common thickness of only 25 ft at El Dorado suggests that erosion and leaching were important in producing the vuggy textures. At El Dorado, additional uplift and reexposure appears to have produced several cycles of leaching and cementation as evidenced by zoned quartz overgrowths (Biederman 1966, p. 55). Calcite, dolomite, and chert cements were also deposited at various times. Each pulsation along the Nemaha ridge probably introduced new fractures that healed with a variety of cementing materials.

Although the Simpson Formation is discussed as a separate reservoir in the chapter that follows, it should be kept in mind that they appear to grade into each other with numerous grains of Simpson Sand floating in the lower part of the Viola carbonate. These two chapters should be considered as a unified whole.

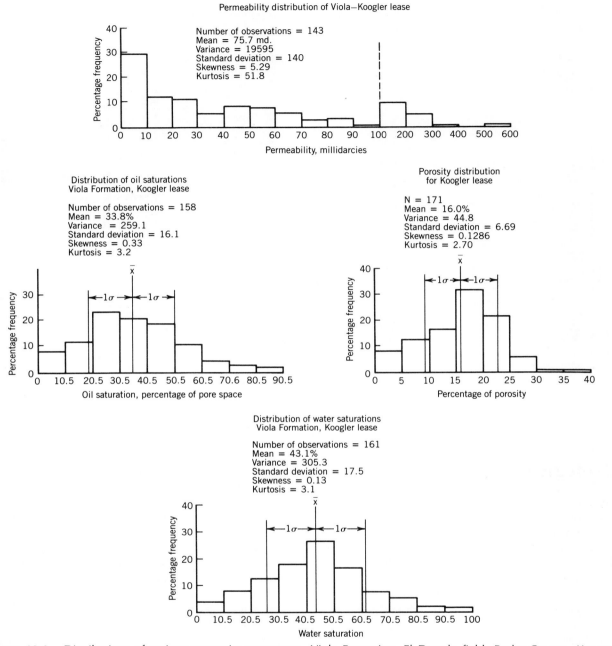

Figure 10.6. Distributions of various reservoirs parameters, Viola Formation, El Dorado field, Butler County, Kansas. (After Biederman 1966, pp. 48, 52.)

Geologic Interpretation

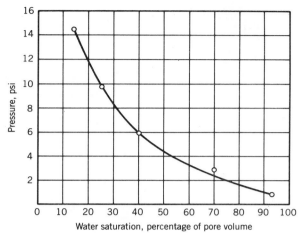

Figure 10.7. Capillary pressure curve, Viola Formation. C.S.O. D. L. Howe 2W, 2523 ft. Air permeability, 23.7 millidarcies; porosity, 17.7%. (After Biederman 1966, p. 52.)

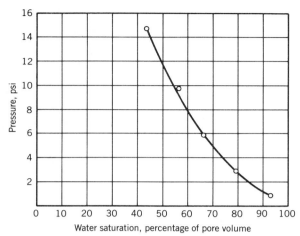

Figure 10.9. Capillary pressure curve, Viola Formation, C.S.O. D. L. Howe 2W, 2526 ft. Air permeability, 4.1 millidarcies; porosity, 6.1%. (After Biederman 1966, p. 54.)

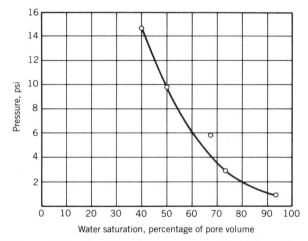

Figure 10.8. Capillary pressure curve, Viola Formation. C.S.O. D. L. Howe 2W, 2525 ft. Air permeability, 1.2 millidarcies; porosity, 9.3%. (After Biederman 1966, p. 53.)

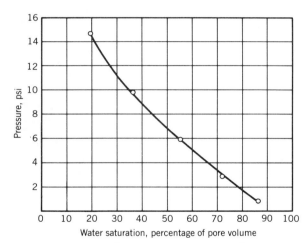

Figure 10.10. Capillary pressure curve, Viola Formation, C.S.O. D. L. Howe 2W, 2526 ft. Air permeability 4.3 millidarcies; porosity, 15%. (After Biederman 1966, p. 56.)

Figure 10.11. Viola Formation, 2501 ft (762.5 m), Cities Service Fulkerson 2W, El Dorado field, Butler County, Kansas (0.44 in. in the photo equals 0.04 mm). View with uncrossed nicols of a medium-grained dolomite showing zoned rhombs with darker interiors and clearer rims. The sample has been impregnated with green plastic to show the pore space and stained red for calcite. Note the angular quartz grain near the center.

Figure 10.12. Viola Formation, 2503 ft (763.1 m), Cities Service Fulkerson 2W, El Dorado Field, Butler County, Kansas (0.44 in. in the photo equals 0.04 mm). Dolomitization has obliterated the organic structures, and much of the pore space has been filled in with larger rhombs; however, the overall reservoir properties are still good.

Figure 10.13. Viola Formation, 2523.8 ft (769 m), Cities Service D. L. Howe 2W, El Dorado field, Butler County, Kansas (0.44 in. in the photo equals 0.08 mm). Vugular dolomite (moldic porosity) where the original organic structure has been dolomitized.

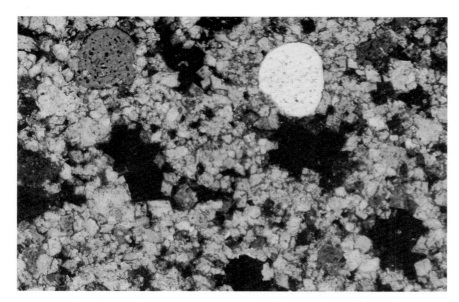

Figure 10.14. Viola Formation, 2526.5 ft (770.3 m), Cities Service D. L. Howe 2W, El Dorado field, Butler County, Kansas (0.44 in. in the photo equals 0.08 mm). Porous dolomite with highly rounded quartz grains (gray and white) suspended in the midst of the dolomite grains. The black areas are pores.

Figure 10.15. Viola Formation, 2531 ft (771.6 m), Cities Service D. L. Howe 2Q, El Dorado field, Butler County, Kansas (0.44 in. in the photo equals 0.08 mm). A porous zone with quartz sand and dolomite rhombs. Quartz overgrowth and dolomite are filling in the pore space and competing with each other. In other cases, dolomite rhombs are surrounded by quartz overgrowths.

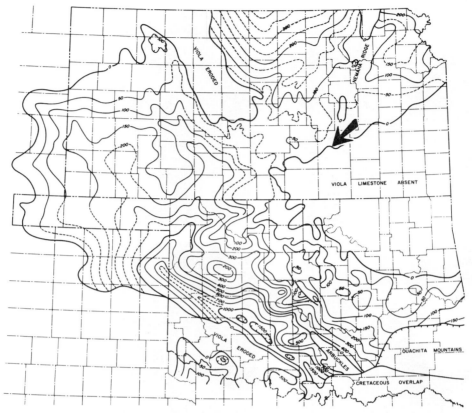

Figure 10.16. Isopach map of Viola, Ferndale, and equivalent rocks, Oklahoma and Kansas. Arrow shows approximate location of El Dorado field. (After Chenoweth 1966, p. 111.)

References

Biederman, E. W., Jr., 1966, The Petrology of the Viola Formation, El Dorado Field, Butler Co., Kansas, in Symposium on the Viola, Ferndale, and Sylvan, *Tulsa Geological Society Digest,* Vol. 34, pp. 41–59.

Chenoweth, P. A., 1966, Viola Oil and Gas Fields of the Mid-Continent, in Symposium on the Viola, Ferndale, and Sylvan, *Tulsa Geological Society Digest,* Vol. 34, pp. 110–118.

Merriam, D. F., 1963, The Geologic History of Kansas, *Kansas Geological Survey Bulletin,* 162, 317 pp.

Reeves, J. R., 1929, *El Dorado Oil Field, Butler County, Kansas,* in Structure of Typical American Oil Fields, *American Association of Petroleum Geologists Bulletin,* Vol. 2, pp. 160–167.

11

El Dorado Oil Field, Simpson Sand Formation, Butler County, Kansas

Geologic Background

Tho Ordovician age Simpson Sand (St. Peter) at El Dorado is remarkable for the degree of rounding of the quartz grains. This feature allows them to be recognized when they occur in other formations such as the Viola Limestone that overlies the Simpson. The cross sections in Figures 10.1 and 10.2 show that in some parts of the El Dorado dome, both the Simpson and Viola have been removed by erosion, whereas in other parts they are present over a considerable area.

The principal reservoir is the Stapleton zone, which consists of eroded truncated edges of Viola, Simpson, and Arbuckle rocks immediately beneath the Pennsylvanian unconformity. The logical question at this point is, do these three reservoir rocks have the same source for their oil. Figure 11.1 shows a plot of the distillation curves for the crude oils obtained from the Viola, Simpson, and Arbuckle reservoirs at El Dorado. It is tempting to conclude that the source rock for all three was the same, namely the overlying Pennsylvanian Cherokee Shale. Baker (1962), in his detailed work on the Cherokee, concluded that the Cherokee shales of Kansas were good source rocks for oil. Therefore, although there has been considerable disagreement concerning the origins of the Viola, Simpson, and Arbuckle oils (Pan 1982, p. 1602), the available data favor a Cherokee Shale source for the Stapleton zone at El Dorado.

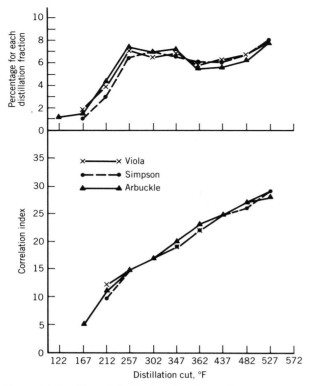

Figure 11.1. Plot of distillation data for Viola versus Simpson crude oils, El Dorado field. Data obtained from Bureau of Mines R.I. 3688. (After Biederman 1966, p. 59.)

The grainsize distributions for the long axes of quartz grains in Figure 11.2 show a relative coarseness that is not characteristic of the oil sands from the Appalachian basin. There is also the hint that there is a contribution from two sources because the sections from 2527 ft (770 m) and 2556 ft (779 m) both have some unusually large grains.

The capillary pressure curve for the Simpson as shown in Figure 11.3 is illustrative of the classical sandstone reservoir where both large and small capillaries contribute. The water saturation in a clean sand such as this is reduced to 48% at 4 lb/in^2.

It is appropriate to recall Fath's description of some of the large wells producing from the Stapleton pay zone, which includes both the Simpson and Viola formations. In his 1921 account, Fath states:

Contrary to the usual condition in similar large wells in other fields, the oil did not gush, in the popular sense of that word, i.e., it did not pour forth in a stream shooting high into the air. This absence of gushing is due, it seems, to a lack of dissolved gas, and in this 14,000 (?) barrel well the flow of oil did not run wild (out of control) but instead flowed a gentle, continuous stream that offered no difficulties whatever in controlling it.

Mineral Composition

The Simpson Sand as observed in its top portion is almost monomineralic. It is a clean quartz sand except for several carbonate-cemented streaks. In other wells penetrating the same reservoir, pyrite, phosphatic plates or nodules, and a few grains of glauconite have been observed. As we shall see, there is another zone in addition to these, and this zone will be discussed later in this section.

Several streaks of green shale are also encountered, and the major clay mineral in this shale is illite as shown by x-ray diffraction studies.

The photomicrographs in Figures 11.4–11.15 are representative of the various differences that occur within the Simpson Sand at El Dorado.

Oolitic Zone

One well that was cored over much of the Simpson section deserves particular mention because of the oolitic zone that was encountered. The Cities Service Entz No. 2 was cored beginning at the top of the Simpson (2487 ft, 758 m) through most of the Simpson Formation as it exists on the flank of the El Dorado structure. The photomicrographs in Figures 11.4–11.15 portray the oolitic zone that extends from 2502 ft (762.8 m) to 2509 ft (764.9 m) plus some of the core beneath this zone.

The fact that we are dealing with a hematite oolite helps in interpreting the environment of deposition; however, it is unfortunate that there are no modern analogies for comparison (Tucker 1981, p. 174). In the case that follows, the environment changes rather rapidly from oxidizing to reducing; therefore, some interesting changes in mineralogy occur and can readily be observed in Figures 11.4–11.15.

Geologic Interpretation

The foregoing pages indicate that the Simpson Sand at El Dorado exhibits a number of changes that ap-

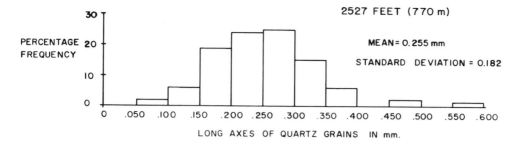
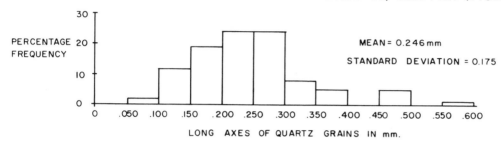
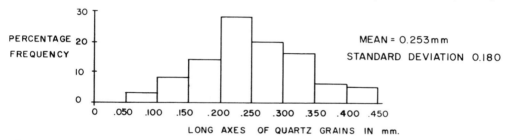

Figure 11.2. Grainsize distributions for long axes of quartz grains, Simpson Formation, Cities Service Fulkerson 2W, El Dorado field, Butler County, Kansas.

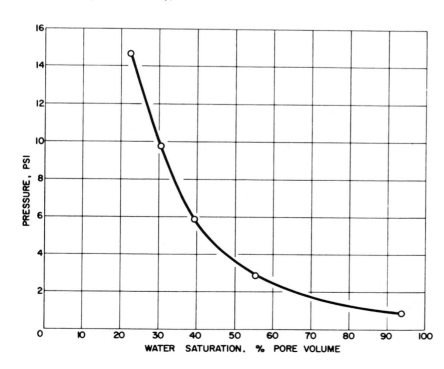

Figure 11.3. Capillary pressure curve, Simpson Sand, Cities Service D. L. Howe 2W, 2539 ft (774 m), El Dorado field, Butler County, Kansas. Air permeability, 330 millidarcies; porosity, 14.5%. (After Biederman 1966, p. 57.)

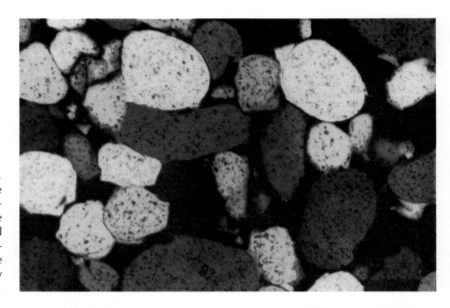

Figure 11.4. Simpson Formation, 2527 ft (770 M), Cities Service Fulkerson 2W, El Dorado field, Butler County, Kansas (0.44 in. in the photo equals 0.17 mm). Rounded quartz grains are cemented with numerous quartz overgrowths. The sand is very clean with very few grains of minerals other than quartz.

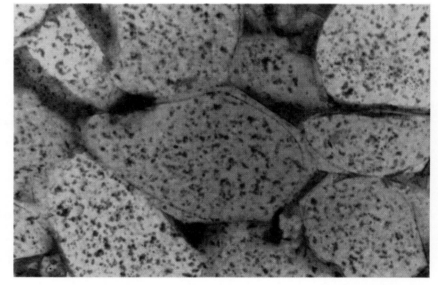

Figure 11.5. Simpson Formation, 2527 ft (770 m), Cities Service Fulkerson 2W, El Dorado field, Butler County, Kansas (0.44 in. in the photo equals 0.08 mm). Sample is photographed in transmitted light to show the pores that are impregnated with green plastic. Most of the quartz grains show quartz overgrowths that cement the rock together and provide straight surfaces for many of the pores.

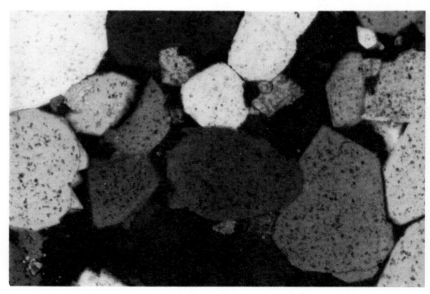

Figure 11.6. Simpson Formation, 2653 ft (809 m), Cities Service Houston No. 32, El Dorado field, Butler County, Kansas (0.44 in. in the photo equals 0.17 mm). A few dolomite grains are growing within the pore space. Numerous quartz overgrowths occur as pyramids jutting into the open pores (center right and far left).

101

Figure 11.7. Simpson Formation, 2653 ft (809 m), Cities Service Houston No. 32, El Dorado field, Butler County, Kansas (0.44 in. in the photo equals 0.17 mm). Small layer is cemented with dolomite. Larger rounded quartz grains "float" in the dolomite along with a population of smaller quartz grains, hinting that there may be two sources for the quartz. The high degree of rounding for most of the quartz grains is a typical feature of the Simpson.

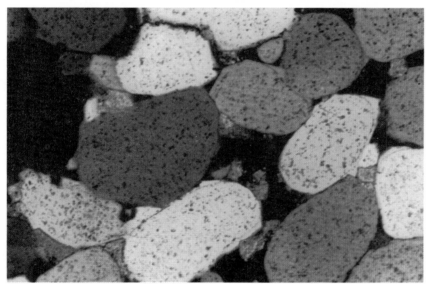

Figure 11.8. Simpson Formation, 2653 ft (809 m), Cities Service Houston No. 32, El Dorado field, Butler County, Kansas (0.44 in. in the photo equals 0.17 mm). Clear rounded quartz grains with numerous overgrowths hold the framework open. Some dolomite has filled in several of the pores; however, the overall communication between pore spaces is good.

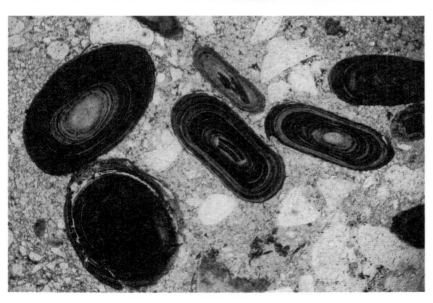

Figure 11.9. Simpson Formation, 2502 ft (762.8 m), Cities Service Entz No. 2, El Dorado field, Butler County, Kansas (0.44 in. in the photo equals 0.17 mm). Hematite oolites floating along with quartz grains in a fine grain matrix. Usually the presence of such oolites indicates oxidizing conditions at the time of deposition.

Figure 11.10. Simpson Formation, 2504 ft (763.4 m), Cities Service Entz No. 2, El Dorado field, Butler County, Kansas (0.44 in. in the photo equals 0.17 mm). Chamosite (iron silicate) ooids and quartz grains. In some cases, the quartz grains have pushed into the sides of the ooids, whereas in others the ooids have been greatly flattened, suggesting that the ooids were originally soft. The mineral chamosite is greenish in color and has a low birefringence. Its composition is variable, but it is related to chlorite and kaolinite (Tucker 1981, p. 278).

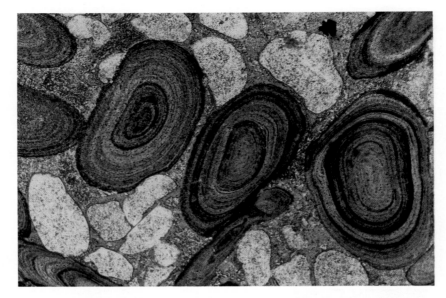

Figure 11.11. Simpson Formation, 2505 ft (763.7 m), Cities Service Entz No. 2, El Dorado field, Butler County, Kansas (0.44 in. in the photo equals 0.17 mm). Chamosite ooids surrounded by a zone of pyrite (black). In some cases, the pyrite crystallization has invaded the chamosite ooids and seems to have followed high-iron zones.

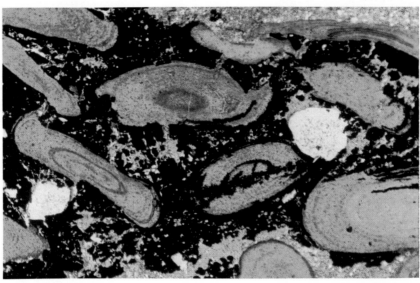

Figure 11.12. Simpson Formation, 2508 ft (764.6 m), Cities Service Entz No. 2, El Dorado field, Butler County, Kansas (0.44 in. in the photo equals 0.17 mm). Hematite ooids surrounded by fine-grained matrix and quartz silt. Some of the ooid cores appear to have begun as fine-grained, hematite-rich mud balls that have become covered with a cortex of hematite.

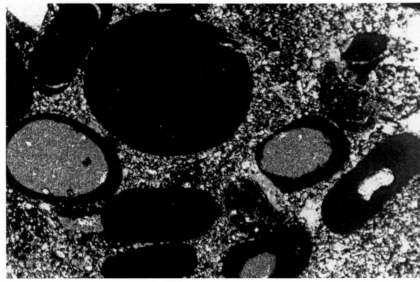

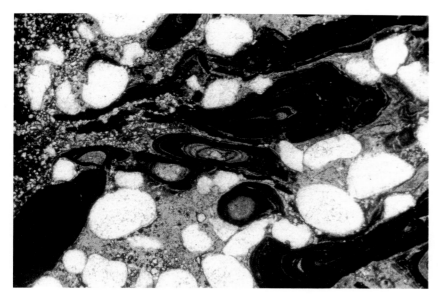

Figure 11.13. Simpson Formation, 2509 ft (764.9 m), Cities Service Entz No. 2, El Dorado field, Butler County, Kansas (0.44 in. in the photo equals 0.17 mm). The original soft chamosite ooids have been deformed into spastoliths (squashed ooids) and subsequently have been replaced by hematite. Quartz grains have been pushed into the ooids in several places. According to Tucker (1981, p. 182), initially, chamosite is precipitated as a mixed gel of $Fe(OH)_3$, $Al(OH)_3$, and $SiO_2 \cdot nH_2O$, which is stable at positive Eh. Conversion of the gel to chamosite takes place after burial within a reducing environment, which is below the interface of water and sediment.

Figure 11.14. Simpson Formation, 2520 ft (765.2 m), Cities Service Entz No. 2, El Dorado field, Butler County, Kansas (0.44 in. in the photo equals 0.17 mm). Well-rounded Simpson Sand exhibiting poikiloptic cementation wherein all the detrital sand grains are held in a single calcite crystal. The occurrence of such textures may be related to the scattered presence of shell debris in the original sediment. (See Scholle 1979, p. 119.)

Figure 11.15. Simpson Formation, 2514 ft (766.5 m) Cities Service Entz No. 2, El Dorado field, Butler County, Kansas (0.44 in. in the photo equals 0.17 mm). Individual quartz grains are more angular and not as well sorted as the usual Simpson Sand. Fine-grained clay matrix (probably illite) is common (lower right) along with numerous glauconite (green) grains, which indicate a distinctly marine environment.

pear to be related to local conditions along the Nemaha uplift. Near the base of the Simpson, the sand reveals an angularity and a relatively large amount of clay mineral matter (probably illite). The same zone shows a considerable amount of glauconite, suggesting a marine environment.

The next horizon appears to provide a beach environment where shell fragments have provided calcite for cementation of the well-rounded quartz. As pointed out previously, the well-rounded quartz grains are probably derived from wind-blown deposits.

The next zone is that of the hematite oolite along with the interlaying of chamosite oolites and spastoliths, suggesting a rapid alternation between the oxidizing (hematite) environment and the reducing (chamosite) environment. The latter also contains zones with pyrite, which further supports the reducing nature of the chamosite oolite environment.

It appears that the Nemaha uplift was active enough to produce a shoreline where the alternating oxidizing and reducing environments could occur.

The top portion of the Simpson provides the clean, well-rounded quartz-cemented sand typical of the producing sand both in Kansas and Oklahoma.

In reviewing this sequence, it is clear that at the base of the sand considerable marine organic activity provided the glauconite. As we move up in the section, the iron-rich environment produces chamosite gels that are sometimes converted to hematite. In other cases, the hematite oolites are formed in an oxidizing environment where the gel stage does not appear to have been important.

It should be mentioned that other cores of the

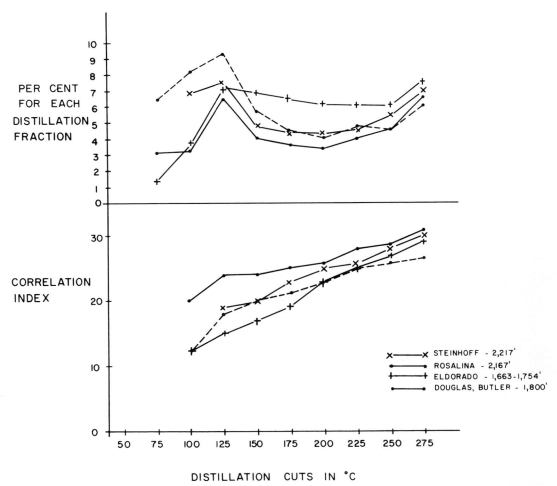

Figure 11.16. Comparison of oils from the Lansing–Kansas City Limestone, four different fields in Butler County, Kansas.

Simpson at El Dorado have also penetrated the hematite oolite zone, hence the occurrence, although probably very limited, is not a totally local phenomenon.

Source of the Oil at El Dorado

Any field that has produced as prolifically as the El Dorado field invites speculation concerning the source rocks that supplied the oil. Of the five reservoir zones, three have been examined in this study; therefore, it is appropriate to ask if the oils from these productive horizons show significant differences in composition. If such differences exist, one is tempted to say that each horizon had its own source bed. On the other hand, if the crude oils from all five zones are very similar, it is then proper to suggest that vertical communication of the reservoir fluids has taken place.

In order to determine the kind of variation one would expect from Lansing–Kansas City reservoirs, a set of correlation index curves from fields in Butler County (the same county where El Dorado is located) is plotted in Figure 11.16).

When the correlation index curves for these Lansing–Kansas City fields are compared with those for the five producing zones at El Dorado (Figure 11.17), it becomes clear that all five zones at El Dorado,

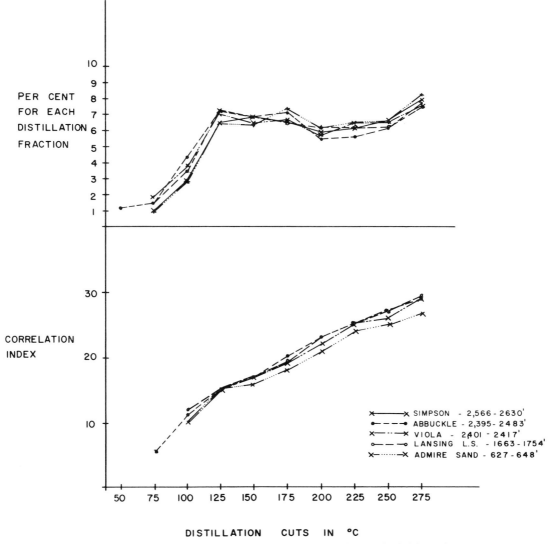

Figure 11.17. Comparison of oils from five producing zones at El Dorado field, Butler County, Kansas.

which range in age from Cambro-Ordovician to Permian, are quite similar with regard to both their paraffinic character and the amount retained in each distillation fraction, whereas the Lansing–Kansas City fields, all from the same county, are quite different.

On the basis of this comparison, it seems likely that most of the oil at El Dorado was generated at one horizon and was spread to porous zones higher in the section through the vertical fractures. It should also be pointed out that seismic activity along the Nemaha anticline is occurring regularly up to this day; hence, a continual shifting and reopening of the cemented fractures has probably taken place, allowing the oil to move vertically.

The source beds that appear to have been in contact with the most prolific zones are the Cherokee shales; therefore, the writer believes that these provided the major source rocks for the reservoirs at El Dorado.

Exploration

The search for other El Dorados along the Nemaha anticline in the relatively early days of the oil business led to the discovery of additional large fields such as Garber, Augusta, Tonkawa, West Edmond, and Oklahoma City. For some reason, there are no large fields north of El Dorado that have the same structural setting and the same type of reservoir rocks. As shown in Figure 11.18, the area north of El Dorado in Chase, Morris, Geary, Riley, Pottawatomie, Marshall, and Nemaha counties is not noteworthy for its Viola production. Yet, as we see from the pre-Pennsylvanian paleogeographic map of eastern Kansas and southeastern Nebraska (Figure 11.19), the Viola–Sylvan and Simpson formations would appear to have possibilities in addition to those at El Dorado. It is with these points in mind that the next field was selected for petrologic study.

Recent Sediment Equivalents

Finding a Recent equivalent of the Viola and Simpson formations is very difficult indeed. In the case of a sand being added to a shallow-water carbonate environment, there is the work by Fryberger et al. (1983), which discusses the "offshore prograding sand sea" of the Saudi Arabian coastline near Dharhran where eolian sand is blown into the shallow-water marine carbonate environment. However, the next step involving the location of an environment where a shallow-water carbonate is being leached and overlain

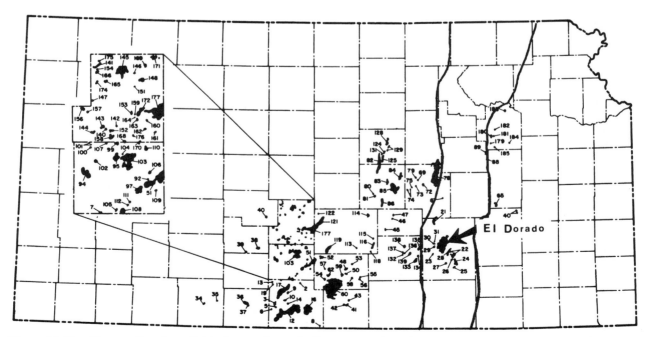

Figure 11.18. Kansas oil and gas fields that produce wholly or in part from the Viola Formation shown in relation to the Nemaha anticline. (Modified from Chenoweth 1966, p. 113.)

Recent Sediment Equivalents

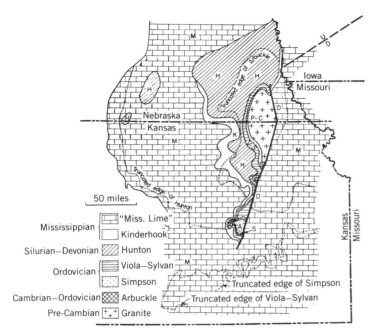

Figure 11.19. Pre-Pennsylvanian paleogeologic map of eastern Kansas and southeastern Nebraska showing the geologic map of the region as it existed at the beginning of Pennsylvanian time. (After Levorsen 1956, p. 587.)

with Pennsylvanian-type sediments is all but impossible. The closest possibility seems to be the Everglades area near the tip of Florida. The type of the Florida peninsula is underlain by the Miami oolite of Pleistocene age. Modern grasses and mangrove swamps are forming peats and soils on top of the consolidated limestone as shown in Figure 11.20. Some of the small solution pits are observable in the photograph. Looking at Florida Bay from Flamingo, some areas of lime mud are in the bay, and shell fragment sands are present at the shoreline, as seen in Figure 11.21. Offshore in the same photograph mangrove-covered islands are observable.

To intimate that these environments duplicate the situation at El Dorado where the Viola was overlain by the Pennsylvanian Cherokee sediments would be

Figure 11.20. Everglade grasses growing on top of Miami oolite limestone, Everglades National Park, Florida.

Figure 11.21. View of Florida Bay from Flamingo. Carbonate sand beach is in the foreground. Mangrove Islands are visible offshore.

misleading; however, some of the same elements are present. Future study of the transition zone between the Pleistocene–Miocene carbonates and the overlaying beds could be rewarding.

References

Baker, D. R., 1962, Organic Geochemistry of Cherokee Group in Southeastern Kansas and Northeastern Oklahoma, *American Association of Petroleum Geologists Bulletin,* Vol. 46, pp. 1621–1642.

Biederman, E. W., Jr., 1966, The Petrology of the Viola Formation, El Dorado Field, Butler County, Kansas, in Symposium on the Viola, Ferndale, and Sylvan, *Tulsa Geological Society Digest,* Vol. 34, pp. 41–59.

Chenoweth, P. E., 1966, Viola Oil and Gas Fields of the Mid-Continent, in Symposium on the Viola, Ferndale, and Sylvan, *Tulsa Geological Society Digest,* Vol. 34, pp. 110–118.

Fath, A. E., 1921, Geology of the El Dorado oil and Gas Field, Butler Co., Kansas, *Kansas State Geological Survey Bulletin,* Vol. 7, 187 pp.

Fryberger, S. G., A. M. Al-Sari and T. J. Clisham, 1983, Eolian Dune, Interdune, Sand Sheet, and Siliciclastic Sabkha Sediments of an Offshore Prograding Sand Sea, Dhahran Area, Saudi Arabia, *American Association of Petroleum Geologists Bulletin,* Vol. 67, No. 2, pp. 280–312.

Lane, E. C. and E. L. Garton, 1943, Analyses of Crude Oil from Some Fields in Kansas, *Report of Investigations No. 3688,* U.S. Dept. of Interior—Bureau of Mines, 95 pp.

Levorsen, A. I., 1956, *Geology of Petroleum,* W. H. Freeman and Company, San Francisco, 703 pp.

Pan, C. H., 1982, Petroleum in Basement Rocks, *American Association of Petroleum Geologists Bulletin,* Vol. 66, No. 10, pp. 1597–1643.

Scholle, P. A., 1979, A Color Illustrated Guide to Constituents, Textures, Cements, and Porosities of Sandstones and Associated Rocks, *American Association of Petroleum Geologists,* Memoir 28, 201 pp.

Tucker, M. E. 1981, *Sedimentary Petrology, An Introduction,* John Wiley & Sons, New York, 252 pp.

12

Yeage Field, Hunton Formation, Riley County, Kansas

Geologic Background

Examination of the oil production map for Kansas suggests that there are areas which for some reason have not contained as much oil as one would expect. The Nemaha anticline north of El Dorado provides this kind of puzzle. In order to examine the problem, it is appropriate to examine the Yeage field in Riley County, Kansas. According to the literature, this field produces from a carbonate of Hunton age. The fact that the field is located on the Nemaha anticline and is approximately 90 miles north of El Dorado makes it of particular interest.

As in several of the previously covered fields, the amount of information available to the writer is small indeed; however, because the Yeage field's location (Figure 9.1 gives the location in Riley County.) and the fact that cores were obtained from the Hunton Formation make it worth describing.

The distribution of the Siluro-Devonian age Hunton as it occurs beneath the Pennsylvanian is shown in Figure 11.13. From this illustration, it is clear that a fairly large area of Hunton was exposed and then covered by Pennsylvanian sediments. The Yeage field provides another good example of the type of porosity that was created by leaching and fracturing of the carbonate surface.

Examination of the Hunton cores and thin sections at the Yeage field leaves one with the impression that the moldic porosity reflects what was once a

Figure 12.1. Hunton Formation, 1718 ft (523.8 m), Cities Service Yeage No. 2, Yeage field, Riley County, Kansas (0.44 in. in the photo equals 0.17 mm). Moldic porosity (upper right) has had several episodes of fluid flow and leaching. Initially, coarse dolomite rhombs, some of which are zoned, grew into the open pores. During the next episode, sparry calcite filled in some of the major pores (center).

Figure 12.2. Hunton Formation, 1718 ft (523.8 m), Cities Service Yeage No. 2, Yeage field, Riley County, Kansas (0.44 in. in the photo equals 0.17 mm). Again, the initial changes favored the growth of large dolomite rhombs. After this stage, the fracture was filled with spar, leaving the rock relatively impermeable.

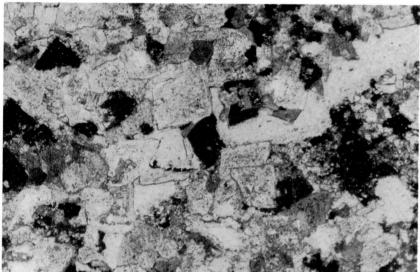

Figure 12.3. Hunton Formation, 1718 ft (523.8 m), Cities Service Yeage No. 2, Yeage field, Riley County, Kansas (0.44 in. in the photo equals 0.17 mm). The coarse dolomite rhombs initially grew protruding into the open pore space. A sudden change in the fluids brought about the deposition of sparry calcite. Several partial dolomite rhombs can be seen floating in the sparry calcite. For some reason, finer-grained dolomite lines some of the pore openings that are not filled.

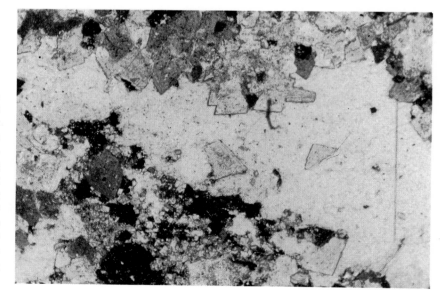

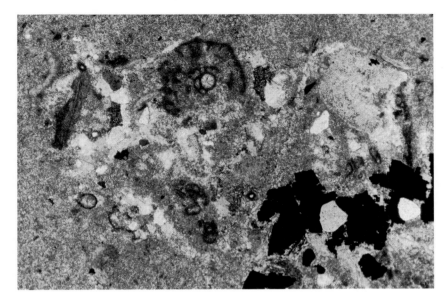

Figure 12.4. Hunton Formation, 1605.5 ft, (489.5 m), Cities Service Yeage No. 3, Yeage field, Riley County, Kansas (0.44 in. in the photo equals 0.17 mm). Fossiliferous zone with crinoid columnal and other debris. Black material at the lower right is pyrite. The porosity has been filled in with coarse carbonate crystals. Several quartz grains show as grains included within the pyrite.

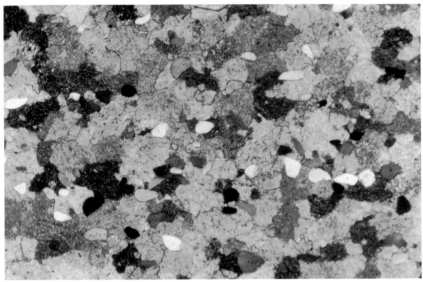

Figure 12.5. Hunton Formation, 1696 ft (517.0 m), Cities Service Yeage No. 4, Yeage field, Riley County, Kansas (0.44 in. in the photo equals 0.17 mm). Coarse-grained, dense dolomite with occasional small quartz grains floating in the dolomite. Several sets of quartz grains show layering. The size and position of these grains suggests that they may have been deposited by wind.

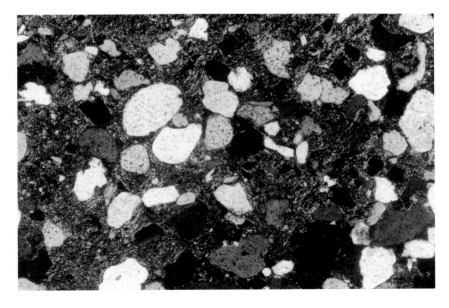

Figure 12.6. Hunton Formation, 1681 ft (512.5 m), Cities Service Yeage No. 4, Yeage field, Riley County, Kansas (0.44 in. in the photo equals 0.17 mm). Clay layering around quartz grains shows some crushing and bending effects. Rounding of some of the grains suggests that Simpson Sand is being eroded and redeposited nearby.

carbonate bank full of shell fragments and crinoid debris. The dolomitization that has taken place effectively removed the traces of most of the organic structure as shown in Figures 12.1–12.6. Even the coarse dolomite itself, with the zoned crystals, indicates a number of changes in the fluid composition that migrated through the reservoir.

The last set of fluids apparently filled much of the available pore space with spar, and what is left has an average porosity of around 8% with only streaks of high permeability.

The frequency of changes in the fluids traversing the porous zones is similar to that observed for the Viola Formation at the El Dorado field.

Exploration

From the explorationists' point of view, the puzzle as to why there are relatively few fields along the Nemaha uplift north of El Dorado remains. The fracturing and leaching conditions that helped produce the reservoir conditions at El Dorado appear to have been somewhat the same at the Yeage field. The source rocks that supplied the oil for the Yeage field must be capable of filling other reservoirs along the uplift; therefore, additional discoveries along the flanks of this structure north of El Dorado will probably be made in the future.

13

Oklahoma City Field, Wilcox Sandstone, Oklahoma County, Oklahoma

Geologic Background

The Oklahoma City field is a classic case of a giant oil field that has produced over a billion barrels of oil and oil-equivalent gas. The Wilcox Sandstone has produced roughly half of the oil in place and therefore deserves to be included in any discussion of reservoir rocks. The fact that it is located at the southern end of the Nemaha anticline makes it particularly appropriate to this study because it allows us to compare samples of the Simpson from two locations that are on the same structure, but 175 miles apart.

As shown in Figure 13.1, the Oklahoma City field is located on top of the southern end of the Nemaha uplift. Growth of the Oklahoma City field structure probably began in Cambrian time and continued through the Ordovician up through the Early Pennsylvanian. In this regard, it closely resembles the history of the El Dorado field. The oil is trapped in a faulted anticline as illustrated in Figure 13.2.

According to Gatewood (1970, p. 226), there are at least three important stages in the structural development of the anticline, folding, faulting, and truncation. The fault that forms the eastern boundary had its beginnings long before erosion removed the sequence of sediments on the upthrown side. It is a normal, almost vertical fault with a maximum displacement at the level of the

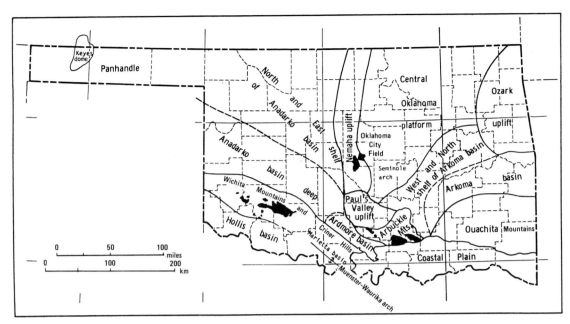

Figure 13.1. Tectonic map of Oklahoma showing the relationship of the Nemaha uplift and the Oklahoma City field. (Modified after Landes, Mackey, and Bowles 1963.)

Arbuckle of 2400 ft (731.7 m). At the Simpson level, this displacement is approximately 2000 ft (609.8 m). According to Gatewood, the fault was a significant factor in the accumulation of hydrocarbons on the high side of the structure. In other words, oil and gas migrated into the structure while the fault was developing. Repeated movements along the fault are indicated in beds up to the surface strata of Permian age. (See Figure 13.3 for stratigraphic section.) The Cherokee sediments at the base of the Pennsylvanian lie in contact with the Ordovician Arbuckle on the top of the structure and cover progressively younger Simpson sediments proceeding west on the flank. At the high point of the structure, erosion of several hundred feet of Cherokee strata has occurred.

There are also indications that quantities of pre-Pennsylvanian oil may have escaped during erosion of the Simpson and Arbuckle groups.

The field that developed as the result of these numerous geologic events is 12 miles (19.3 km) long, 4½ miles (7.2 km) wide, and its productive area encloses 32 square miles. The producing closure is 1000 ft, and 1810 wells have been drilled to tap 20 different significant zones (Gatewood 1970, p. 229).

Sample Location

In this case, the specific well from which at least one of the samples is taken is of historical as well as petrologic interest. The sample from which the photomicrograph in Figure 13.6 was taken was bailed from the famous ITIO No. 1 Mary Sudik, the discovery well for the Ordovician Wilcox zone. (See Figure 13.4 for location of Sudik lease.) Gatewood's (1970, p. 239) description of this discovery gives a feeling for the event as follows:

Development drilling was extended southward into Sec 31, T11N, R2W and on March 26, 1930, as the crew was pulling drill pipe out of the hole of the ITIO No. 1 Mary Sudik, it blew in with a roar, carrying 20 joints of heavy drill pipe into the derrick. The crew had failed to keep the hole full of mud as they withdrew the drill pipe. [See Figure 13.5.] The Mary Sudik was brought under control after 11 days, but in the meantime, carried by a strong north wind, it had sprayed 200 million ft^3 of gas and 20,000 bbl of oil per day as far south as the university town of Norman, 12 miles away. . . .

Sample Location

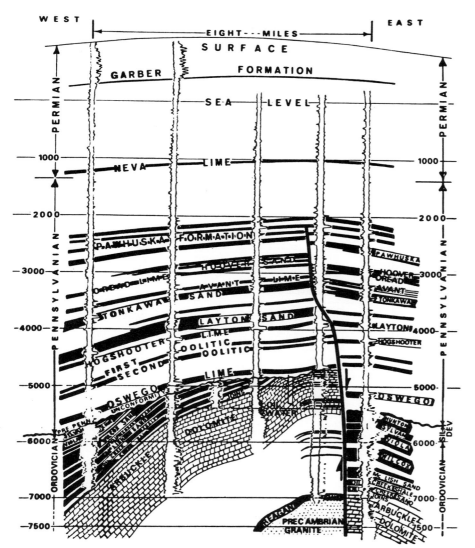

Figure 13.2. Structural cross section of the Oklahoma City field showing the change in fault displacement with depth. (After Gatewood 1970, p. 233.)

One of the main problems encountered in controlling the Sudik well and others like it producing from the Wilcox at Oklahoma City was the loose sand that was produced with the gas. In some cases, the abrasive action of the sand was such that it would cut through 1-in.-thick mild steel in 10 minutes. The middle 100 ft of the Wilcox is the most porous. In the Mary Sudik well, the average porosity as measured from the recovered core materials was 22.5%; however, this figure is probably low because the most heavily cemented parts are those that are recovered by coring. This same observation holds true for the consolidated sand samples bailed from the Sudik lease and used for the thin sections in this work. Also for this reason, mineralogic point count data are misleading and were not carried out for the samples pictured. For the most part, however, the Wilcox is a very clean, monomineralic rock that is a classic orthoquartzite.

It is hoped that the photomicrographs in Figures 13.6–13.11 will do justice to this famous reservoir rock.

Generalized Stratigraphic Section

SYSTEM	SERIES	GROUP		FORMATION OR RESERVOIR ROCK NAME (Asterisk designates rock unit producing oil or gas)	
T.				Ogallala. Late Tertiary continental deposits in western Oklahoma	
CRET.	Gulfian	Colorado		Occurs both in Cimarron Co. in panhandle and in southeast Oklahoma.	
	Comanchean			Dakota Sandstone in Cimarron Co., Paluxy Sandstone in southeast; Bilbo sand*	
J.				Cimarron Co.	
TR.		Dockum		Panhandle only	
PERMIAN	Ochoan			Outcropping red beds in western Oklahoma	
	Guadalupean	Whitehorse El Reno			
	Leonardian			Local sands* in lower part produce in southeast Anadarko basin	
	Wolfcampian	Chase	Pontotoc*	Gas reservoirs* in Hugoton field	
		Council Grove			
		Admire			
PENNSYLVANIAN	Virgilian	Wabaunsee to Douglas	Hoxbar*	"Granite wash" * Hoover sand * Endicott sand * Tonkawa (Stalnaker) sand*	
	Missourian	Lansing		Perry sand *	
		Kansas City		Cottage Grove sand Hogshooter limestone * Layton sand* County Line lime *	
		Pleasanton		Checkerboard lime * Cleveland sand *	
	Desmoinesian	Marmaton	Deese*	Wayside sand* Peru sand* Oswego Limestone*	Lone Grove sands*
		"Cherokee"		Prue sand *; Squirrel sand * Skinner sand * Red Fork sand *; Burbank sand *; Earlsboro sand *; Dora sand* Bartlesville sand *; Glenn sand* Booch sand * Burgess sand*	
	Atokan	U. Dornick Hills		Dutcher Sand*; Gilcrease Sand*	
	Morrowan	L. Dornick Hills		Wapanucka Limestone Union Valley Formation: Cromwell sand *; Primrose sand*	
	Springeran			Springer sand *	
SIL.-DEV.-MISS.	Chesterian Maramecian			Goddard Shale Caney Shale; "Chester Lime" *; Manning Limestone Sycamore Limestone	
	Osagean Kinderhookian			Boone Chert; "Mississippi chat" *; "Mississippi lime" *; Arkansas Woodford Shale; Misener sand* Novaculite	
	Oriskanian to Albion	Hunton*		Bois d' Arc Limestone *	
ORDOVICIAN	Cincinnatian Trentonian			Sylvan Shale Viola Limestone	
	Black Riverian	Simpson*		Bromide sand *; "Wilcox sand" * ←	
	Chazyian			McLish sand * Oil Creek sand *	
	Canadian	Arbuckle*		"Siliceous lime" *; Turkey Mt. sand*	
CAM-BRIAN	Croixian				
		Timbered Hills		Reagan sand *	
	Precambrian	crystallines		Granites, porphyries, basic igneous rocks, metasediments	

Figure 13.3. Stratigraphic section for Oklahoma. (After Landes 1970, p. 127.)

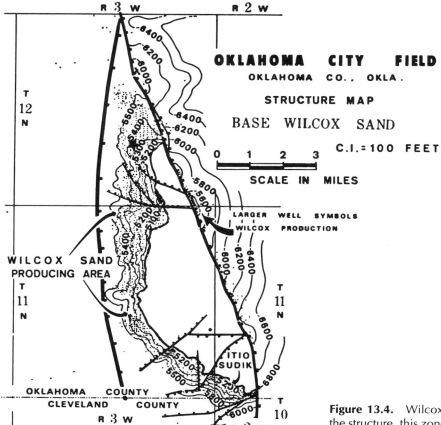

Figure 13.4. Wilcox structure map. Although lowest on the structure, this zone was the most extensive and prolific. Note the location of the ITIO Sudik lease. (After Gatewood 1970, p. 242.)

Figure 13.5. Discovery well for the Wilcox zone in the Oklahoma City field, the "Wild Mary Sudik #1." (After Gatewood 1970, p. 241.)

Figure 13.6. Second Wilcox (Simpson Formation), bailed from ITIO No. 1 Mary Sudik, Oklahoma City field, Oklahoma County, Oklahoma (0.44 in. in the photo equals 0.17 mm). The sample is partially cemented with carbonate, which explains why it remained intact. The part in the upper right has very little cement and is more representative of the loose sand characteristic of the blowout of the Wild Mary Sudik well. Larger quartz grains floating among the others are also characteristic. As mentioned previously, the porosities measured from the Mary Sudik core ranged between 20 and 25%; however, the sand was so loosely cemented that core recovery was far from complete.

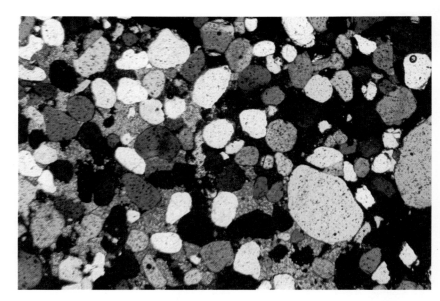

Figure 13.7. Second Wilcox (Simpson Formation) bailed from Cities Service Sudik No. 13, Oklahoma City field, Oklahoma County, Oklahoma (0.44 in. in the photo equals 0.17 mm). Well-rounded but highly cemented clean quartz sand. The cementing material in this case is quartz, and this fact accounts for its recovery. Most of the quartz grains show no undulose extinction, suggesting that no strain existed in the source material from which the quartz was eroded. Also there are no bubble trains crossing grain boundaries, as was observed in the Oriskany Sand in West Virginia.

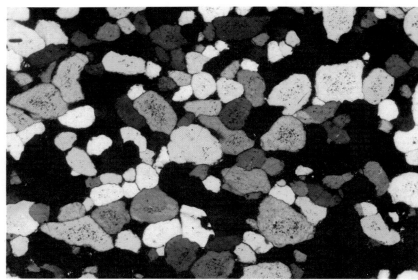

Figure 13.8. Second Wilcox (Simpson Formation) bailed from Cities Service Sudik No. 13, Oklahoma City field, Oklahoma County, Oklahoma (0.44 in. in the photo equals 0.04 mm). Quartz grains are well cemented, leaving decreased pore space. Clay mineral linings are visible as thin yellow streaks at several grain boundaries.

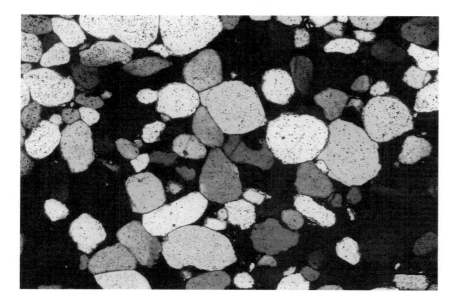

Figure 13.9. Second Wilcox (Simpson Formation) bailed from Cities Service Wickline No. 4, Oklahoma City field, Oklahoma County, Oklahoma, (0.44 in. in the photo equals 0.17 mm). The sand is quite clean and in places shows cementing of very rounded grains with quartz overgrowths. The sorting of the quartz suggests that there may be a mix of two quartz grainsize populations.

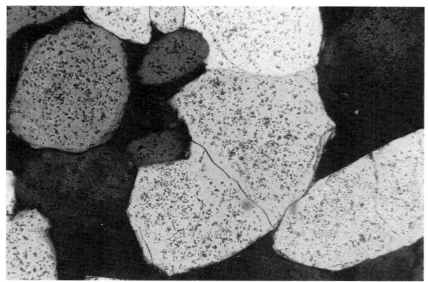

Figure 13.10. Second Wilcox (Simpson Formation) bailed from Cities Service Wickline No. 4, Oklahoma City field, Oklahoma City, Oklahoma (0.44 in. in the photo equals 0.08 mm). Higher magnification showing quartz cement. Note quartz overgrowths on top of rounded surfaces (white grain at top) in contrast to the rather angular grains with cemented boundaries (center and lower right).

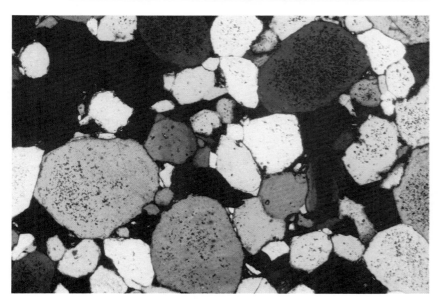

Figure 13.11. Second Wilcox (Simpson Formation) bailed from Cities Service Wickline No. 4, Oklahoma City field, Oklahoma County, Oklahoma (0.44 in. in the photo equals 0.17 mm). Shows large variation in grainsize, suggesting that two populations are involved. Quartz overgrowths (cement) provide a number of straight-edged grain boundaries on top of well-rounded grain surfaces.

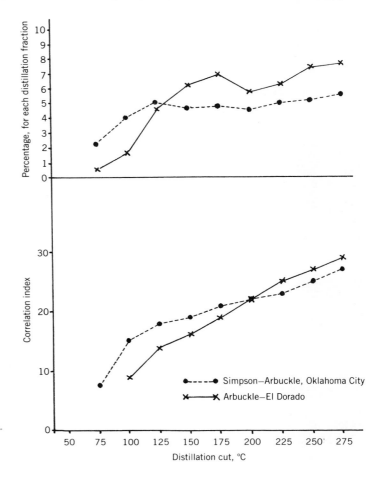

Figure 13.12. Plot of distillation data for oils from the Arbuckle–Simpson reservoir at Oklahoma City and the Arbuckle at El Dorado (McKinney and Garton 1957, pp. 80, 164).

Source of the Oil

In trying to assess where the oil in the Oklahoma City field came from, it is helpful to recall that the Simpson Group is entirely eroded from part of the crest so that Cherokee Shale (Pennsylvanian) rests on top of eroded Arbuckle. On the west flank of the structure, the Cherokee lies on top of Simpson strata. One well on top of the structure has no Cherokee Shale; therefore, we can conclude that an island existed at that point while the Cherokee Shale was being deposited elsewhere (Newman 1965, p. 216).

The fact that the Cherokee Shale acts as the cap rock for the Arbuckle and Simpson reservoirs has suggested that the Cherokee is also the source rock. On the other hand, Newman (1966, p. 217) states:

Most of the oil at the Oklahoma City pool is produced from eroded Simpson sands which have been made very porous and permeable by leaching of the cementing material. These beds were at the surface in early Pennsylvanian time. It appears probable that there was an accumulation of oil in the Simpson sands before Pennsylvanian erosion, as there is some solid asphaltic residue in these sands.

If we assume that the pre-Pennsylvanian materials that produced the asphaltic residues continued to provide oil to the Wilcox, then it appears probable that the Oklahoma City oil has both a Cherokee Shale component and a pre-Pennsylvanian component.

If the oil from the Simpson–Arbuckle at El Dorado is compared with that from the Simpson at Oklahoma City, it appears that the oils are different. The plots of the correlation indexes for the two crude oils are provided in Figure 13.12.

Recent Equivalents

It seems doubtful that any good Recent equivalents of the Oklahoma City field second Wilcox sand are available for study. The highly rounded sand and the monomineralic nature of the sand make the finding of a modern example very difficult, and the writer cannot suggest an appropriate case.

Pettijohn, Potter, and Siever (1972, p. 221) make the following comments concerning the Recent equivalents of orthoquartzites (quartz aernites) in general:

Because quartz arenites are by definition essentially quartz sands with less than 5% other constituents, they are exceptional sands. They are exceptional, that is, in the sense, that sands of this character seem not to be forming today. Modern sands of this type, if they exist at all, are small accumulations derived from or formed by redeposition of an older quartz arenite. Such seems to be the case with the pure quartz sands of the Libyan Desert in Egypt that contain 95 or more percent quartz (Mizutani and Suwa, 1966) which probably were derived from disintegration of the Nubian sandstone.

Geologic Interpretation

At this point, two occurrences of Simpson Sand along the Nemaha anticline (or uplift) have been examined. The fact that relatively little difference exists in the Simpson over a distance of 175 miles (282 km) from Oklahoma City to the El Dorado field is remarkable enough; however, when the whole extent of the Simpson and its equivalents are considered, the question of how such a continuous and uniform sand came into being always arises.

As Pettijohn (1957, p. 299) observed, orthoquartzitic sandstones or quartz arenites are "the end product of protracted and profound weathering and abrasion. In order that there be sufficient time to achieve these results, it is imperative that the source area and site of deposition be tectonically very stable or that the sand go through several cycles of sedimentation."

Again according to Pettijohn (1957, p. 300): "Only locally, as near granitic monoadnocks or near the underlying grainite basement, would any feldspar appear in the sandstone."

It may well be asked, what causes the difference between the granite high that contributes to a Simpson-like sediment and a granite high that produces a "granite wash"? Later in Chapter 15 the case of the granite wash will be examined as it is associated with the Amarillo uplift.

According to Kuenen (1960), the only environment in which sand can be effectively rounded is that of the desert dune. This suggests that the highly rounded grains that form much of the Simpson sand experienced a desert environment before being deposited in the Cambro-Ordovician sea.

It should be recalled that land plants do not become significant features until the Silurian; therefore, the deep erosion and weathering of the land surface along with the formation of extensive deserts could proceed unhindered by stabilizing vegetation.

As other orthoquartzitic reservoirs are examined, it will become clear that none have the highly rounded grains observed in the Simpson.

References

Gatewood, L. E., 1970, Oklahoma City Field—Anatomy of a Giant, in *Geology of Giant Petroleum Fields,* M. T. Halbouty, Ed., American Association of Petroleum Geologists, Memoir 14, pp. 223–254.

Kuenen, P. H., 1960, Experimental Abrasion 4, Eolian Action, *Journal of Geology,* Vol. 68, pp. 427–449.

Landes, K. K., 1970, *Petroleum Geology of the United States,* Wiley-Interscience, New York, 571 pp.

Mackey, F. L. and J. P. F. Bowles, Jr., 1963, Oil and Gas Exploration Developments in Oklahoma During 1962, *American Association of Petroleum Geologists Bulletin,* Vol. 47, No. 6, pp. 1013–1034.

McKinney, C. M. and E. L. Garton, 1957, *Analyses of Crude Oils from 470 Important Oilfields in the United States,* Bureau of Mines Report of Investigations 5376, 276 pp.

Newmann, M. L., 1965, The Simpson Play—Some Notes and Reminiscences, in Symposium on the Simpson Group, *Tulsa Geological Society Digest,* Vol. 33, pp. 212–218.

Pettijohn, F. J., 1957, *Sedimentary Rocks,* 2nd ed., Harper & Brothers, New York, 718 pp.

Pettijohn, F. J., P. E. Potter, and R. Siever, 1972, *Sand and Sandstone,* Springer-Verlag, New York, 618 pp.

14

Victory Oil Field, Lansing–Kansas City Group, Haskell County, Kansas

Geologic Background

The Lansing–Kansas City Group of Middle Pennsylvanian age (see Figure 14.1) is one of the major oil and gas horizons within Kansas and produces in more fields than any other reservoir group. The carbonates are usually quite porous and fossiliferous.

In the case of the Victory field in Haskell County, it exhibits some rather unusual textures that deserve the attention of both the explorationist and the reservoir engineer. As is the case in several other discussions of reservoirs in this work, much of the information concerning, for example, well logs and structure contour maps is missing. Nevertheless, the petrologic detail deserves to be covered.

The Victory field is located in the southwestern corner of Haskell County, Kansas, and geologically, it is within the structural unit called the Hugoton embayment (see Figure 14.2). The field itself appears to be a highly fossiliferous reef or bank that, at least for short periods of time, was probably exposed to the atmosphere. A portion of the petrologic discussion is devoted to the diagenetic effects observed in thin section. In this instance, it is the oolitic zones that show the most marked effects. The reason for this appears to be that oolites in their initial uncemented stages are essentially well sorted sands

AGE	ROCK UNIT	THICKNESS	LITHOLOGY	RESERVOIRS	
Permian		5000			Overlying rocks consist, in ascending order, chiefly of upper Permian red beds and evaporites, Cretaceous shales, sandstones, limestones, and chalky shales, and the late Tertiary Ogallala Formation of continental origin. Top of reservoir section (as of 1968).
Permian		335		○○○	Chase Group. The principal gas reservoir in the Hugoton embayment.
Permian		320		○	Council Grove Group. Gas reservoir in Hugoton embayment. Permian rocks crop out in central Kansas.
Permian		130		●	Admire Group. Shallow reservoir on Nemaha anticline.
Pennsylvanian	Wabaun-see	500		●	Wabaunsee is the uppermost group in the Pennsylvanian. Pennsylvanian rocks from the Cherokee to the Wabaunsee crop out in eastern Kansas.
Pennsylvanian		325		●	Shawnee Group. Includes Oread and other producing limestones in western Kansas.
Pennsylvanian		340		●	Douglas-Pedee Groups.
Pennsylvanian	Lansing-Kansas City	430		●	Limestone reservoirs within the Lansing-Kansas City Groups produce in more fields than any other reservoir group. ←
Pennsylvanian		100		●	Pleasanton Group. Much thinner in northwestern Kansas.
Pennsylvanian		250		●	Marmaton Group. Both limestone and sandstone reservoirs may be present.
Pennsylvanian	Cher-okee	450		●	Cherokee Group. Contains Bartlesville and other sands. Absent across Central Kansas uplift.
Pennsylvanian		200		●	Atokan Group. Hugoton embayment only.
Pennsylvanian		180		●	Morrowan Group. Hugoton embayment only. Pennsylvanian basal sand or conglomerate on western uplifts.
Miss. Mississippian		~30 650		●	"Mississippi Lime". Also "chat". Absent across top of Central Kansas uplift. The oldest rock to crop out in Kansas (in southeastern corner of Kansas, where it directly underlies the Cherokee Shale). Also reservoirs in the Chester and Meramecian (Upper Mississippian) and Osage and Kinderhook (Lower Mississippian) Series.
Siluro-Dev. Dev.		105		●	Chattanooga Shale. Eastern two-thirds of Kansas only. Misener sand in Sedgwick basin only.
Siluro-Dev.		200		●	"Hunton" Limestone. Salina, Sedgwick, and Forest City basins.
Ordo-vician		40			Maquoketa Shale. Not everywhere present.
Ordo-vician		160		●	Viola or Kimmswick Limestone
Ordo-vician		90		●	Simpson sand. Also St. Peter, "Wilcox".
Cambro-Ordovician		570			"Siliceous Lime".
Cambro-Ordovician		~30			Reagan or Lamotte Sandstone.
				●	Precambrian granite, quartzite, and other crystallines.

STRATIGRAPHIC SECTION
Not to scale

Figure 14.1. Stratigraphic section for Kansas showing the relative position of the Lansing–Kansas City Reservoir. (Moore, R. C., Frye, J. C., Jewett, J. M., Lee, Wallace, and O'Connor, H. G., 1951. The Kansas Rock Column: Kansas Geological Survey, Bulletin 89, 132 p.)

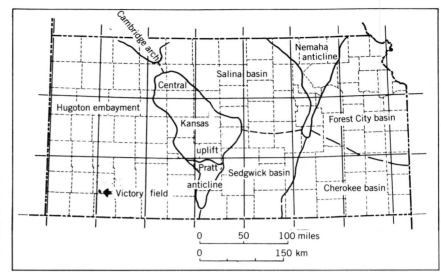

Figure 14.2. Tectonic map of Kansas showing the location of The Victory Field in Haskell County, (Hilpman, P. L., 1958, Producing Zones of Kansas Oil and Gas Fields: Kansas Geological Survey, Oil and Gas Investigations No. 16, map.)

through which fluids flow with relative ease. It is these fluids that cause important changes in the reservoir properties.

Photomicrographs and Description

When sampling a highly fossiliferous core, it is difficult to obtain samples that are truly representative. In this case, the photomicrographs (Figures 14.3–14.8) were chosen to provide a look at some of the more interesting zones as well as some shots representative of the major part of the core. The following discussion focuses upon the oolitic zone where the porosity and permeability values shown in Table 14.1 were obtained.

When porosity values are as high as 32% (Table 14.1, 4174 ft, 1272.6 m), a permeability value of 5.3 millidarcies is anomalous. Illing's (1954) study of consolidated carbonates that compose the islands of the Bahama platforms postulated that the blocky calcite enclosing the rounded pores is characteristic of leached carbonates that have been exposed to percolating rain water containing organic acids. More recent work by Jomas and Choquette (1984) indicates that the blocky calcite is a phreatic cement where the pores are always filled with water and crystals grow unimpeded except by intercrystalline competition. The cement rinds left surrounding the dissolved ooids were probably formed under normal marine conditions which became more brackish as the sediments were exposed to meteoric waters.

The reservoir engineer should take note; high porosities with low permeabilities (such as those observed above) usually raise questions about the accuracy of the data. It is at this point that it pays to look at the thin sections.

Exploration

Zones that exhibit subaerial leaching commonly occur on the landward side of modern reefs. The occurrence of the leached texture with the clear interlock-

TABLE 14.1 Examples of Porosities and Permeabilities from Leached Oolite Zones in Cities Service Conover 1C Well (Biederman 1966, p. 174)

Depth (ft)[a]	Porosity (%)	Permeability (millidarcies)
4163.5 (1269.4)	28.4	5.2
4166 (1270.1)	12.3	0.92
4174 (1272.6)	32.0	5.3
4175 (1272.9)	21.2	1.0
4188 (1276.8)	19.9	0.57
4190 (1277.4)	18.6	0.46
4194 (1278.7)	14.1	0.05
Average	20.9	1.93

[a] Numbers in parentheses in meters.

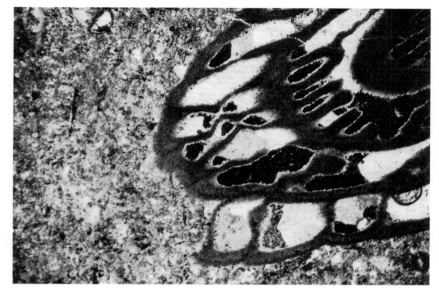

Figure 14.3. Lansing–Kansas City Formation, Cities Service Conover D-2, 4147 ft (1264.3 m), Victory field, Haskell County, Kansas (0.44 in. in the photo equals 0.42 mm). Numerous fusulinid fragments with considerable pore space inside the fusilinids (upper right corner) are common. Coarse dolomite fills some of the pores. Fine-grained lime mud shows evidence of organic fragments and unidentifiable fossil hash. Permeability, 20 millidarcies; porosity, 12%; oil, 2.6%; total water, 63.0%.

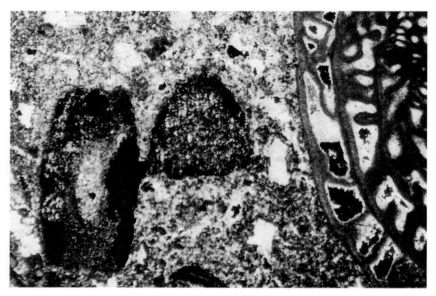

Figure 14.4. Lansing–Kansas City Formation, Cities Service Conover D-2, 4148 ft (1264.6 m), Victory field, Haskell County, Kansas (0.44 in. in the photo equals 0.42 mm). Lime mud contains numerous fossil fragments—mainly crinoid debris and fusilinids. Some larger dolomite rhombs are filling in the pores.

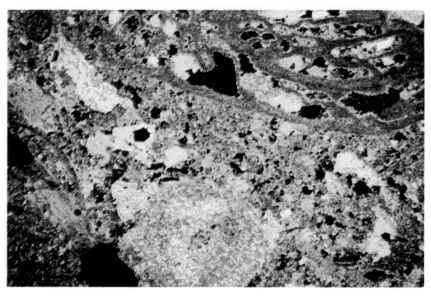

Figure 14.5. Lansing–Kansas City Formation, Cities Service Conover D-2, 4152 ft (1265.9 m), Victory field, Haskell County, Kansas (0.44 in. in the photo equals 0.42 mm). Skeletal sand debris with fusilinids (TOP), crinoid columnals (colored fragment near bottom center). Porosity appears to be partially leached. Permeability, 12.10 millidarcies; porosity, 14.5%; oil, 4.7%; total water, 54.0%.

Figure 14.6. Lansing–Kansas City Formation, Cities Service Conover D-2, 4155 ft (1266.8 m), Victory field, Haskell County Kansas (0.44 in. in the photo equals 0.42 mm). Lime mud with coral fragment floating in it. Other fragments present include chertified crinoid columnals and a few bryozoans. Permeability, <0.1 millidarcies; porosity, 11.2%; oil, 0.0%; total water, 74.9%.

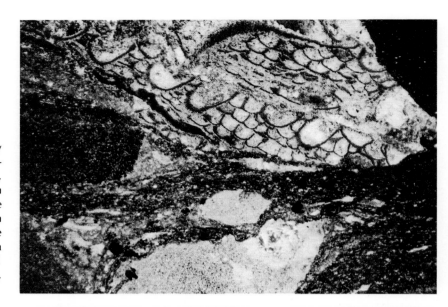

Figure 14.7. Lansing–Kansas City Formation, Cities Service Conover 2-D, 4157 ft (1267.4 m), Victory field, Haskell County, Kansas (0.44 in. in the photo equals 0.30 mm). Lime mud with occasional bryozoan and a chertified crinoid columnal. Some pores and fractures (not visible) have been filled with dolomite. Permeability, <0.1 millidarcies; porosity, 7.0%; oil, 2.7%; total water, 75.1%.

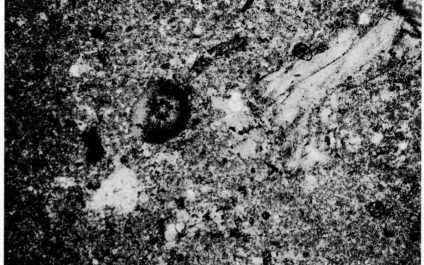

Figure 14.8. Lansing–Kansas City Formation, Cities Service Conover 1C, 4168 ft (1270.9 m), Victory field, Haskell County, Kansas (0.44 in. in the photo equals 0.30 mm). Leached oolites with clear interlocking granular calcite characteristic of subaerial deposition of the Lansing–Kansas City Formation. The black circular voids are surrounded by a variety of calcite that is relatively impermeable. Therefore, the rock may exhibit unusual porosities and permeabilities as shown in Table 13.1.

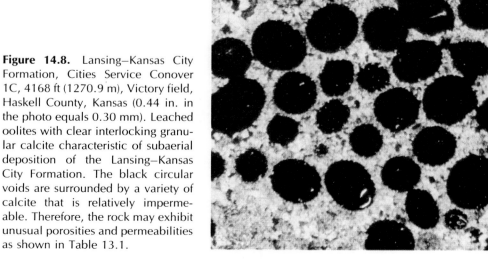

ing granular calcite should provide the explorationist with a reasonable idea of the location in the ancient environment and where the major part of the reef may be found.

Recent Equivalents

Locating a present-day situation that closely resembles the Lansing–Kansas City as it occurs in the Victory field is not easy. Present-day reef materials involve different organisms and different overall settings.

The closest U.S. environment involving reefs is in Florida. At first glance, we would be tempted to conclude that the Pennsylvanian reefs are so far removed from today's reef that nothing could be learned from making the comparison. However, the fact that part of the Lansing–Kansas City has been raised above sea level and leached by both freshwater rain and sea spray suggests that some effort to examine the Recent reefs under somewhat similar circumstances might be instructive (Figures 14.9–14.11).

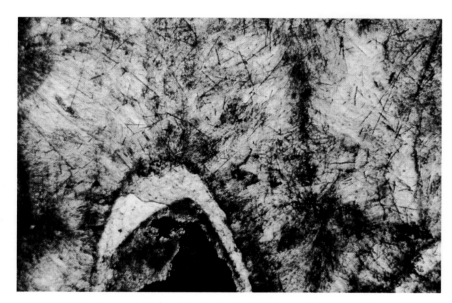

Figure 14.9. Thin section view of coral (C. F. Diploria) with calcite and fine mud filling the open pore (0.44 in. in the photo equals 0.30 mm). In some instances, almost no change in terms of cementation has taken place, and the pores are almost entirely open, providing a highly porous framework. An example of this open framework is shown in Figure 14.10.

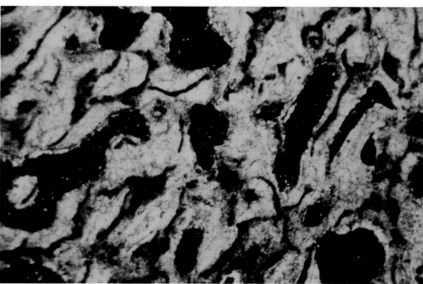

Figure 14.10. Thin section of a recent coral fragment collected from beach detritus on lower Matecombe Key, Florida (0.44 in. in the photo equals 0.42 mm). Much of the debris associated with a reef is highly porous and yet can have considerable rigidity. A typical sample of reef detritus associated with the coral heads of the Key Largo Limestone is illustrated in Figure 14.11.

128

Figure 14.11. Thin section showing typical fragmental detritus (*Halimeda* Sp. and Foraminifera) associated with Diploria framework in Key Largo Limestone, Key Largo, Florida (0.44 in. in the photo equals, 0.63 mm). It is interesting to note that the fine-grained mud so commonly observed in many of the ancient reservoirs appears to be absent in these samples of reef rock. In Florida Bay, reef detritus and lime muds are present, but the reef itself is strikingly clean.

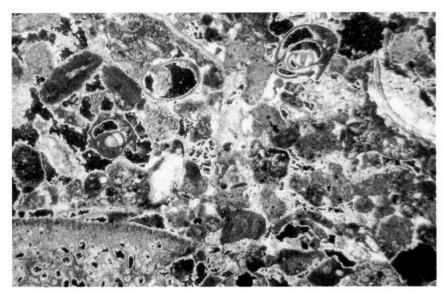

Figure 14.12. Leached surface of the Key Largo Limestone, Atlantic Ocean side of Key Largo.

Figure 14.13. Close view of leached Key Largo Limestone showing the structure of a coral (C. F. Diploria). Thin sections of the coral structures indicate that calcite and other fine-grained organic-rich sediment have begun to fill the open pores (Figure 14.9).

The photographs in Figures 14.12 and 14.13 were obtained from areas along the shores of the Florida Keys. Here the reef rock of Pleistocene age has been exposed to the atmosphere and suffered the degradation caused by both fresh and salt water as well as that caused by tropical vegetation. A typical view of the leached and storm-battered surface of the Key Largo Limestone reef is shown in Figure 14.12. In the partially cemented reef rock, the structures of corals are clearly visible, as illustrated in Figure 14.13.

On the Atlantic side, work by Cloud (1962) indicated that hydrous mica and kaolinite were abundant in the Straits of Florida.

In looking for leaching effects, it is clear that the modern equivalent of a leached oolite is more likely in a Bahama Island setting.

References

Biederman, E. W., Jr., 1966, Leached Oolites—Oil Finders' Key, *World Oil,* Vol. 163, No. 1, p. 174.

Cloud, P. E., Jr., 1962, *Environment of Calcium Carbonate Deposition West of Andros Island Bahamas,* Geological Survey Professional Paper 350, p. 138.

Hilpman, P. L., 1958, Producing Zones of Kansas Oil and Gas Fields, *Kansas Geological Survey,* Oil and Gas Investigations No. 16, Map.

Illing, L. V., 1954, Bahaman Calcareous Sands, *American Association of Petroleum Geologists Bulletin,* Vol. 38, No. 1, pp. 1–95.

James, N. P. and P. W. Choquette, 1984, Diagensis 9—Limestones—The Meteoric Diagenetic Environment, *Geoscience Canada,* Vol. 11, No. 4, pp. 161–194.

15

Panhandle Field Reservoirs of Texas

The Panhandle–Hugoton field is large by any measure. It is 275 miles (442.5 km) long and ranges in width from 8 miles (12.9 km) to 57 miles (91.7 km) and contains within its boundaries approximately 5 million acres. Figure 15.1 shows its area distribution and those places where oil is produced. Since its discovery, more than 20,000 wells have been drilled in the Panhandle field and as of January 1, 1967, there were 11,827 still producing oil or gas (Pippin 1970, p. 206).

This study focuses on three reservoirs from this giant field and these are the Granite Wash, Brown Dolomite, and Red Cave formations. The first two reservoirs are productive of oil whereas the last produces gas only. The areas within the field that are discussed are in Hutchinson and Carson counties, Texas. Where applicable, some additional photomicrographs of the basement rocks in these areas are provided.

A stratigraphic column of the Panhandle–Hugoton field is illustrated in Figure 15.2 with the three producing zones discussed in this work marked by arrows.

Geologic Background

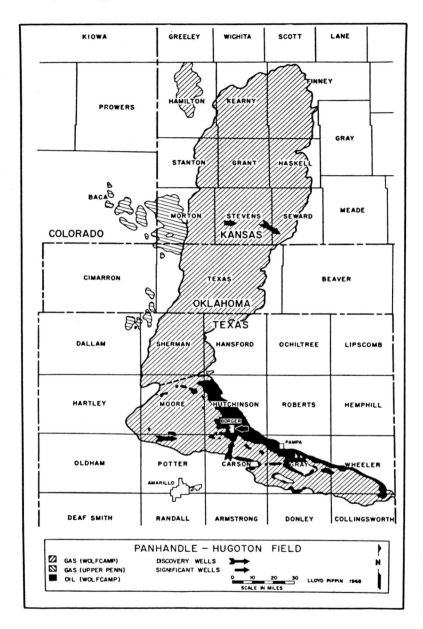

Figure 15.1. Location of the Panhandle–Hugoton field in Texas, Oklahoma, and Kansas showing oil- and gas-bearing areas. (After Pippin 1970, p. 205.)

Panhandle Field, Granite Wash Formation, Empire Granite Wash, Unit 4, Well 14, Carson County, Texas

Geologic Background

The Granite Wash of the Panhandle field is closely associated with a narrow granite mountain uplift several hundred miles long. The producing sediments were deposited in the Pennsylvanian and Early Permian seas mostly along the north flank. The granite core was finally buried, and the area was folded, faulted, and finally eroded. The sedimentary strata that resulted include an interconnected zone of porosity that extends for 125 miles (Levorsen 1956; p. 63).

Figure 15.3 shows a general cross section of the Amarillo uplift. The cross section in Figure 15.3 cuts across both Carson and Hutchinson counties and

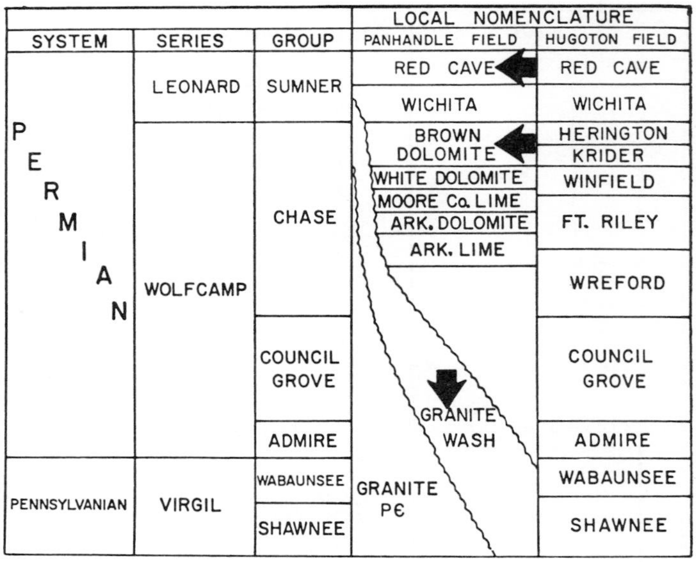

Figure 15.2. Stratigraphic column the Panhandle–Hugoton field showing relative positions of the Red Cave, Brown Dolomite, and Granite Wash reservoirs. (After Pippin 1970, p. 210.)

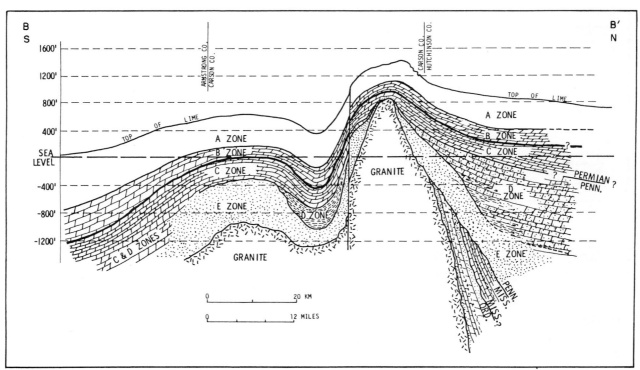

Figure 15.3. South-to-north section across the Amarillo uplift and the Panhandle field, Carson and Hutchinson counties, Texas. (After Landes 1970, p. 318).

Mineral Composition—Amorphous Materials

In attempting to analyze the clay minerals from the Granite Wash sediments, it became clear that in addition to kaolinite there was what has been called an "amorphous hump" on the x-ray pattern at 4-Ås spacing. This hump is characteristic of the clay mineral fraction of the Granite Wash in several locations. Some preliminary studies suggested that this was amorphous silica of some variety and that when the amorphous hump was large, the water-flooding properties were poor.

In terms of environmental interpretation, the fact that the muds associated with the Granite Wash are particularly rich in this type of material suggests that chemical weathering has been very important in the breakdown of the granitic materials. Dutton and Land (1985, p. 31) indicate that amorphous aluminosilicates from the Granite Wash probably contributed to the formation of authigenic chlorite in the reservoir rocks of the Mobeetie Field in Wheeler County, Texas.

Although the literature concerning the occurrence of amorphous materials in sediments is not large, some observations worth mentioning have been made. Velde (1977, p. 27), discusses the case of intense weathering as follows:

Under conditions of intense weathering, silica is unstable in the crystalline form. Mature bauxites, soils representing the most intense weathering conditions, contain no quartz and little combined silica. High rates of water influx remove SiO_2 at low solution concentrations. Normal ground water and streams carry about 17 ppm SiO_2 and less in high rainfall areas (Davis, 1964). In some weathering profiles silicification or deposition of silica has been observed. Most often the form of the phase deposited is crypto-crystalline, either opal or chalcedony.

Velde's (1977, p. 29) concluding remarks are worth noting with respect to the occurrence of amorphous silica in such ancient sediments as the Granite Wash. He states:

It is evident then that quartz is largely inert in many weathering or sedimentary environments. A study of Mizutani (1970) indicates that the transformation amorphous silica → cristobalite (opal, chalcedony) → quartz takes 10^9 years at 0 °C, 10^6 at 100 °C at 100 bars. This insures that amorphous silica will be present for significant periods of time in sedimentary rocks after its deposition from solution.

Sampling

Attempting to sample the Granite Wash for the purposes of porosity and permeability measurements presents problems because the cores frequently fall apart while the plugs are being drilled. Furthermore, grainsize measurements in thin section have a very wide range, which reflects the large granite fragments and the fine-grained clays all mixed together. In any case, Granite Wash-type sediments should be placed in a class by themselves. The thin-section photomicrographs in Figures 15.4–15.8 illustrate the problem. Comparison of the permeability measurements on these sediments with those in Figures 15.9 and 15.10 can cause difficulties in interpretation if the actual core is not examined in detail.

The Granite Wash-type reservoir can provide many problems for the personnel attempting either secondary or tertiary recovery projects. The major problems, which are inherent, involve the water sensitivity of the fine-grained clays and amorphous materials along with the distribution of the cleaner sands.

Water sensitivity and sampling can be related in that cores drilled for porosity and permeability measurements may fall apart so that the only values recorded are those where cementation of one kind or another is strong enough to hold the rock together. Permeability and porosity values on only the cemented samples can provide a misleading picture, and it is important to know whether or not such sampling problems have occurred.

Exploration

In terms of the major structural and stratigraphic possibilities, it is likely that most of the options have been explored rather thoroughly. It should be mentioned, however, that wells that produce from fractures in the granite have probably not been promoted because of

Figure 15.4. Granite Wash Formation, 3222 ft (982.3 m), Cities Service unit 4, Well No. 14, Empire Granite Wash, Panhandle field, Carson County, Texas (0.44 in. in the photo equals 0.17 mm). Angular granite fragments along with weathered perthite feldspar. Fine-grained, iron-stained clay matrix fills in some of the porosity. A small carbonate shell fragment (upper left corner) suggests that some biologic activity was taking place as the sediments were deposited.

Figure 15.5. Granite Wash Formation, 3222 ft (982.3 m), Cities Service Unit 4, Well No. 14, Empire Granite Wash, Panhandle field, Carson County, Texas (0.44 in. in the photo equals 0.17 mm). More angular granite rock fragments along with some weathered feldspar. Fine-grained, iron-stained clay matrix agains fills in much of the available pore space. Most of the feldspar appears to be microcline and perthite.

Figure 15.6. Granite Wash Formation, 3231 ft (985.1 m), Cities Service Unit 4, Well No. 14, Empire Granite Wash, Panhandle field, Carson County, Texas (0.44 in. in the photo equals 0.17 mm). Granite rock fragments showing varying degrees of weathering. One plagioclase feldspar grain (gray and black stripes) appears to be more fresh than the remaining granite fragments.

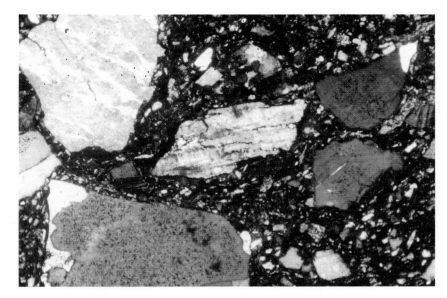

Figure 15.7. Granite Wash Formation, 3233 ft (985.7 m), Cities Service Unit 4, Well No. 14, Empire Granite Wash, Panhandle field, Carson County, Texas (0.44 in. in the photo equals 0.17 mm). Iron-stained red clay matrix becomes a dominant feature with larger weathered granite fragments, perthite feldspars, and quartz floating in the clay matrix. Specimens such as this are more likely to show the amorphous hump on the x-ray patterns.

Figure 15.8. Granite Wash Formation, 3252 ft (991.5 m), Cities Service Unit 4, Well No. 14, Empire Granite Wash, Panhandle field, Carson County, Texas (0.44 in. in the photo equals 0.17 mm). Photograph is taken without crossed polarizers to show the black opaque minerals and their distribution. The opaque mineral appears to be titaniferous magnetite. Weathered feldspars show up in various stages of light brown.

Figure 15.9. Granite Wash Formation, 3317 ft (1011.3 m), Cities Service Unit 4, Well No. 14, Empire Granite Wash, Panhandle field, Carson County, Texas (0.44 in. in the photo equals 0.17 mm). Relatively fine grained Granite Wash located near the basement shows orientation of grains and a texture that indicates some degree of low-grade metamorphism has taken place.

Figure 15.10. Basement Granite, 3332 ft (1015.8 m), Cities Service Unit 4, 23LL No. 14, Empire Granite Wash, Panhandle field, Carson County, Texas (0.44 in. in the photo equals 0.17 mm). Fracture zone in the basement granite. Note the numerous small fractures. Some of the basement rock production in the Panhandle field is produced from a fracture system such as this in the granite basement. The granite appears similar to that described in the outcrops of the Wichita Mountains in southwestern Oklahoma. These latter rocks have been described as "perthite leucogranites" (Ham et al. 1964, p. 22). Levorsen (1956, p. 117) discusses the production from this type of reservoir as follows: "Several wells produce from fresh basement granite in the Amarillo field, in the Texas Panhandle. One such well alone has produced over a million barrels of oil. Such production is undoubtedly the result of fractures in the granite, since nearby wells were either dry or nearly so. Presumably the oil has entered the granite from the sediments that were deposited along the flanks of the buried mountain range along which the field runs."

Figure 15.11. Basement diabase, depth unknown, Cities Service Burnett Ranch C-19, Panhandle field, Carson County, Texas (0.44 in. in the photo equals 0.17 mm). In many areas of the Wichita Mountains in Oklahoma, diabase occurs as dikes that cut the Wichita granite. Diabase has also been encountered in at least five wells in southwestern Oklahoma (Ham et al. 1964, p. 78), and it is only logical that this rock type should be encountered in the Amarillo uplift area, an extension of the Wichita Mountains. Plaigoclase (striped, gray) and pyroxene (blue and yellow) are the two most important minerals. The plagioclase is typically labradorite and the pyroxene is augite. The remaining orthopyroxene is hypersthene. Opaque grains (not shown) are also present, and these are probably titaniferous magnetite as is the case in the Oklahoma diabases. (Ham et al. 1964, p. 79.)

the high risk. Some of the newer seismic approaches may be able to provide information concerning these fracture systems so that future prospects of this kind can be expected to have reasonable payouts.

Dutton and Land (1985) show that the ground water recharge probably occurred in the coarse-grained Granite Wash sediments in the proximal fans at the foot of the Amarillo Uplift. It appears that the circulating waters flowed outward into sands such as those of the Mobeetie Field and produced a number of diagenetic changes resulting in the formation of chlorite, calcite of various types, quartz cement, feldspar cement, ankerite, anhydrite, and celestite.

In some wells, diabase has been encountered rather than granite as shown in Figure 15.11. Similar basement rocks have been drilled in a number of other wells along the Amarillo trend and probably contributed some of the ilmenite and magnetite observed in the Granite Wash.

References

Dutton, S. P., and L. S. Land, 1985, Meteoric Burial Diagenesis of Pennsylvanian Arkosic Sandstones, Southwestern Anadarko Basin, Texas, *American Association of Petroleum Geologists Bulletin,* Vol. 69, No. 1, pp. 22–38.

Ham, W. E., R. E. Denison, and C. E. Merritt, 1964, Basement Rocks and Structural Evolution of Southern Oklahoma, *Oklahoma Geological Survey Bulletin,* Vol. 95, 302 pp.

Landes, K. K., 1970, *Petroleum Geology of the United States,* Wiley-Interscience, New York, 571 pp.

Levorsen, A. I., 1956, *Geology of Petroleum,* W. H. Freeman, San Francisco, 703 pp.

Pippin, L., 1970, Panhandle–Hugoton Field, Texas–Oklahoma–Kansas—, the First Fifty Years, in *Geology of Giant Petroleum Fields,* M. T. Halbouty, Ed., American Association of Petroleum Geologists, Memoir 14, pp. 204–222.

Velde, B., 1977, *Developments in Sedimentary Clays and Clay Minerals in Natural and Synthetic Systems,* Elsevier Scientific Publishing Co., Amsterdam, Oxford, and New York, 218 pp.

16

Panhandle Field, The Brown Dolomite, Hutchinson County, Texas

Geologic Background

The most productive reservoir in the Panhandle field is probably the oolitic zone in the Brown Dolomite. According to Pippin (1970, p. 213), production from this single zone in Hutchinson County has been about 10,000 bbl/acre.

As will be seen from the photomicrographs in this chapter, much depends on the amount of anhydrite crystallized in the available pores and fractures. A brief glance at the oil pay map in Figure 16.1 shows the importance of this reservoir rock in terms of real distribution.

Later in this text, it will be observed that the reservoir rocks from the San Andres Formation of west Texas and New Mexico resemble those of the Brown Dolomite.

Geologic Interpretation

The Brown Dolomite illustrates the highly stratified nature of the oolite-type deposits. The situation is complicated by the fact that vertical fractures connect some, but not all, of the porous zones. As illustrated in the photomicrographs (Figures 16.2–16.7), the fractures are filled and ultimately sealed with varying

Geologic Interpretation

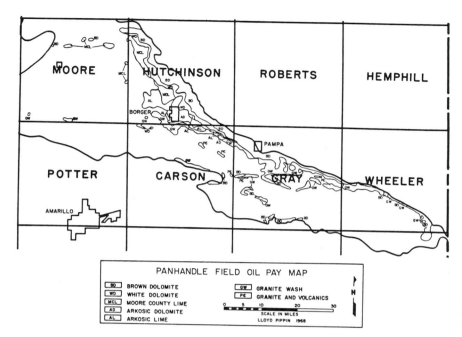

Figure 16.1. Oil pay map of the Panhandle field showing location of Brown Dolomite, Granite Wash, and basement production. (After Pippin 1970, p. 215.)

types of cementing material; therefore, timing the fracture with respect to the emplacement of the oil is important.

It is also quite clear that the composition of the circulating fluids changes along with the cementing materials. With these points in mind, it can be hypothesized that the Brown Dolomite reservoirs closest to the granite core will experience more fracturing and hence more vertical communication between porous zones. Those reservoirs further away from the granite ridge should have more discrete zones that show little or no fluid communication.

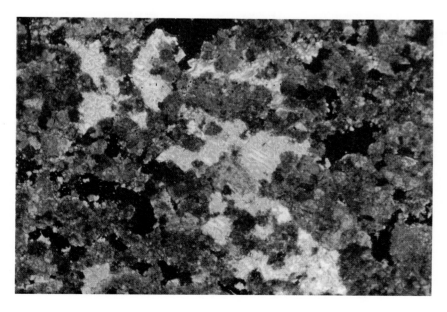

Figure 16.2. Brown Dolomite Formation, 3119 ft (950.9 m), Cities Service Starnes F, No. 18, Panhandle field, Hutchinson County, Texas (0.44 in. in the photo equals 0.17 mm). The most prolific part of the Brown Dolomite is the "dolomitized oolite zone shown above." Several oolite ghosts can be observed, although the dolomitization has masked the original sediment carbonate shell fragments of various kinds were also probably present. The initial porosity was quite high. The anhydrite filling (brightly colored) is patchy and continues to allow the flow of fluids. The environment appears to have been an "ooid shoal" (Longacre, 1980, p. 111).

140

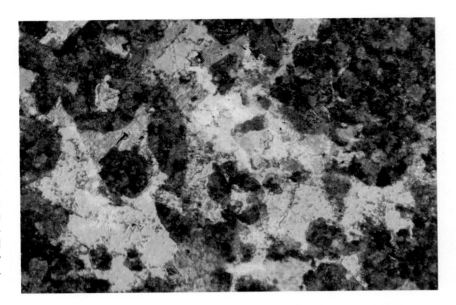

Figure 16.3. Brown Dolomite Formation, 3127 ft (953.4 m), Cities Service Starnes F, No. 18, Panhandle field, Hutchinson County, Texas (0.44 in. in the photo equals 0.17 mm). In this view, anhydrite has filled most of the available pore space. The round dolomite blobs reflect both dolomitized oolites and pellets. Other regular shapes appear to reflect dolomitized shell fragments.

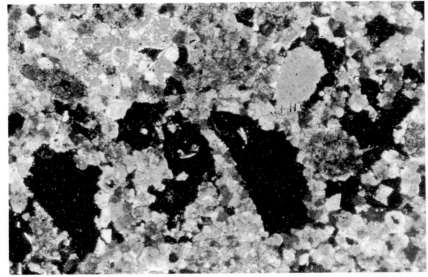

Figure 16.4. Brown Dolomite Formation, 3127 ft (953.4 m), Cities Service Starnes F, No. 18, Panhandle field, Hutchinson County, Texas (0.44 in. in the photo equals 0.17 mm). Large (moldic) pores with only minor infilling of anhydrite indicate good reservoir properties. Coarser dolomite is observed, with finer-grained dolomite revealing ghosts of dolomitized ooids. Some of the dolomite appears to be of the interlocking granular variety, suggesting subaerial exposure of the ooid shoal.

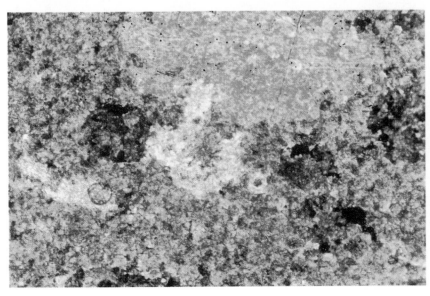

Figure 16.5. Brown Dolomite Formation, 3134 ft (955.5 m), Cities Service Starnes F, No. 18, Panhandle field, Hutchinson County, Texas (0.44 in. in the photo equals 0.17 mm). Finer-grained dolomite with anhydrite filling in almost all the pore space. Slightly darker rounded shapes suggest former ooids that have dolomitized.

Figure 16.6. Brown Dolomite Formation, 3173 ft (967.4 m), Cities Service Starnes F, No. 18, Panhandle field, Hutchinson County, Texas (0.44 in. in the photo equals 0.17 mm). Much finer-grained dolomite with a small amount of anhydrite lining a large void that was subsequently filled with "zebraic" chalcedony. This association of zebraic chalcedony with evaporites has been described by McBride (1977).

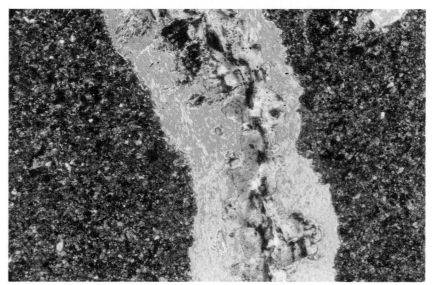

Figure 16.7. Brown Dolomite Formation, 3177 ft (968.6 m), Cities Service Starnes F, No. 18, Panhandle field, Hutchinson County, Texas (0.44 in. in the photo equals 0.17 mm). Fracture in fine-grained dolomite has first been lined with anhydrite and then finally filled with chalcedony. After these two episodes of deposition, little porosity or permeability remain.

References

Longacre, S. A., 1980, Dolomite Reservoirs from Permian Biomicrites, in *Carbonate Reservoir Rocks,* Notes for SEPM Core Workshop No. 1, R. B. Halley and R. G. Toucks, Eds., Society of Economic Paleontologists and Mineralogists, pp. 105–117.

McBride, E. F., 1977, Secondary Porosity—Importance in Sandstone Reservoirs in Texas, *Gulf Coast Association of Geological Societies, Transactions,* Vol. 27, pp. 1199–1208.

Pippin, L., 1970, Panhandle–Hugoton Field, Texas–Oklahoma–Kansas, the First Fifty Years, in *Geology of Giant Petroleum Fields,* M. T. Halbouty, Ed., American Association of Petroleum Geologists, Memoir 14, pp. 204–222.

17

Panhandle Field, The Red Cave Formation, Carson County, Texas

Geologic Background

The Red Cave reservoir is composed mainly of fine-grained sandstone with streaks of red shale and siltstone. The field limits are determined by the presence or absence of porosity. The photomicrographs (Figures 17.2–17.7) are more representative of the reservoir at the perimeter of the field where anhydrite forms patches of impermeable rock. Anhydrite-cemented sandstones are somewhat rare; however, they are characteristic of the Red Cave reservoir. The overall distribution of Red Cave gas production in the Panhandle field is shown in Figure 17.1.

The lack of porosity in contemporaneous strata around the field indicates that the gas did not migrate laterally into the field, but probably migrated vertically from the Wolfcamp below, through fractures. This hypothesis is supported by the fact that formation pressure originally was the same in both (the) Red Cave and the Wolfcamp.

This statement by Pippin (1970, p. 220) appears to be correct.

Petrologically, the Red Cave involves various sand influxes that are fine grained and are associated with either fine-grained carbonates or clay mate-

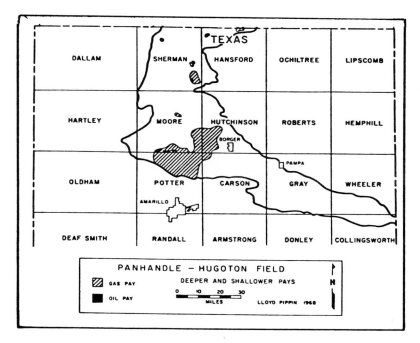

Figure 17.1. Gas pay areas reflect production from the Red Cave Formation. (After Pippin 1970, p. 221.)

rials. The photomicrographs in Figures 17.2–17.7 illustrate these associations and provide a good review of how anhydrite fills available pore space.

Source of Oil for the Panhandle Field

The location of the major oil production in the Panhandle field with respect to the Amarillo uplift indicates the direction from which the oil migrated. As was illustrated in Figure 15.1, the updip migration clearly favored the north side of the Amarillo uplift. In other words, the source rocks and the oil-generating conditions were far more favorable in the Anadarko basin than they were in the Palo Duro basin to the south. The paleogeographic map (Figure 17.8) of the mid-continent during Late Wolfcampian time shows the Panhandle field in relation to the marine sea to the north, which appears to have been key to oil formation.

Figure 17.9 shows the basement structure and the relationship between the Wichita uplift and the Amarillo uplift. In attempting to find a recent equivalent, the problem of locating even part of the situation that helped create the Granite Wash is not easy. A general feeling for what must have occurred can be obtained by climbing to the top of Mount Scott in the Wichita Wildlife Refuge near Lawton, Oklahoma, and considering what the terrane would look like if a sea surrounded the base of the granite mountains. At least, the mineralogy of the sediments produced today should be somewhat similar and the structural setting close to that of the Amarillo uplift.

In studying the available data concerning the oil composition of the Panhandle field oils, one can say very little other than that the oil from the Moore County Limestone appears to be different from what is called the Permian and Pennsylvanian oil as shown by the plates in Figure 17.10. One might expect local differences between the Granite Wash oils and those of the Brown Dolomite or the Moore County Lime on the basis of the active nature of the fine-grained sediments in the Granite Wash. However, this hypothesis will have to be tested elsewhere.

In terms of future exploration, it seems unlikely that there are undiscovered fields within the continental United States that are similar to the Panhandle field.

144

Figure 17.2. Red Cave Formation, 1772 ft (540.2 m), Cities Service Gas Company Deahl No. 1-R, Panhandle field, Carson County, Texas (0.44 in. in the photo equals 0.08 mm). Fine-grained sandstone with anhydrite filling in the most porous zone. Clay matrix is also abundant. Sand grains are angular, and in some cases they appear to have been replaced by the anhydrite cement.

Figure 17.3. Red Cave Formation, 1774 ft (540.9 m), Cities Service Gas Company Deahl No. 1-R, Panhandle field, Carson County, Texas (0.44 in. in the photo equals 0.08 mm). Coarser sandstone with anhydrite cement filling in the pores and reducing the permeability.

Figure 17.4. Red Cave Formation, 1780 ft (542.7 m), Cities Service Gas Company Deahl No. 1-R, Panhandle field, Carson County, Texas (0.44 in. in the photo equals 0.08 mm). Fine-grained dolomite surrounds the sand material. Perthite feldspar is present (middle). Sand appears to be angular and not very well sorted.

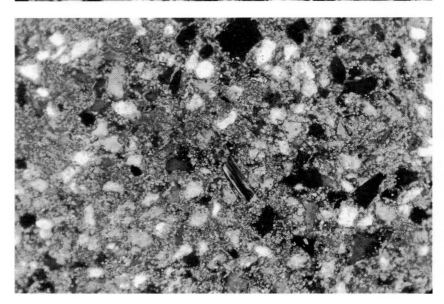

Figure 17.5. Red Cave Formation, 1782.5 ft (543.5 m), Cities Service Gas Company Deahl No. 1-R, Panhandle field, Carson County, Texas (0.44 in. in the photo equals 0.08 mm). Red clay layer still contains numerous angular quartz grains. Clay minerals (probably kaolinite) appear to be associated with hematite.

Figure 17.6. Red Cave Formation, 1795 ft (543.3 m), Cities Service Gas Company Deahl No. 1-R, Panhandle field, Carson County, Texas (0.44 in. in the photo equals 0.08 mm). Coarser sandstone with anhydrite cement. Many of the quartz grains show evidence of replacement by anhydrite.

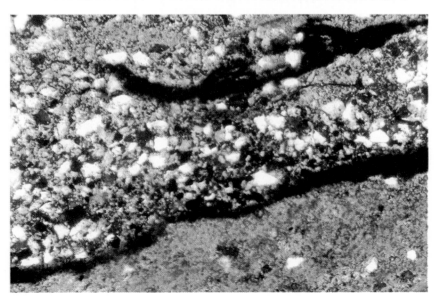

Figure 17.7. Red Cave Formation, 1847.5 ft (563.3 m), Cities Service Gas Company Deahl No. 1-R, Panhandle field, Carson County, Texas (0.44 in. in the photo equals 0.08 mm). Fine-grained silt layer bounded by two organic rich zones.

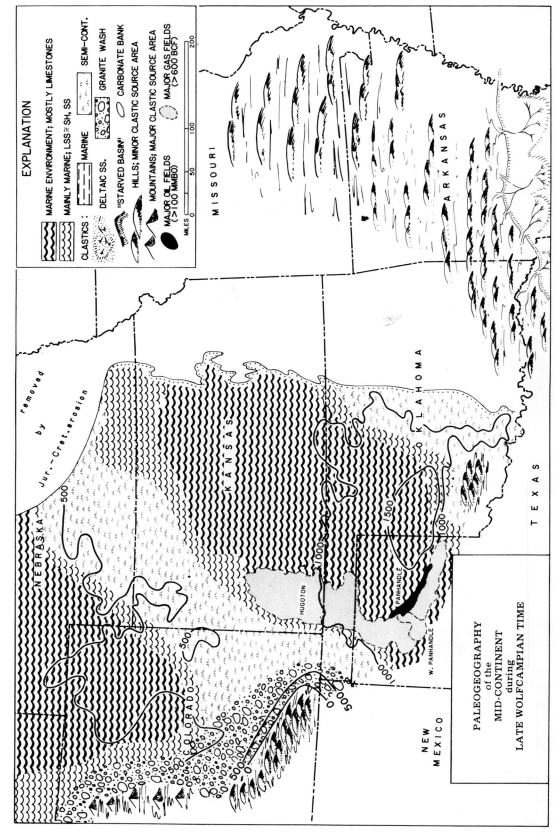

Figure 17.8. Paleogeography and isopachs of Wolfcampian strata for the mid-continent. The oil and gas accumulations in the Panhandle field are related to the Amarillo uplift. (After Rascoe and Adler 1983, p. 999.)

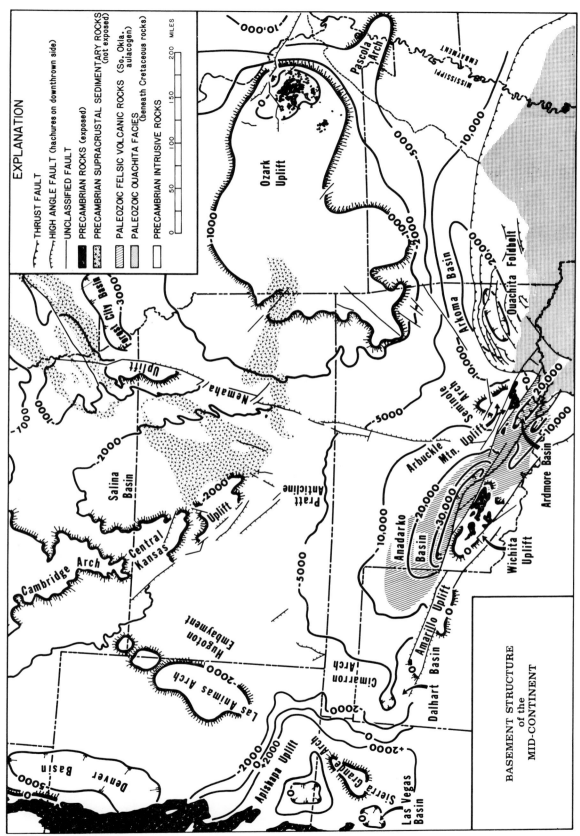

Figure 17.9. Basement structure of the mid-continent. (After Adler et al. 1971, Figure 3.)

147

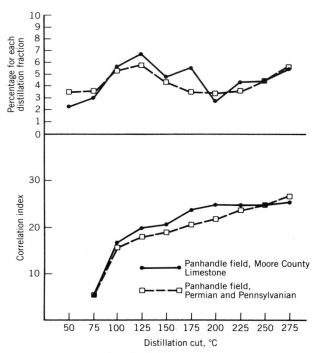

Figure 17.10. Panhandle field, Hutchinson County, Texas (McKinney and Garton 1957, McKinney et al. 1966).

References

Adler, F. J., M. W. Caplan, M. P. Carlson, E. D. Goebel, H. T. Henslee, I. C. Hicks, T. G. Larson, M. H. McCracken, M. C. Parker, B. Rascoe, Jr., M. W. Schramm, Jr., and J. S. Wells, 1971, Future Petroleum Provinces of the Mid-Continent Region, in *Future Petroleum Provinces of the United States—Their Geology and Potential,* American Association of Petroleum Geologists, Memoir 15, Vol. 2, pp. 985–1120.

McKinney, C. M. and E. L. Garton, 1957, *Analyses of Crude Oils from 470 Important Oil Fields in the United States,* Bureau of Mines Report of Investigations 5376, 276 pp.

McKinney, C. M., E. P. Ferrero, and W. J. Wenger, 1966, *Analyses of Crude Oils from 546 Important Oil Fields in the United States,* Bureau of Mines Report of Investigations 6819, 345 pp.

Pippin, L., 1970, Panhandle–Hugoton Field, Texas–Oklahoma–Kansas, the First Fifty Years, in *Geology of Giant Petroleum Fields,* M. T. Halbouty, Ed., American Association of Petroleum Geologists, Memoir 14, pp. 204–222.

Rascoe, B., Jr., and F. J. Adler, 1983, Permo-Carboniferous Hydrocarbon Accumulations, Mid-Continent, U.S.A., *American Association of Petroleum Geologists Bulletin,* Vol. 67, No. 6, pp. 979–1001.

18

West Texas and Eastern New Mexico

One of the most productive areas of oil and gas in the world is encompassed by the region that includes the Midland basin and Central basin platform of west Texas and southeastern New Mexico. The reservoir types are so numerous and varied that only a very few can be mentioned in a volume of this kind. Fortunately, a great deal has been written about this prolific province and those who care to dig deeper into the literature concerning various fields and their geologic setting can get a good start by consulting the guidebooks published by the West Texas Geological Society and the American Association of Petroleum Geologists.

There is a logical tendency to separate this province from that that includes the Panhandle field discussed in Chapter 17, but as these reservoirs are examined, it becomes clear that there are similarities in the dolomite reservoirs. As illustrated in Figure 18.1, the next large structural element south of the Amarillo uplift is the Palo Duro basin. The fact that this basin has been disappointing in terms of oil and gas production has helped in setting apart the west Texas production from the continuous sequence of sediments that cover the whole area. Moving south over the Matador arch, there is a whole spectrum of oil- and gas-producing reservoirs covering almost all of the stratigraphic units shown in Figure 18.2. Why the Palo Duro basin has not been as highly productive appears to be a reflection of a lack of appropriate source rocks.

The particular reservoir rocks discussed were not selected on the basis of the unusual features they show but rather on the availability of core materials and thin sections.

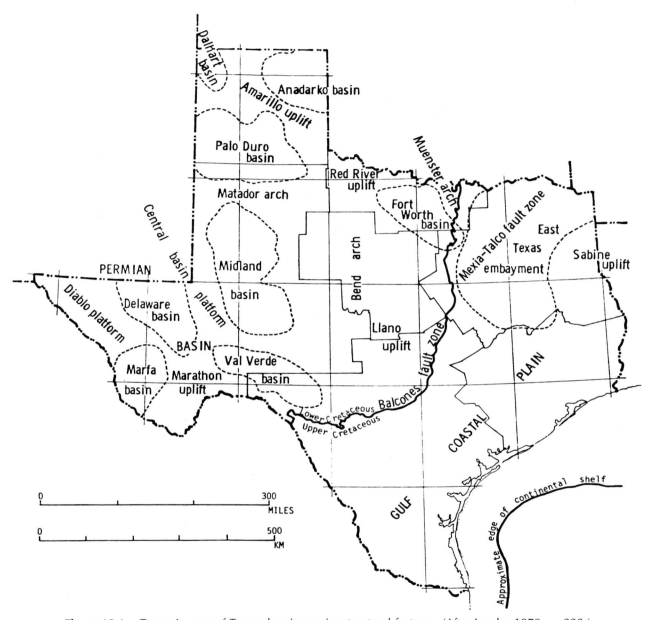

Figure 18.1. Tectonic map of Texas showing major structural features. (After Landes 1970, p. 222.)

General Geologic Setting

The limiting structural elements of the large area known as the Permian basin are the Bend arch on the east, the Amarillo–Wichita uplift on the northeast, the Pedernal massif on the northwest, the Diablo platform on the southwest, and the Marathon folded belt on the south. The major structural features within this area are the Palo Duro basin, the Matador arch, the Northwestern shelf, the Delaware basin, the Central basin platform, the Midland basin, the Eastern shelf, the Ozona platform, and the Val Verde and Marfa basins (see Figures 18.1 and 18.3).

Most of the oil and gas production is located to the south of the Palo Duro basin and south of the Matador arch. It is this highly productive area that will be the focus of the studies in this volume.

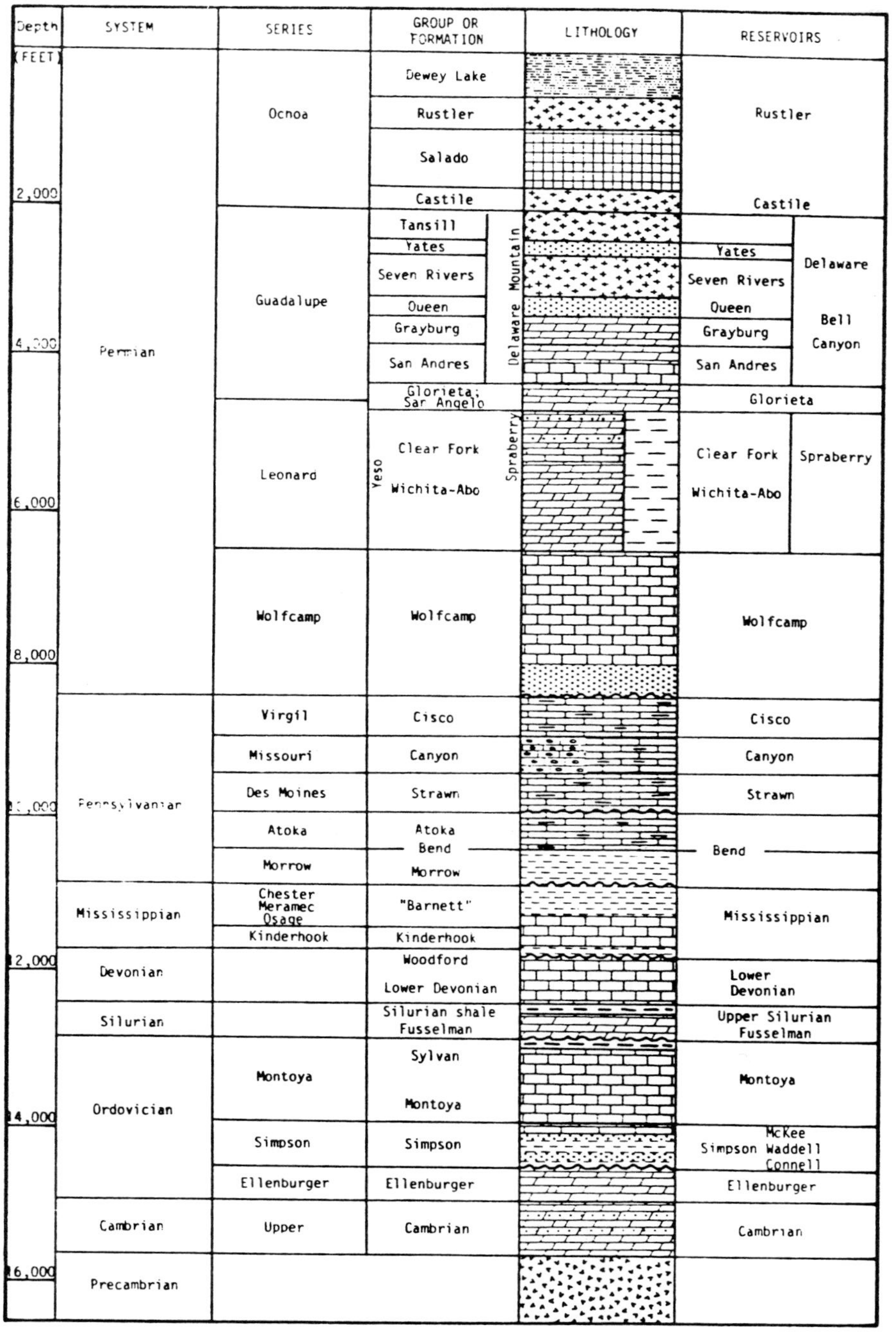

General Geologic Setting

Depth (FEET)	SYSTEM	SERIES	GROUP OR FORMATION	LITHOLOGY	RESERVOIRS	
	Permian	Ochoa	Dewey Lake			
			Rustler		Rustler	
2,000			Salado			
			Castile		Castile	
		Guadalupe	Tansill			
			Yates		Yates	Delaware
			Seven Rivers		Seven Rivers	
			Queen		Queen	
4,000			Grayburg		Grayburg	Bell Canyon
			San Andres		San Andres	
			Glorieta; San Angelo		Glorieta	
		Leonard	Clear Fork		Clear Fork	Spraberry
6,000			Wichita-Abo		Wichita-Abo	
		Wolfcamp	Wolfcamp		Wolfcamp	
8,000						
	Pennsylvanian	Virgil	Cisco		Cisco	
		Missouri	Canyon		Canyon	
		Des Moines	Strawn		Strawn	
10,000		Atoka	Atoka — Bend		— Bend —	
		Morrow	Morrow			
	Mississippian	Chester Meramec Osage	"Barnett"		Mississippian	
		Kinderhook	Kinderhook			
12,000	Devonian		Woodford		Lower Devonian	
			Lower Devonian			
	Silurian		Silurian shale Fusselman		Upper Silurian Fusselman	
	Ordovician	Montoya	Sylvan		Montoya	
			Montoya			
14,000		Simpson	Simpson		Simpson McKee Waddell Connell	
		Ellenburger	Ellenburger		Ellenburger	
	Cambrian	Upper	Cambrian		Cambrian	
16,000	Precambrian					

Figure 18.2. Generalized stratigraphic column for west Texas and southeastern New Mexico. The principal reservoirs are shown at right. (After Landes 1970, p. 328.)

Central Basin Platform

The structural feature that includes the most oil and gas productive area within the Permian basin is the Central basin platform. As illustrated in Figure 18.3, it is a north–northwest trending uplift roughly 170 miles long and 50 miles wide. Its northern end extends into southeastern New Mexico and its southern end turns east and combines with a small structure called the Ozona platform.

The arched foundation of the Central basin platform contains folded and faulted sedimentary rocks ranging in age from Cambro-Ordovician to Pennsylvanian. The Cambro-Ordovician sequence is divided into the Ellenburger, Simpson, and Montoya formations. Above this is the Silurian Fusselman Dolomite, the Upper Silurian Shale, the Lower Devonian carbonates, the Mississippian carbonates, and the dark Barnett Shale. The Pennsylvanian is divided into the Bend, Strawn, Canyon, and Cisco groups, which exhibit varying lithologies.

Lying unconformably on top of this foundation is a thick sequence of Permian sediments. Most of the lower half of this sequence is composed of dolomite interbedded with a few limestones and some sandstones and shales in the central part. Important subdivisions of this dolomitic half are the Wichita, Clear Fork, San Angelo (Glorieta), San Andres, and Grayburg formations.

Above the Grayburg is a thick section composed mainly of salt and anhydrite with a few interbedded sandstones. Recognizable formations in this sequence are the Queen, Seven Rivers, Yates, Tansill, Salado, Rustler, and Dewey Lake as shown in Figure 18.2.

The general relationship of the Central basin platform to the Delaware basin on the west and the Midland basin to the east is illustrated in Figures 18.3 and 18.4.

For the purpose of this work, another structural feature of particular interest is the Midland basin. This basin parallels the east flank of the Central basin platform and is limited to the north by the Matador arch, to the east by the so-called Eastern shelf, and to the south by the Ozona platform. The Cambro-Ordovician, Silurian–Devonian, and Mississippian sediments in the Midland basin are relatively thin and pinch out or are truncated on the east side of the basin. The Pennsylvanian strata composed of the Bend (mostly shale) and Strawn (limestone) cover most of the basin. In the northern Midland basin and on the western flank of the Eastern shelf, prolific reefs occur. The famous Scurry reef and the Horseshoe atoll are present in this sequence.

In the deeper parts of the Midland basin, we en-

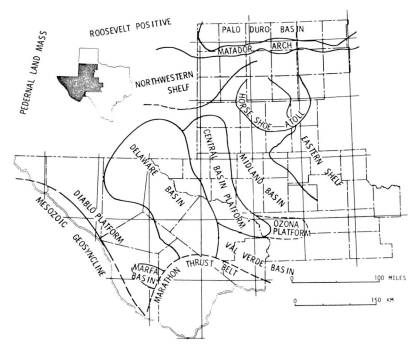

Figure 18.3. Tectonic map of west Texas and southeastern New Mexico. (Modified after Sax 1967, pp. 1056–1057.)

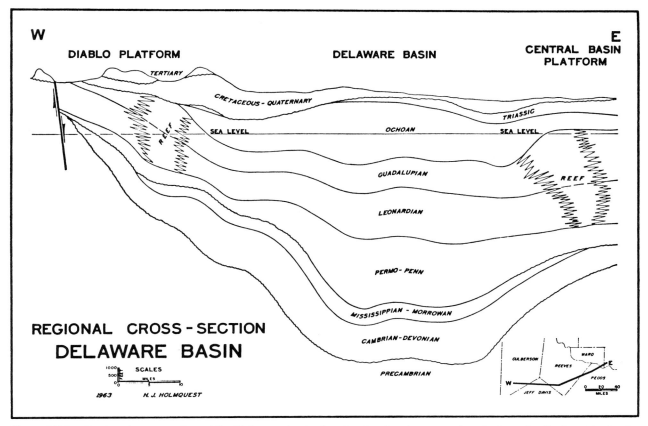

Figure 18.4. Regional cross section of the Delaware basin showing the development of reefs along the flanks of the basin. (After Holmquest 1965, p. 261.)

counter the Permian sediments, which include the "Dean Sand" and the Spraberry Sand. This latter sand will be examined along with two San Andres reservoirs.

Delaware Basin

The Delaware basin is bounded by the Northwestern shelf on the north and northwest, the Central basin platform on the east, the Diablo platform on the southwest, and the Val Verde basin on the southeast. Its main axis trends northwest for about 150 miles and its greatest width is about 100 miles. The Pennsylvanian Strawn, Canyon, and Cisco formations, which are mostly carbonates on and around the folded core of the Central basin platform, change into shale and sandstone in the Delaware basin. The dolomitic Permian sediments of the Central basin platform grade into a thick series of sandstones, shales, and limy shales in the Delaware basin. Formations composing this 8000 ft of section are the Wolfcamp, Leonard (Bone Spring), and the Guadalupe (Delaware Mountain) Group.

Reefs in the Guadalupe (Delaware Mountain) such as the famous Capitan reef outline a large part of the Delaware basin.

Other relatively small basins included within what is called the Permian basin are the Val Verde basin and the Marfa basin. The fields mentioned in this work are not close enough to these features to warrant a discussion of their stratigraphy.

Reservoirs from the following formations will be covered in this work:

1. the Ellenburger Dolomite (Lower Ordovician, TXL field),
2. the Spraberry Sand (Lower Permian, Spraberry Trend),

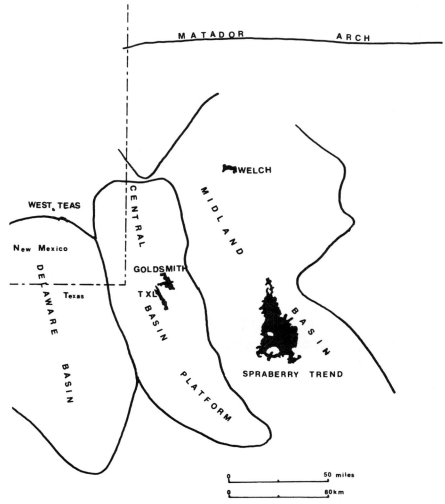

Figure 18.5. Map of west Texas showing locations of the TXL, Welch, Goldsmith West Texas fields and the Spraberry Trend. (After Cooper and Ferris 1957, p. ix.)

3. the San Andres Dolomite (Middle Permian, Welch, Goldsmith fields), and
4. the Yates Sand (Upper Middle Permian, west Texas field).

(See Figure 18.5 for the locations of these fields.)

Two examples will be drawn from the San Andres Formation because of the different settings involved and the overall importance of San Andres production to the west Texas regions.

The Ellenburger Formation TXL Field, Ector County, Texas

Geologic Background

As mentioned earlier, the Central basin platform is one of the most highly oil productive areas in Texas. The TXL field is located close to the western edge of Ector County and not far from the Goldsmith field, which is a short distance to the east (3 miles). Both of these fields are discussed; however, in the case of the Goldsmith field, the discussion focuses on the San Andres (Permian age) Formation and the Ordovician age Ellenburger in the case of the TXL field.

In the TXL field, Permian rocks lie unconformably on truncated pre-Permian strata of the TXL anticline. The major unconformity is expressed in the form of a detrital zone composed of a Devonian chert (David 1946, p. 118).

The TXL field produces from a number of zones; according to Landes (1970, p. 345), it has a history of oil and gas production from 11 reservoirs. (See Figure 18.6.) The leading productive zones are the Ellenburger Dolomite and the Devonian Chert.

The trapping mechanism is a long narrow anticline (see Figure 18.7). The productive area of the TXL field is roughly 9 miles long and $2\frac{1}{2}$ miles wide with the southern end narrowed by a fault that cuts off production on the eastern side of the anticline. The nature of the cutoff is shown in the cross section C–C' in Figure 18.8.

The Ellenburger is brown to gray crystalline dolomite that lacks porosity in many places; therefore, the oil is stored and released from numerous microfractures. Examples of these microfractures are illustrated in Figure 18.12.

According to Jones and Smith (1965, p. 118), the Ellenburger rocks form a single sedimentary sequence that was deposited on a broad southeastward dipping shelf. The overall thickness ranges from zero in the northwest to 2000 ft on the southeast. The Ellenburger dolomite covers the whole area and contains most of the oil.

The sequence of events that resulted in the oil reservoir at the TXL field began at the end of Ellenburger time, when the shelf was elevated above sea level and eroded. This erosion caused joints and fractures to be widened by solution. Subsidence at the beginning of Simpson time introduced clastics, which covered the Ellenburger in part of the area.

The fact that the Ellenburger is the major producer in the Cambrian–Lower Ordovician sequence focuses attention on the environment of deposition. Jones and Smith (1965, p. 118) state that the Ellenburger was deposited in a shallow, warm, well-lighted, oxygenated environment. Subsequently, it underwent little compaction and was eroded.

In looking for the source of the oil, it appears that there are two possible options. The first possibility includes the shaly Cambrian and Ordovician rocks in the deeper basin south of the area and the second option is the Simpson Group overlying the Ellenburger.

Of the two options, Jones and Smith (1965, p. 119) favored the Simpson origin. In the particular case of the TXL oil, it is believed to have migrated across faults, along fault planes and other fractures, and in some cases laterally through the Ellenburger under hydrodynamic forces. The correlation index profile for the Ellenburger oil at TXL (Figure 18.9) closely resembles those for several of the oils from the Simpson; therefore, it seems likely that they have the same origin.

As will be observed, the Ellenburger samples at our disposal are particularly tight, and fracture porosity appears to have been important in this reservoir.

Another feature of the TXL oil that should be observed is its relatively high paraffinicity (low correlation index). The feature should be kept in mind as additional oils from reservoirs higher in the section (Permian age) are studied.

Photomicrographs and Description

It is unfortunate that only two samples from roughly the same depth were available for study; however, what we have does provide an idea of the type of dolomite commonly encountered. The rock is almost totally dolomite, and hence, mineralogic point counts are not helpful. Furthermore, dolomitization has been so complete that there are no recognizable structures left as shown in Figures 18.10–18.13. Those who desire to know more about the whole spectrum of rock types that can be encountered in the Ellenburger can consult the work of Loucks and Anderson in their detailed study of the Ellenburger in the Puckett field in Pecos County, Texas (1980, pp. 1–31).

Comments

Finding a present-day equivalent to the Ellenburger Dolomite as it occurs on the Central basin platform of west Texas is a tall order. We can learn much from the recent carbonate sequences in and around the Gulf of Arabia; however, the overall geologic setting

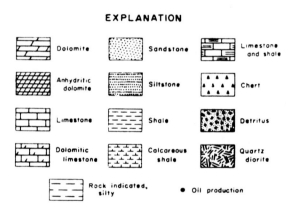
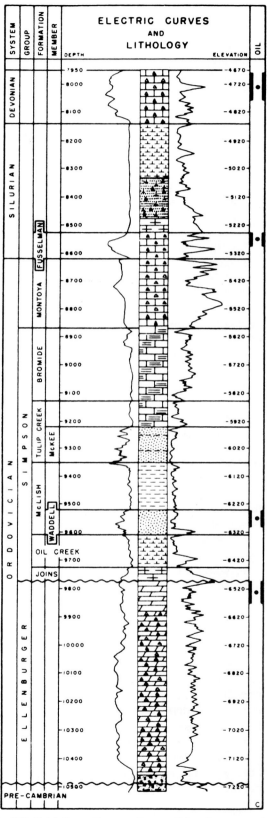

Figure 18.6. Typical section of pre-Mississippian rocks penetrated in the TXL field, Ector County, Texas. (After Cooper and Ferris 1957, p. 360.)

Comments

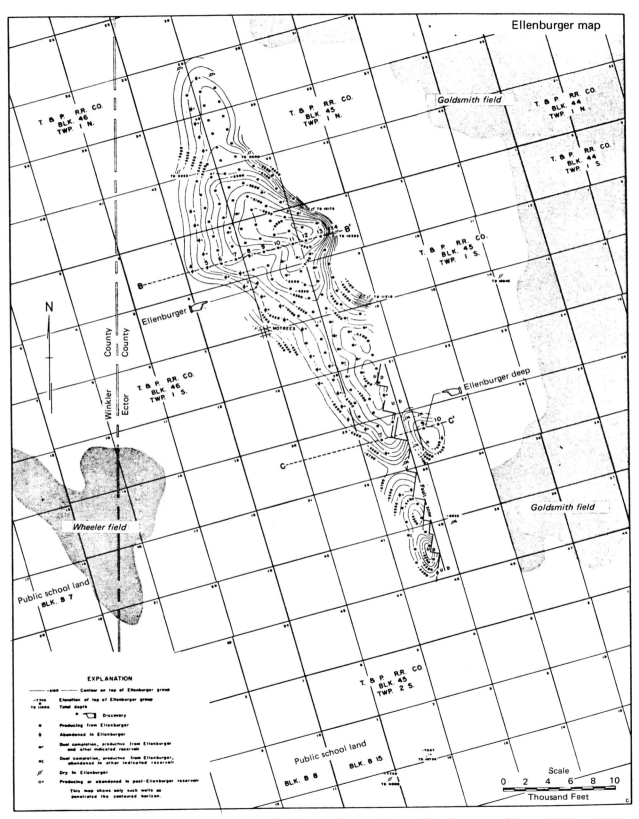

Figure 18.7. Structure contour map of the TXL field, Ector County, Texas. (After Cooper and Ferris 1957, p. 367.)

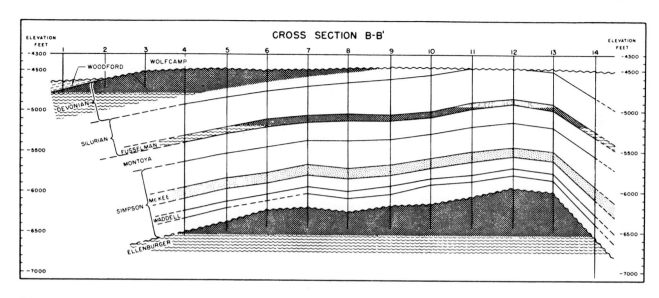
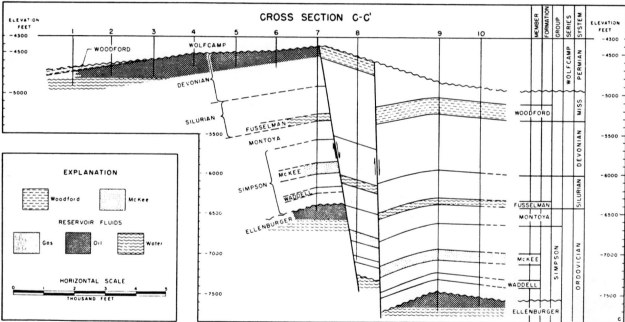

Figure 18.8. Cross sections of the TXL field, Ector County, Texas, showing structural position of the Ellenburger Dolomite. (After Cooper and Ferris 1957, p. 361.)

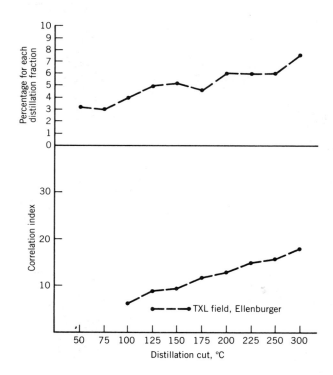

Figure 18.9. Correlation index profile of crude oil from Ellenburger Dolomite, 10,790–10,975 ft, TXL field, Ector County, Texas, Bureau of Mines R. I. 5378, p. 251. (McKinney and Garton 1957.)

Figure 18.10. Ellenburger Formation, 9000 ft (2743.9 m), Cities Service Cummins J-2, TXL field, Ector County, Texas (0.44 in. in the photo equals 0.17 mm). Coarse-grained interlocking dolomite rhombs. Dark (black) areas represent grains that are at optical extinction and are not pores.

Figure 18.11. Ellenburger Formation, 9000 ft (2743.9 m), Cities Service Cummins J-2, TXL field, Ector County, Texas (0.44 in. in the photo equals 0.08 mm). Higher magnification shows zoned dolomite with microinclusions in the cores of the dolomite rhombs. Dolomitization has almost completely filled in any primary porosity.

Figure 18.12. Ellenburger Formation, 9000 ft (2743.9 m), Cities Service Cummins J-2, TXL field, Ector County, Texas (0.44 in. in the photo equals 0.08 mm). Coarse-grained dolomite exhibits fractures across the grains. These microfractures suggest that this mechanism is at least partially responsible for the flow of fluids in the reservoir.

Figure 18.13. Ellenburger Formation, 9000 ft (2743.9 m), Cities Service Cummins J-2, TXL field, Ector County, Texas (0.44 in. in the photo equals 0.08 mm). Coarse-grained dolomite at higher magnification showing small quartz grain in the center. These fine-grained sand grains are probably introduced by the wind as the dolomite was being deposited.

of the Ellenburger does not lend itself to modern conditions.

In terms of future exploration, it is safe to say that exploration for Ellenburger reservoirs in west Texas is in its mature stage. New oil will certainly be found in the deeper parts of the basins around the Central basin platform; however, it will be expensive in terms of exploration, drilling, and production.

References

Cooper, C. G. and B. J. Ferris, 1957, TXL Field, in *Occurrence of Oil and Gas in West Texas,* University of Texas Publication 5716, pp. 358–368.

David, M., 1946, Devonian (?) Producing Zone, TXL Pool, Ector County, Texas, *American Association of Petroleum Geologists Bulletin,* Vol. 30. No. 1, pp. 118–119.

Holmquest, H. J., 1965, Deep Pays in Delaware and Val Verde Basins, in *Fluids in Subsurface Environments,* American Association of Petroleum Geologists, Memoir 4, pp. 257–279.

Jones, T. S. and H. M. Smith, 1965, Relationships of Oil Composition and Stratigraphy in the Permian Basin of West Texas and New Mexico, in *Fluids in Subsurface Environments,* Memoir 4, American Association of Petroleum Geologists, Tulsa OK, pp. 101–224.

Landes, K. K., 1970, *Petroleum Geology of the United States,* Wiley-Interscience, New York 571 pp.

Loucks, R. G. and J. H. Anderson, 1980, Depositional Facies and Porosity Development in Lower Ordovician Ellenburger Dolomite, Puckett Field, Pecos County, Texas, in *Notes for SEPM Core Workshop No. 1,* R. B. Halley and R. G. Loucks, Eds., Society of Economic Paleontologists and Mineralogists, pp. 1–31.

McKinney, C. M. and E. L. Garton, 1957, *Analyses of Crude Oils form 470 Important Oilfields in the United States,* Bureau of Mines Report of Investigations 5376, 276 pp.

Sax, N., 1967, Developments in West Texas and Southeastern New Mexico in 1966, *American Association of Petroleum Geologists Bulletin,* Vol. 51, No. 6, pp. 1053–1061.

19

Spraberry Field, The Spraberry Formation, Reagan County, Texas

The most important field within the confines of the Midland basin itself is the Spraberry field, or Spraberry "trend." The area contains a number of "pools" near the middle of the basin that produce from a fine-grained sandstone in the Permian Spraberry Formation.

As Landes (1970, p. 337) notes, the remarkable thing about the Spraberry trend is its large area and the difficulty with which oil is recovered from the reservoir rock. The trend appears to underlie a minimum of 1,000,000 acres spread among parts of 12 west Texas counties. The productive area is approximately 150 miles long and is at least 75 miles wide (Levorsen, 1956, p. 118).

The Spraberry section averages about 1000 ft in thickness and is underlain by the so-called Dean Sand (Figure 19.1). This latter unit is actually composed of approximately 300 ft of fine-grained sand, dark shale, and shaley limestone. On top of the Spraberry lies the Clearfork Formation, locally called the Tubb Sand, which is composed of shales and crystalline limestones.

According to Warn and Sidwell (1953, p. 67), the Spraberry Formation "was deposited slowly and intermittently in a subsiding basin." The Midland basin in Spraberry time occupied part of the Texas foreland and was almost surrounded by shelves and platforms such as the Central basin platform, which was discussed in the Chapter 18. Although there has been some difference of opinion concerning the source of the Spraberry sands, Schmitt (1954, p. 1977)

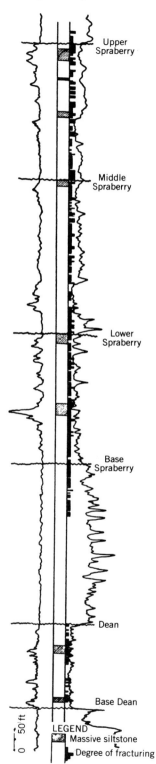

Figure 19.1. Typical electric log profile through the Spraberry Formation, Midland basin, Texas. Diagonal shading shows sandstone and siltstone. Solid black shows degree of fracturing. (After Wilkinson 1953, p. 254.)

indicated that he favored a source to the north or northwest. His reasoning is based on the following: Particle size decreases to the south over a broad area, and the total sand thickness decreases southward, as illustrated in Figure 19.2. This author feels that this latter interpretation is a reasonable one.

Photomicrographs and Description

As pointed out by several authors (Warn and Sidwell 1953; Wilkinson 1953), the major component of the Spraberry sand is fine-grained quartz. The next most significant mineral is carbonate cement. Table 19.1 provides the average mineral composition for the thin sections pictured in this chapter.

The grainsize distribution for the long axes of quartz grains is provided in Figure 19.3. Compared to most other reservoirs, the Spraberry is very fine grained, and it is surprising that significant amounts of oil can be produced at all. The photomicrographs in Figures 19.4 and 19.5 make one wonder about the flow of fluids through such a system. In fact, the clastic ratio map shown in Figure 19.6 suggests that much of field is in the shale end of the classification triangle.

Fractures in the Spraberry

Wilkinson (1953) studied the fractures in the Spraberry reservoir and concluded that fracturing was the important feature in determining the production characteristics. His concept (p. 261) was as follows:

The matrix siltstone which serves as the main reservoir rock has an average permeability of 0.50 millidarcy and an average porosity of 8 percent. With this type of permeability in the reservoir rock, it becomes obvious that the fractures serve mainly as "feeder lines" to conduct oil to the well bore. . . .

Wilkinson's initial production potential map (Figure 19.7) of the Tex–Harvey pool shows a strong north–northeast pattern that parallels and reflects the fracture system. Any attempts at secondary or tertiary recovery from this type of reservoir must take the fracture system into account.

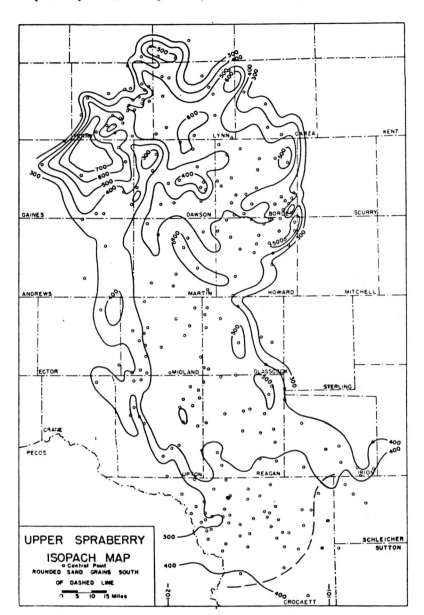

Figure 19.2. Isopach map of the upper Spraberry unit showing increasing thickness to the north end of the Midland basin. (After Schmitt 1954, p. 1972.)

TABLE 19.1 Average Mineral Composition, Upper Spraberry Sand, 6567.7–6568.5 ft (2002.4–2002.6 m), Phillips Oil Co., No. 2 Malone, Spraberry Field, Reagan County, Texas

Quartz	Carbonate	Feldspar	Clay	Heavy Mineral	Chert
74.3%	17.0%	2.3%	5.0%	1.0%	0.4%

Fractures in the Spraberry

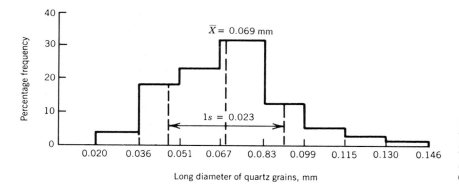

Figure 19.3. Grainsize distribution for long axes of quartz grains, upper Spraberry Sand, Phillips Oil Co., No. 2 Malone, Spraberry field, Reagan County, Texas.

Figure 19.4. Spraberry Formation (upper sand) 6567.7–6568.5 ft (2002.4–2002.6 m), Phillips Oil Co., No. 2 Malone, Spraberry field, Reagan County, Texas (0.44 in. in the photo equals 0.19 mm). Very fine grained quartz sandstone. Cementing materials include carbonate (light brown), which is scattered about suggesting fragments of shell material. Quartz cement also appears to be very important.

Figure 19.5. Spraberry Formation (upper sand), 6567.7–6568.5 ft (2002.4–2002.6 m), Phillips Oil Co., No. 2 Malone, Spraberry field, Reagan County, Texas (0.44 in. in the photo equals 0.19 mm). Very fine grained, angular, cemented sandstone. Carbonate cementing probably reflects shell fragments of various organisms. Silica cement is also important. The feldspar (striped mineral) does not appear to have suffered much weathering. Some pyrite is present, reflecting reducing conditions at the time of deposition.

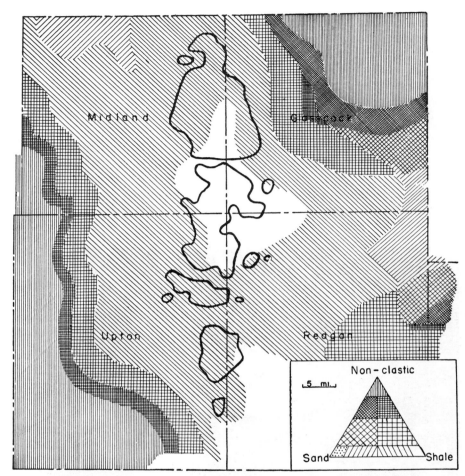

Figure 19.6. Clastic ratio map showing the Spraberry Formation of west Texas. The Spraberry field is outlined in black. (After Levorsen 1956, p. 584.)

The studies of both the San Andres and Yates formations will also show the influence of fractures.

Levorsen (1956, p. 118) states the following with reference to the Spraberry reservoirs:

Porosity is generally less than 10 percent, and the average permeability is ½ md. Oil pools are unusual in reservoir rocks of such low porosity, and low permeability. The effective permeability, therefore, is almost altogether in fractures and conchoidal openings, which extend through the finer materials in all directions, but chiefly vertically.

Exploration

Once again, locating a recent equivalent is very nearly impossible. Perhaps the basic lesson from the geologists' point of view is that where good source beds are in contact with an appropriate fracture system, oil is likely to be found at fracture junctions.

Source of Spraberry Oil

Jones and Smith, (1965, p. 153) indicate that the plots of the correlation indices for the Spraberry oils show no unusual characteristics (Figure 19.8). Examination of the environment of deposition shows that the sands are associated with dark shales (gray to black) of the restricted basin type. The fact that pyrite and dark gray to black limestone and dolomite are present further supports the concept of a restricted basin and reducing conditions. According to Schmitt (1954, p. 1977), the sea was probably very shallow on the sur-

Source of Spraberry Oil

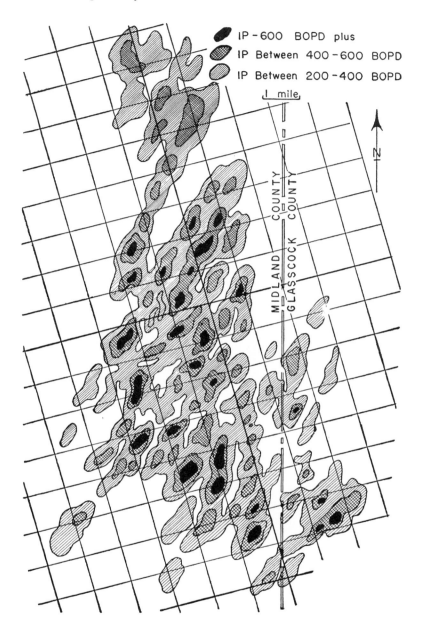

Figure 19.7. Isopotential map of the Tex–Harvey pool in the Spraberry field of west Texas. The linear pattern is a reflection of the fracture system. Where fractures are more numerous, greater well productivity occurs. (After Levorsen 1956, p. 596.)

rounding shelves and was less than 600 ft in the basin itself during most of Spraberry time. Schmitt (1954, p. 1978) goes on to state the following:

The oil was probably formed from the bituminous shale which apparently was rich in the necessary constituents for the formation of petroleum. Because of the low permeability of the rocks, the generated petroleum did not have an opportunity to escape or migrate any great distance other than into the coarser beds associated with the Spraberry section.

This author feels the above concept is correct. If the oil has not moved far from its source, then its correlation index profile (Figure 19.8) can be used as an end member against which other oils in the Permian can be compared.

Table 19.2 outlines the reserves and productive capacity of the Spraberry field.

TABLE 19.2 Summary Report of Reserves and Productive Capacity, Natural Gas, December 31, 1974, Spraberry Trend Field[a,b]

	Crude Oil (MMbbl)	Lease Condensate (MMbbl)	Associative (BCF)	Nonassociative (BCF)	Liquids (MMbbl)
		(Wet Basis)			
Hydrocarbons originally in place	7771.9	NA	3885.9	NA	—
Proved ultimate recovery	469.0	NA	1261.2	NA	—
Cumulative production	378.8	NA	990.7	NA	—
Proved reserves	90.2	NA	270.5	NA	—
		(Dry Basis)			
Proved reserves[c]	—	—	230.0	NA	21.3
Reserves in shut-in reservoirs	0	NA	0	NA	0
Indicated secondary and tertiary reserves	75.0	—	128.0	—	—
Production					
1973 (total)	19.8	NA	41.7	NA	4.2
1974 (total)	17.6	NA	42.0	NA	4.2
Long-term projection of production (annual total)					
1975	13.8	NA	33.9	NA	—
1976	11.0	NA	28.0	NA	—
1977	9.2	NA	24.3	NA	—
1978	7.8	NA	21.1	NA	—
1979	6.8	NA	18.3	NA	—
1980	5.9	NA	15.9	NA	—
1981	5.1	NA	13.8	NA	—
1982	4.5	NA	12.0	NA	—
1983	3.9	NA	10.4	NA	—
1984	3.4	NA	9.0	NA	—
	(Mbbl)	(MMbl)	(MMCF)	(MMCF)	(Mbbl)
Daily Averages					
December 1974 production	45.3	NA	105	NA	10.5
Short-term productive capacity (60-day basis)	45.3	NA	105	NA	—

[a] Does not include the Clearfork, Wolfcamp, and Devonian zones, which would increase the data by about 1–2%.

[b] Table obtained from a report by the Office of Policy and Analyses and Office of Energy Resource Development, 1975, Final Report, Vol. II, Oil and Gas Resources, Reserves, and Productive Capacities, Federal Energy Administration, Washington, D.C., 151 pp.

[c] Proved reserves do not include 75 MMbbl and 191 BCF of inferred reserves from future extension of field limits.

References

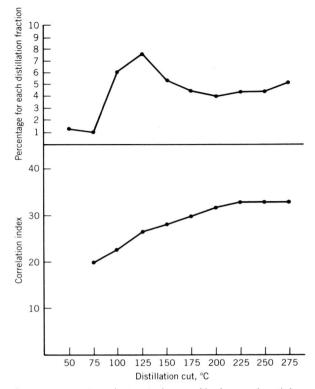

Figure 19.8. Correlation index profile for crude oil from the Spraberry Formation, 7128–7228 ft (2173–2204 m), Spraberry Trend field, Reagen County, Texas (McKinney and Shelton 1966, p. 140).

Landes, K. K., 1970, *Petroleum Geology of the United States*, Wiley-Interscience, New York, 571 pp.

Levorsen, A. I., 1956, *Geology of Petroleum*, W. H. Freeman, San Francisco, 703 pp.

McKinney, C. M. and E. M. Shelton, 1966, *Analyses of Some Crude Oils from Fields in West Texas*, Bureau of Mines Report of Investigations 6752, U. S. Department of Interior, p. 141.

Schmitt, G. T., 1954, Genesis and Depositional History of Spraberry Formation, Midland Basin, Texas, *Bulletin of the American Association of Petroleum Geologists*, Vol. 38, No. 9, pp. 1957–1978.

Warn, G. F. and R. Sidwell, 1953, Petrology of the Spraberry Sands of West Texas, *Journal of Sedimentary Petrology*, Vol. 23, No. 2, pp. 67–74.

Wilkinson, W. M., 1953, Fracturing in Spraberry Reservoir, West Texas, *Bulletin of the American Association of Petroleum Geologists*, Vol. 37, No. 2, pp. 250–265.

20

Welch Field, The San Andres Formation, Dawson County, Texas

The San Andres Formation is a predominantly carbonate sequence that extends from central Texas all the way to Arizona and Utah. The San Andres Formation is particularly important as an oil producer, and more than 80% of the oil production in the Northern shelf of the Midland basin is from lower San Andres reservoirs (Ramondetta, 1982, p. 1).

In the Permian basin, the San Andres grades northward, gradually changing into anhydrite, salt, and red beds in the northern Texas panhandle, Oklahoma, and Kansas. The rock unit grades from a deep-water carbonate environment to shallow-water oolite bar deposits, shallow-shelf plus lagoonal carbonates, and ultimately to sabkha, brine pan, and mud-flat deposits, (Todd 1976). Toward the end of San Andres time, open-marine environments no longer prevailed on the Northern shelf, and only nearshore and continental sediments were deposited.

There is evidence that suggests that periodic subaerial exposure of the shelf and platform areas occurred during San Andres time, and some investigators (Hills, 1972; Todd 1976) postulate that large eustatic drops in sea level account for this. Porosity developed in the San Andres, even in the basinal rocks, and appears to be the result of the leaching of unstable carbonates by meteoric waters.

The Welch field is one of the few San Andres fields located in the basinal

Welch Field, The San Andres Formation, Dawson County, Texas

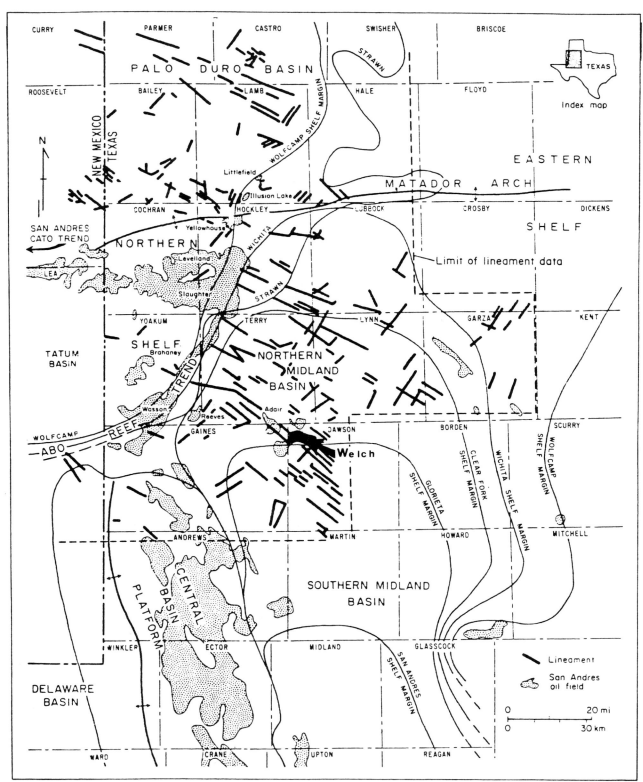

Figure 20.1. Map showing the location of the Welch field in relation to the major structural features and the surface lineaments. (After Ramondetta 1982, p. 2.)

environment. Figure 20.1 shows the area of most of San Andres production in Texas and provides outlines of the major structural features as well as the locations of the surface lineaments.

Concerning the trapping mechanism, Ramondetta (1982, p. 10) states the following:

Porous cyclic San Andres and Clearfork dolomites are overlain by laterally persistent nonporous dolomite that was deposited in an intertidal to supratidal environment. Anhydrite-rich and containing bedded anhydrite, these nonporous beds form an effective seal for the underlying reservoir, which in turn maintains lateral continuity among the oils of the Northern Shelf.

Looking at the trapping mechanisms in some detail, it becomes clear that much of the oil is trapped by porosity pinchouts updip from a structural front such as a shelf edge where reef buildups occur. In fields such as the Levelland–Slaughter trend, the porosity pinchout is produced by secondary anhydrite, which plugs up the pores. In the Welch field, the anhydrite pore filling is observed in a number of zones; however, subsequent fracturing has repeatedly broken the anhydrite seal, allowing oil to have access to a number of additional porous zones.

Organic Matter

According to Ramondetta (1982, p. 16), samples from San Andres wells in Dawson County "exhibit a strong dominance of amorphous sapropel and algal debris reflecting deeper water conditions." One of the problems encountered in the study is that the organic matter associated is not sufficiently matured to have generated the large amounts of hydrocarbons observed (Figure 20.2) (Ramondetta 1983, p. 17).

The geothermal gradient of the Northern shelf is approximately 1.1°F/100 ft (Dutton 1980a) and with

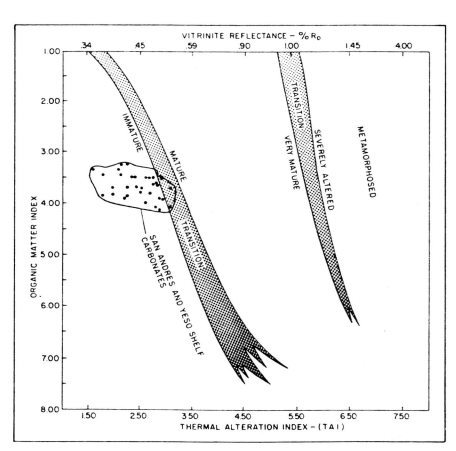

Figure 20.2. Cross plot of maturation indices for organic matter contained in the San Andres carbonates. (After Ramondetta 1982, p. 5.)

this gradient a depth of 7700 ft is theoretically required for oil formation (catagenesis). From this, it is clear that the San Andres oil probably did not form in place. Detailed work by Ramondetta (1982, p. 21) on the gas chromatographic spectrum suggests that the source materials were both algal and terrigenous.

Authors such as Jones and Smith (1965), among others, have favored the concept of oil generated in deep subsiding basins being squeezed into the more porous sediments of the surrounding shelf margins. As was pointed out in the discussion of the Spraberry reservoir (Chapter 19), there is adequate source rock potential in the shales and carbonates of Spraberry age to supply much of the oil in surrounding structural highs. Wolfcampian basinal sediments, which range from about 1000 to 2000 ft thick, have an average total carbon content of 2.8% and have excellent source rock potential Dutton (1980b), Ramondetta (1982, p. 24) postulates that differential shale–carbonate compaction leads to fracturing and the development of a hingeline such as that along the Abo Reef trend. The San Andres Dolomite overlies this reef–shale transition with its associated continuing compaction. The continuance of the compaction of the basinal shales and the relative incompatibility of the carbonate reef resulted in fractures within the San Andres and the entrapment of oil where the dip was maximum.

When the correlation index curves for the Welch field crude and the Spraberry crude are compared (Figure 20.3), there appears to be more of an aromatic hump in the fractions from 100 to 150°C in the crude from the Welch field. Such an increase is generally interpreted as reflecting the effect of biodegradation. The comparison of the correlation index curves for the Spraberry, Welch, and Goldsmith crude oils (Figure 20.3) shows that the differences are not large, suggesting that, indeed, they may have a common source. It must be recalled, however, that the Spraberry crudes from the southern Midland basin are sweet crudes, rich in naphthenes and low in sulfur. San Andres oils from the Northern shelf are contrastingly sour, highly aromatic, and sulfurous.

The shallower oils on the shelf probably experienced greater biodegradation and therefore have more sulfur. Landward from the shelf margins, the San Andres Formation shows larger amounts of anhydrite, which also contributes to the sulfur available.

In the next section, the Yates Formation is examined as it occurs on the Northwestern shelf in Lea County, New Mexico. The sediment variety and the overall mineralogic changes reflect the very shallow water conditions where biodegradation and groundwater recharge are at a maximum.

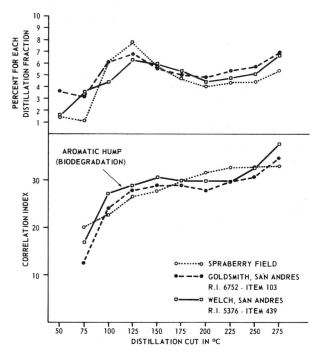

Figure 20.3. Correlation index profiles, crude oils from the San Andres Formation, Welch field, Dawson County, Texas, and Goldsmith field, Ector County, Texas.

Photomicrographs and Description

In the thin sections of the Welch field (Figures 20.4–20.9) some of the anhydrite blockage and filling is observed. Figure 20.10 shows highly leached dolomite with very porous and open framework.

Reservoir Engineering Problems

The occurrence of anhydrite in a dolomite reservoir causes a number of problems for the reservoir engineer who is concerned with a water-flooding or a tertiary recovery project. As can be seen from the photograph of a typical core through the reservoir of the Welch field (Figure 20.11), the anhydrite (white patches) is very irregularly distributed. Point counts of thin sections provide a very poor idea of the configu-

Figure 20.4. San Andres Formation, 4883 ft (1488.7 m), Cities Service Kirkpatrick No. 15, Welch field, Dawson County, Texas (0.44 in. in the photo equals 0.17 mm). Fine-grained dolomite with anhydrite (blue) filling in some of the moldic porosity. Competing for space in the lower pore are (black) pyrite cubes and framboids. Some pores remain free of anhydrite (extreme left).

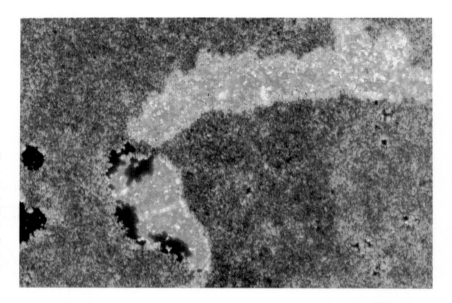

Figure 20.5. San Andres Formation, 4885 ft (1489.3 m), Cities Service Kirkpatrick No. 15, Welch Field, Dawson County, Texas (0.44 in. in the photo equals 0.17 mm). Fine-grained dolomite with complete anhydrite pore filling. Original moldic porosity was high; however, very little pore space is left.

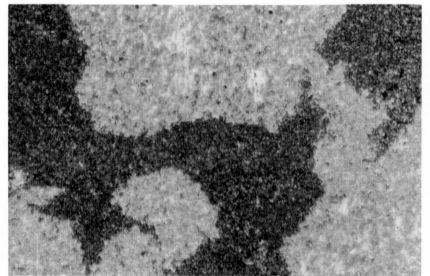

Figure 20.6. San Andres Formation, 4893 ft (1491.8 m), Cities Service Kirkpatrick No. 15, Welch field, Dawson County, Texas (0.44 in. in the photo equals 0.17 mm). Fine-grained dolomite has the appearance of having been leached to produce a highly porous zone. Blue in upper left is anhydrite.

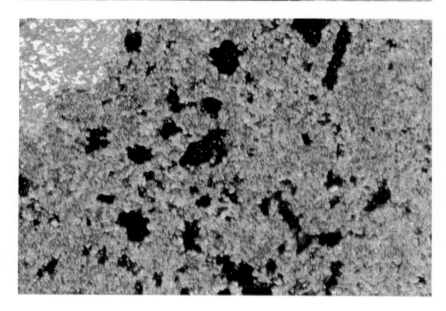

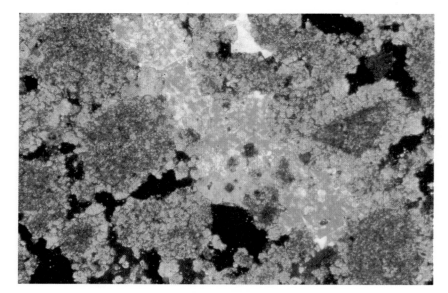

Figure 20.7. San Andres Formation, 4906 ft (1495.7 m), Cities Service Kirkpatrick No. 15, Welch field, Dawson County, Texas (0.44 in. in the photo equals 0.17 mm). Dolomite has become coarser grained and appears to have formed on the surfaces of fine-grained dolomite pellets and fragments. The latter pellets formed an open and highly porous framework that is only partly filled with anhydrite.

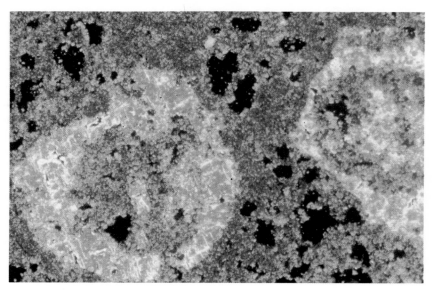

Figure 20.8. San Andres Formation, 4919 ft (1499.7 m), Cities Service Kirkpatrick No. 15, Welch field, Dawson County, Texas (0.44 in. in the photo equals 0.17 mm). Some of the coarser moldic pores are filled in with anhydrite, leaving the smaller pores available for the storage and movement of fluids.

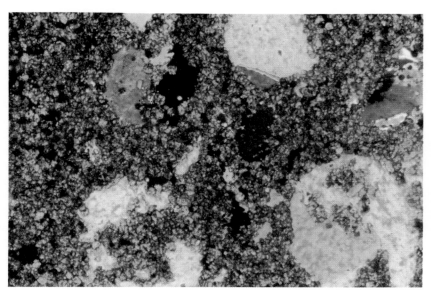

Figure 20.9. San Andres Formation, 4915 ft (1509.5 m), Cities Service Kirkpatrick No. 15, Welch field, Dawson County, Texas (0.44 in. in the photo equals 0.17 mm). Dolomite has become coarser grained, and many of the moldic pores have been filled with anhydrite. Some pyrite is present within the anhydrite inside the filled pores.

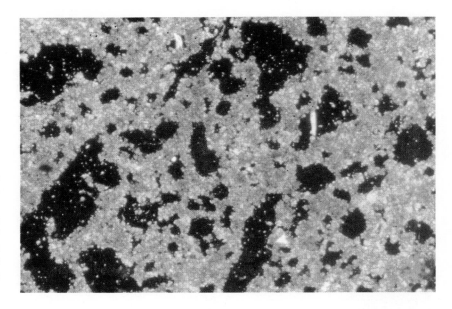

Figure 20.10. San Andres Formation, 4919 ft (1499.7 m), Cities Service Kirkpatrick No. 15, Welch field, Dawson County, Texas (0.44 in. in the photo equals 0.17 mm). Highly leached dolomite with very porous and open framework. The moldic pores appear to have resulted from subaerial exposure.

ration and overall occurrence of anhydrite in this type of reservoir.

In attempting to recover more oil by water flooding, it becomes obvious that one of the major difficulties is the increasing amount of gypsum deposited in the well bores. Ultimately, these gypsum deposits plug up the well bore and must be removed. The cost of "cleaning up" these wells every 5–8 months is excessive; yet, although many approaches to the gypsum problem have been tried, none has proved to be totally satisfactory.

The photomicrographs in Figures 20.12–20.14 were obtained from thin sections of well wall materials that show the boundary between the minerals in the reservoir rock and the gypsum formed in the well bore.

Permeability and Porosity

When anhydrite enters into the reservoir–caprock trapping mechanism as illustrated in Figures 20.12–20.14, and the photograph of the core in Figure 20.11, the relationship expected between porosity and permeability decreases. As has been observed, many of the large pores are plugged with anhydrite whereas other pores remain open. The thin channels between pores can be lined with crystals of dolomite or blocked with anhydrite. Figure 20.15 shows the core analysis data for the Cities Service Aynesworth No. 4, which is typical of many of the wells in the Welch field. Although the porosity profile ranges up to 20%, the permeability values are quite low and do not seem to correspond at all with the maxima in porosity.

Figure 20.11. Core of the San Andres Formation, Welch field, Dawson County, Texas.

Figure 20.12. San Andres Formation, depth unknown, Cities Service 48-25, Welch field, Dawson County, Texas (0.44 in. in the photo equals 0.30 mm). In cleaning out gypsum from producing wells, sometimes samples are retrieved from the bore hole walls. The above sample shows the anhydrite (blue and yellow) filled the pore space prior to the growth of the gypsum (gray). It is clear that the boundary between the reservoir rock (brown dolomite) and the gypsum is sharp. Although gypsum is the stable phase in the bore hole, there is no change observed in the anhydrite.

Figure 20.13. San Andres Formation, depth unknown, Cities Service 48-25, Welch field, Dawson County, Texas (0.44 in. in the photo equals 0.42 mm). From a section across the long axes of the gypsum crystals. The zoning within the crystals reflects the successive stages of crystal growth. The zoning indicates that the periods of growth occur irregularly.

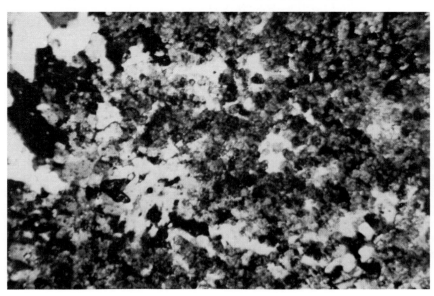

Figure 20.14. San Andres Formation, depth unknown, Cities Service 48-25, Welch field, Dawson County, Texas (0.44 in. in the photo equals 0.30 mm). Where the porosity is particularly good and open to the well bore, gypsum (various shades of gray) growth penetrates into the reservoir rock. In cases where such growth has occurred, treating the well bore alone may not restore production satisfactorily.

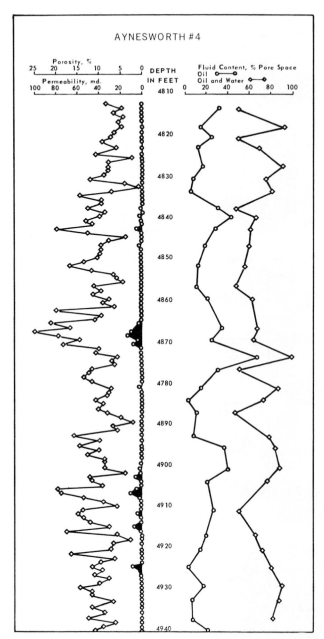

Figure 20.15. Typical core analysis profile, Welch field, Dawson County, Texas.

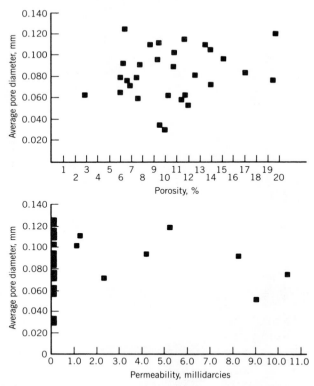

Figure 20.16. Average pore diameter versus porosity and permeability, San Andres Formation, Welch field, Dawson County, Texas.

A further study of the relationship between the pore diameters and the porosity values was undertaken to see what relationship remained, if any. With so many of the most accessible pores filled with anhydrite, it was reasoned that the larger open pores left might be those that are not effective and therefore show little or no relationship. Thin sections were examined on a foot-by-foot basis over the interval from 4904 ft (1495.1 m) to 4940 ft (1506.1 m), and 100 pore diameters were measured for each thin section. The average value for each thin section is plotted against the measured porosity and permeability for that footage in Figure 20.16. It is obvious that there is no relationship. Looking at the permeabilities in general, it is surprising that economic production is obtained at all. As in the case of the Spraberry, the fracture system must be taken into account. Unfortunately, the data presently available to this author do not include fracture studies; however, the fact that they exist and are important is clear from the water-flooding experience where tracers in the input water showed up at producing wells in very short time spans, on the order of a week. It seems likely that the fracture system was also important in the emplacement of the oil in that it provided the pathway from the rich shales and silts of Wolfcampian age. Ramondetta's (1982, p. 27) discussion of the emplacement of San Anders oils along the shelf margin hingeline also applies to the Welch field; therefore, it is included as follows:

Compositional relationships among the oils of the Abo Reef trend may best be explained by the timing of oil expulsion from the basin. Expulsion of the oil can occur only after temperature sufficient for maturation is reached during burial. Continued burial will then aid expulsion. Wolfcampian shales in the northern Midland Basin reached this stage before overlying Leonardian source beds did; hence, emplacement of Wolfcampian oils in equivalent Wolfcamp or Wichita–Albany shelf-margin reservoirs occurred first. Oil migration up vertical fractures along the shelf-margin hingeline into the younger Clearfork and San Andres reservoirs also occurred at this time.

Figure 20.3, which provides the correlation index profiles for crude oils from the Spraberry trend and the Welch field shows that these oils are quite similar. Again, as we look at the San Andres oils produced from the Goldsmith field in Chapter 21, it will be observed that the similarity persists (Figure 20.3).

Exploration

Once more, the attempt to locate a present-day equivalent appears doomed to failure. It is therefore best to consider what the reservoir means in terms of locating others like it. The obvious lesson seems to be that future searches should consider those areas where fracture systems appear to have developed early. The detailed surface lineaments as shown in Figure 20.1 may provide useful hints concerning where the trends are close together and, hence, where oil may have found a pathway to porous zones within the dolomite.

References

Dutton, S. P., 1980a, *Depositional systems and Hydrocarbon Resource Potential of the Pennsylvanian System, Palo Duro and Dalhart Basins, Texas Panhandle;* The University of Texas at Austin, Bureau of Economic Geology Geological Circular 80–8, 49 pp.

Dutton, S. P., 1980b, *Petroleum Source Rock Potential and Thermal Maturity, Palo Duro Basin, Texas;* The University of Texas at Austin, Bureau of Economic Geology Geological Circular 80–10, 48 pp.

Hills, J. M., 1972, Late Paleozoic Sedimentation in West Texas Permian Basin, *American Association of Petroleum Geologists Bulletin,* Vol. 54, No. 10, pp. 1809–1927.

Jones, T. S. and H. M. Smith, 1965, *Relationships of Oil Composition and Stratigraphy in the Permian Basin of West Texas and New Mexico,* American Association of Petroleum Geologists, Memoir 4, pp. 101–224.

McKinney, C. M. and E. L. Garton, 1957 *Analyses of Crude Oils from 470 Important Oilfields in the United States,* Bureau of Mines Report of Investigations 5376, 276 pp.

McKinney, C. M. and E. M. Shelton, 1966, *Analyses of Some Crude Oils from Fields in West Texas,* Bureau of Mines Report of Investigations 6752, 163 pp.

Ramondetta, P. J., 1982, *Genesis and Emplacement of Oil in the San Andres Formation, Northern Shelf of the Midland Basin, Texas,* Bureau of Economic Geology, Report of Investigations No. 116, University of Texas at Austin, 39 pp.

Todd, R. G., 1976, Oolite Bar Progradation, San Andres Formation, Midland Basin, *American Association of Petroleum Geologists Bulletin,* Vol. 50, No. 6, pp. 907–925.

21

Goldsmith Field, The San Andres Formation, Ector County, Texas

As discussed in connection with the TXL field, the Central basin platform is a north–south structural uplifted area between the Delaware basin on the west and the Midland basin on the east. The San Andres Formation is almost the lowermost unit of the Guadalupe strata, as illustrated in Figure 21.1, and is one of the most important producers on the Central basin platform. As Young (1965, p. 280) indicates, by the beginning of Permian deposition (Wolfcampian age), the pre-Permian sediments had been uplifted, folded, and faulted and then eroded to a surface close to sea level. This surface was, however, several hundred feet higher than the neighboring Midland and Delaware basins. It is on this elevated surface that the building of carbonate reefs of San Andres age was initiated.

The Permian strata above the Grayburg (see Figure 21.1) represent a back-reef facies of interbedded salt, anhydrite, and red beds (Young 1965 p. 281).

The member of the San Andres productive at Goldsmith is the so-called upper San Andres, which is approximately 600 ft thick. Along the producing trend that extends from the Penwell field to the Means field and includes Goldsmith, the reservoir rock is a brown, finely crystalline dolomite that includes relatively small amounts of anhydrite and chert. To the west, anhydrite filling becomes more prominent. On the east, the porous reef grades into a facies containing limestone, shale, and tight sandstone, all of which are described as "nearly non porous" (Young 1965, p. 283).

Source of Oil at Goldsmith Field

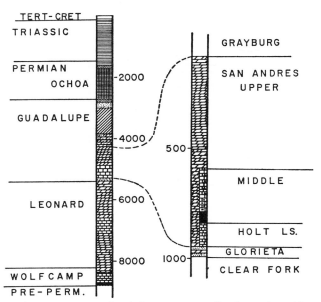

Figure 21.1. On the left is a generalized stratigraphic column that includes the Permian. On the right is the stratigraphic column detail of the San Andres Formation. (After Young 1965, p. 282.)

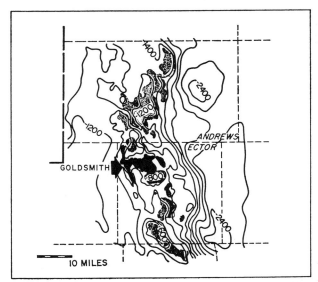

Figure 21.2. Structure contour map of the Penwell-to-Means trend, contoured on the top of the San Andres Formation, showing the structural position of the Goldsmith field. (After Young 1965, p. 284.)

The upper surface of the Central basin platform is composed of a series of low, broad anticlines, most of which have a northwest-to-southeast orientation (see Figure 21.2).

As illustrated in Figures 21.3 and 21.4, the Goldsmith field is situated on a structural high. Figure 21.4, the stratigraphic section, illustrates that dolomite formed where the water was most shallow and the sand and shale facies are associated with the deeper waters of the Midland basin. It should also be observed that the porosity developed in the eastern part of the dolomite facies (Young 1965, p. 287). In other words, the shallow-water environment on the eastern platform edge was favorable for reef growth and better porosity. By contrast, the back-reef area to the west and the basin sediments in the Midland basin to the east are relatively poor in terms of porosity development.

Photo-micrographs and Description

The thin sections in Figures 21.5–21.10 illustrate some of the various kinds of porosity represented in the Goldsmith field. It should be noted that the mineral composition of this reservoir rock is almost totally dolomite with very minor amounts of anhydrite.

Source of Oil at Goldsmith Field

As Ramondetta (1982, p. 29) observes, "Migration apparently occurred also between source beds in the southern Midland Basin and reservoirs in the Central Basin Platform. However, differences in the basinal source rocks and varying amounts of biodegradation that occurred on the Central Basin Platform complicate comparisons." Fortunately, the correlation index profile for the Goldsmith crude is quite smaller to that of the Spraberry.

The regional plot of the API gravities for San Andres crude oils (Stenzel 1965, p. 249) reveals that Goldsmith is included in a suite of oils that extend out into the deep part of the Midland basin adjacent to the eastern edge of the Central basin platform (Figure 21.11). The inference is that although mixing of various oils may have occurred, these oils appear to have a common origin, namely, the Wolfcampian and Clearfork sediments of the northern Midland or Permian basin.

At this point, it makes sense to return to the Wolfcamp strata as they occur in the northern Permian

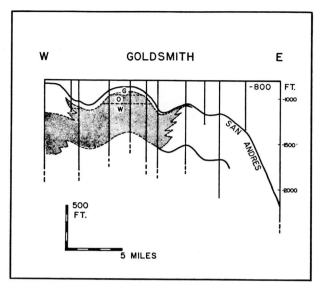

Figure 21.3. West-to-east structural cross section through the Goldsmith field. Porous zone of the Upper San Andres is shaded. Gas–oil and oil–water contacts are shown by horizontal dashed lines. (After Young 1965, p. 289.)

Figure 21.4. West-to-east stratigraphic section through the Goldsmith field, adjusted to the top of the Yoakum member of the Queen Formation. Porous zones of the Upper San Andres are outlined by dashed lines. (After Young 1965, p. 288.)

basin and look for evidence that these beds along with those of the Wichita do in fact contain significant amounts of hydrocarbons. Some data are available. However, first it is necessary to discuss how the samples were taken and how the measurements were made.

Studies covering many counties, in this case 23, could not by definition dwell on the small details. In other words, only one to three wells were analyzed per county. Similarly, the sampling interval in each well was of necessity large, that is, 100 ft. In spite of all these drawbacks, analyses of the hydrocarbons contained in cuttings samples provide a fairly clear picture of where the major source sediments are located and how they probably supplied the prolific reservoirs of the Eastern part of the Central basin platform.

The analytical technique used is discussed in detail in U.S. patent number 3,149,068 by Biederman and Heinze (1964). Briefly, the procedure used was as follows: Cutting samples of 1 g size were pulverized to less than 100 mesh and extracted with a 1 : 1 mixture of toluene and isopropanol. A measured amount of the extract was then evaporated on a corner of 1-in-square chromatographic paper. Two-dimensional chromatography was employed whereby the oxygenated organics were migrated with a polar solvent such as methanol, and the more hydrocarbonlike organics were migrated (at right angles to the first solvent) with heptane. The ultraviolet fluorescence of these two bands was then compared with known quantities of crude oil that had been treated in the same way. The advantage of this technique is that it includes both the mobility and the solubility of the extracted organic material in one measurement. It also can be carried out rather easily in the field without bulky equipment.

At this point, is should be mentioned that much more elegant and sophisticated evaluations of source rock are being used which involve at least 11 different measurements. Dembick (1984, p. 2641) found that in his comparison studies of standard rock samples analyzed by 19 different laboratories, each laboratory was consistent internally; however, important differences were noted in results between laboratories. Caution should be exercised in making comparisons of source rock data.

Two types of maps are readily produced from the two-dimensional chromtographic measurements. The first map (Figure 21.12) shows the maximum heptane value in terms of milliliters per kilogram of rock for the Wolfcamp interval in each well. The contoured map shown in Figure 21.12 reveals that the high contour of 10 ml oil/kg rock or above includes

183

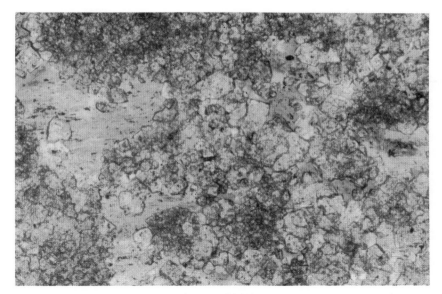

Figure 21.5. San Andres Formation, depth unknown, Cities Service Cummins D19W, Goldsmith field, Ector County, Texas (0.44 in. in the photo equals, 0.17 mm). Coarse-grained dolomite that has been impregnated with blue plastic to show the pore distribution. Taken with transmitted light with uncrossed polarizers so that the blue color is clearly observable. The light brown mineral filling other pore space is anhydrite.

Figure 21.6. San Andres Formation, depth unknown, Cities Service Cummins D19W, Goldsmith field, Ector County, Texas (0.44 in. in the photo equals 0.17 mm). Coarse-grained dolomite growing into the open pores. The finer-grained dolomite acts as a framework on which the coarser crystals grow. The lack of anhydrite filling helps to provide good reservoir characteristics.

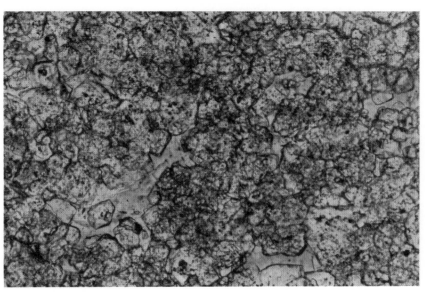

Figure 21.7. San Andres Formation, depth unknown, Cities Service Cummins D19W, Goldsmith field, Ector County, Texas (0.44 in. in the photo equals 0.08 mm). Higher magnification shows small, plastic-filled pores (greenish). Dolomite is relatively coarse grained. There appear to be no pore lining materials other than the dolomite. Taken with transmitted light with uncrossed polarizers.

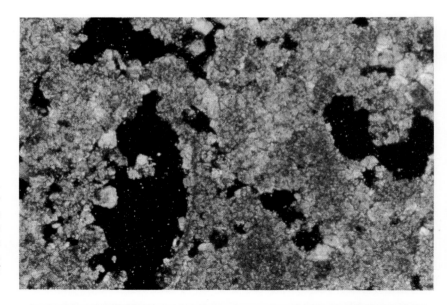

Figure 21.8. San Andres Formation, depth unknown, Cities Service Cummins D19W, Goldsmith field, Ector County, Texas (0.44 in. in the photo equals 0.17 mm). Moldic porosity is dominant. Shell fragments appear to have been leached out, leaving a highly porous structure typical of good reef reservoir rock.

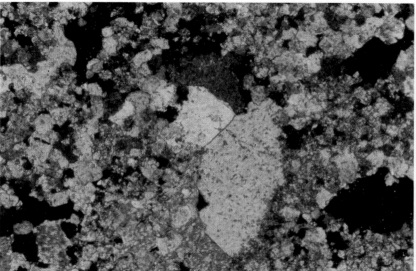

Figure 21.9. San Andres Formation, depth unknown, Cities Service Cummins D19W, Goldsmith field, Ector County, Texas (0.44 in. in the photo equals 0.17 mm). Moldic porosity dominates. Shows one filled-in shell fragment that has not been leached away.

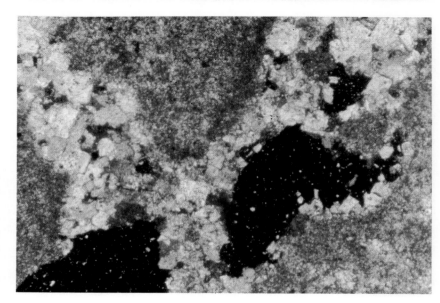

Figure 21.10. San Andres Formation, depth unknown, Cities Service Cummins D19W, Goldsmith field, Ector County, Texas (0.44 in. in the photo equals 0.17 mm). Again the highly porous reef structure is observed. Moldic porosity is common; however, coarse-grained dolomite is growing into the open pore space.

185

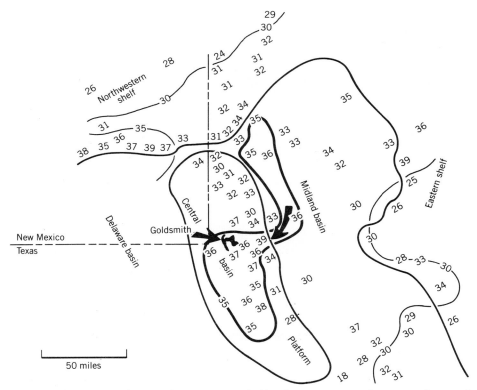

Figure 21.11. Map of regional pattern provided by the API gravities for San Andres crude oils showing the location of the Goldsmith field in relation to a proposed migration path from the Midland basin source rocks. (After Stenzel 1965, p. 249.)

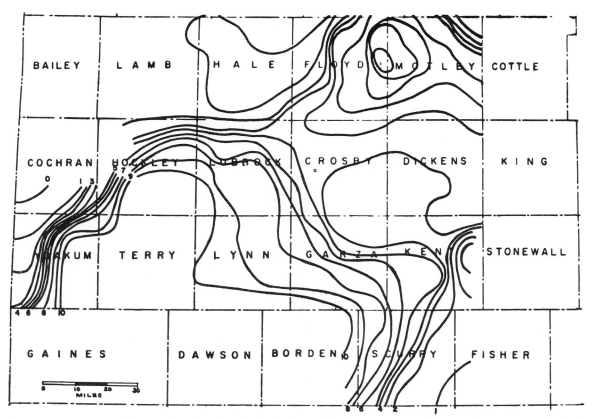

Figure 21.12. Map of the northern Permian basin in Texas showing contours of the maximum values for hydrocarbons extracted from well cuttings of the Wolfcampian interval contoured in terms of milliliters of oil per kilogram of rock.

Terry, Gaines, Dawson, and most of Borden counties.

The second type of map is based on the sum of the heptane values analyzed within the whole Wolfcampian section. The contours again reveal high areas in Dawson and Borden counties and a small area at the junction of Hockley, Lubbock, Terry, and Lynn counties (Figures 21.13). The major message would appear to be that the Wolfcamp strata do indeed contain significant amounts of hydrocarbons that could easily have supplied the San Andres fields surrounding the basin.

Figures 21.14 and 21.15 show maps of the extracted hydrocarbons from the Wichita interval. The contours essentially parallel those observed for the Wolfcamp and strongly suggest that both of these stratigraphic units provided the source rocks for the prolific production that occurs along the margins of the northern Midland (Permian) basin.

It might well be asked if the fracturing mechanisms discussed in previous sections apply to the Goldsmith field. The answer appears to be yes.

The actual measurements available all apply to one core from the Goldsmith reservoir; however, these can provide a good idea of the probability that fracture channels connected the source rock with the reservoir. In this instance a core of the Cities Service 19W Cummins "D" well was slabbed with a diamond saw and each of the flat surfaces was examined in detail for

1. vertical fractures,
2. horizontal fractures,
3. angular fractures, and
4. stylolites.

The above classes were further subdivided into major and minor categories. A major fracture or stylolite was defined as one that extended the whole length of the core fragment. Minor fractures or stylolites were defined as those that terminated within the core fragment. In cases where one stylolite split into several hairline irregular stylolites that were discon-

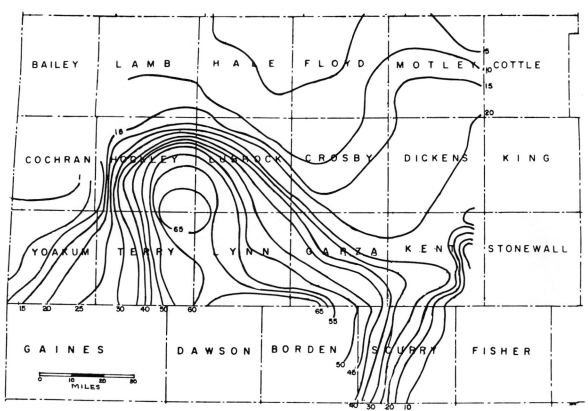

Figure 21.13. Map of the northern Permian basin in Texas showing contours of total value for hydrocarbons extracted from well cuttings of the Wolfcampian interval contoured in terms of milliliters of oil per kilogram of rock summed over the whole Wolfcampian section.

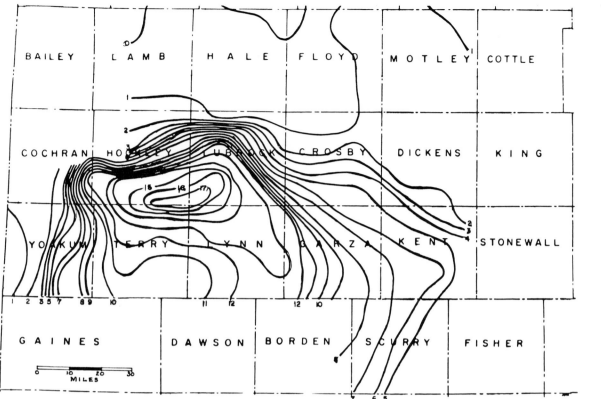

Figure 21.14. Map of the northern Permian basin showing contours of the maximum values for hydrocarbons extracted from well cuttings of the Wichita interval. Contours are in terms of milliliters of oil per kilogram of rock.

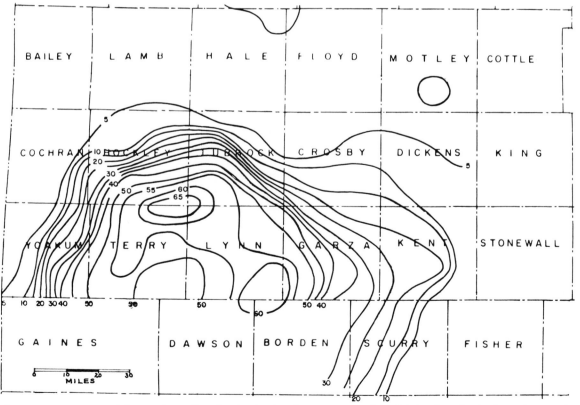

Figure 21.15. Map of the northern Permian basin showing contours of the total values for hydrocarbons extracted from well cuttings of the Wichita interval. Contours are in terms of milliliters of oil per kilogram of rock summed over the Wichita section.

TABLE 21.1 Number of Microfractures and Stylolites Observed In Core Materials From the CSO 19W Cummins D

Depth (ft)	Core Area (in.²)	Vertical Fractures		Horizontal Stylolites		Horizontal Fractures		Angular Fractues	
		Major	Minor	Major	Minor	Major	Minor	Major	Minor
4160.8–4171.2	235	9	18	12	7	1	5	1	—
4171.2–4181.0	264	6	7	12	11	2	—	—	3
4181.0–4191.3	175	4	1	6	—	—	1	1	—
4191.3–4201.4	182	—	9	2	6	2	5	—	—
4201.4–4211.0	244	—	8	3	7	—	1	1	—
4211.0–4221.3	208	3	8	3	2	1	6	—	1
4211.3–4231.8	253	1	12	8	13	1	13	1	5
4231.9–4241.0	236	6	5	1	5	—	12	1	—
4241.0–4251.1	260	1	8	7	9	—	5	—	1
4251.5–4261.4	236	2	4	10	9	1	2	—	4
4261.4–4271.0	213	—	—	1	8	1	3	—	—
4271.0–4281.2	259	—	6	1	7	3	1	—	4
4281.2–4291.2	251	—	—	1	5	1	7	—	—
4291.6–4301.2	262	—	1	2	7	—	10	—	—
4301.5–4311.3	299	1	1	5	10	—	2	—	—
4311.5–4320.9	250	—	12	6	1	—	6	—	2
4321.8–4332.0	238	2	4	4	3	—	—	—	1
Totals	4065	35	104	84	110	13	79	5	21

Total for all types, 451

tinuous across the core, the occurrence was counted as one minor stylolite. Table 21.1 summarizes the number and types of zones of weakness for 10-ft intervals of core. It is interesting to note that over a vertical distance of 171.2 ft (42.2 m) there were 35 major vertical fractures and 104 minor vertical fractures. Major horizontal fractures were fewer in number, with only 13 observed. Summarizing these observations, it is obvious that vertical fracturing is likely to be significant in the San Andres Formation reservoirs that occur on the Central basin platform. These fractures clearly provide excellent pathways for the migration and emplacement of oil that was squeezed out of the Leonardian and Wolfcampian shales and silts in the Midland basin.

Exploration

It seems likely that the San Andres Formation in Ector County is one of the most highly explored areas in the world. It may be, however, that some of the lessons learned from this area can be used to good advantage as frontier areas such as the Palo Duro basin are drilled more heavily.

References

Biederman, E. W., Jr., and B. Heinze, 1964, *Geochemical Exploration*, U.S. Patent number 3,149,068, September 15, 6 pp.

References

Dembicki, H. Jr., 1984, An Interlaboratory Comparison of Source Rock Data, *Geochemica et Cosmochimica Acta*, Vol. 48, No. 12, pp. 2641–2649.

Ramondetta, P. J., 1982, *Genesis and Emplacement of Oil in the San Andres Formation, Northern Shelf of the Midland Basin, Texas*, Bureau of Economic Geology, Report of Investigations No. 116, University of Texas at Austin, 39 pp.

Stenzel, W. K., 1965, *Times of Migration and Accumulation of Petroleum in Abo Reef of Southeastern New Mexico—A Hypothesis*, American Association of Petroleum Geologists, Memoir 4, pp. 243–256.

Young, A., 1965, *The Penwell-to-Means Upper San Andres Reef of West Texas*, American Association of Petroleum Geologists, Memoir 4, pp. 280–293.

22

West Teas Field, The Yates Formation, Lea County, New Mexico

As illustrated in Figure 22.1, the Yates Formation occurs toward the top of the Permian section. The reservoir rock at West Teas differes considerably from the previous two San Andres reservoirs because it is located on the structural shelf just north of the Delaware basin margin (Figure 22.2). In this setting, the environment changes from subaerial to shallow-water marine; the sediment type switches rather rapidly. Furthermore, we are dealing with a lateral transition zone that probably ranges from low, coastal tidal flats to open marine deposits. Most of the photomicrographs (Figures 22.3–22.8) reflect the marine carbonate or terrigenous sand that has been altered by subaerial exposure and leaching.

Summary

TERTIARY	PLIOCENE	
TRIASSIC		
PERMIAN	OCHOA	
	GUADALUPE	TANSILL
		YATES ◀
		SEVEN RIVERS
		QUEEN
		GRAYBURG
		SAN ANDRES
	LEONARD	Upper / subsurface Glorieta YESO
		Lower
		ABO
	WOLFCAMP	

Figure 22.1. Post-Pennsylvanian stratigraphic section for west Texas and southeastern New Mexico showing the relative position of the Yates Formation. (After Stenzel 1965, p. 245.)

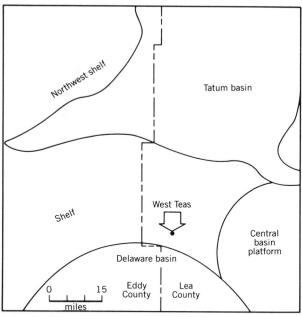

Figure 22.2. Location of the West Teas field in Lea County, New Mexico, in relation to the major structural features.

Recent Equivalents

As pointed out previously, the Florida Keys appear to have environments that closely approximate those of the subaerially exposed shelf in the Yates Formation of southeastern New Mexico. More specifically, the caliche from Big Pine Key appears to be an excellent place to begin a comparison study between a number of the shelf reservoirs in the Yates, Queen, and San Andres formations. The work by Coniglio and Harrison (1983) appears to provide a good base on which such a study can be launched successfully.

Source of Oil at West Teas Field

The fact that the West Teas field is located near the margin of the Delaware basin suggests that the oil could have originated within the basin and been squeezed into the porous zones on the margins. Furthermore, the fact that oil production follows the western margin of the Central basin platform also strongly suggests that the Delaware basin can provide the needed source materials.

Other evidence includes the organic geochemical profiles obtained from deep wells in the Delaware basin. One such well is illustrated in Figure 22.9. In this case, cuttings samples have been extracted with a toluene–isopropanol solvent combination, and the heptane–soluble portion of the extract has been separated with solvent extraction and measured using ultraviolet fluorescence techniques. It becomes clear from this typical profile that the Delaware basin, at least in Culberson and Reeves counties, has sufficient source rock.

Summary

The last three reservoir rock examples (Welch, Goldsmith, and West Teas) have involved Permian age dolomite reservoirs from three different settings. The Welch field represents the mid-basin location; the Goldsmith field provided a good example of the dolo-

Figure 22.3. Yates Formation, 3252 ft (991.5 m), Cities Service state B. F. No. 1, West Teas field, Lea County, New Mexico (0.44 in. in the photo equals 0.17 mm). The area between two large algal pisolites is shown where coarse dolomite crystals are growing into the pore space. These structures are common to the coastal tidal flats and represent low-wave-energy conditions. (Longacre 1980, p. 110.)

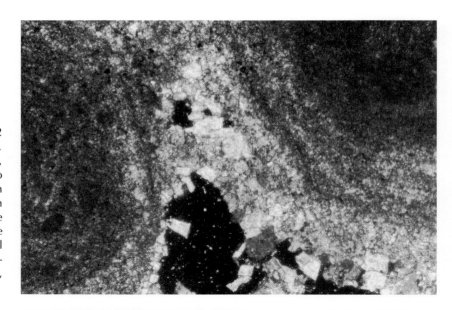

Figure 22.4. Yates Formation, 3256 ft (992.7 m), Cities Service State B. F. No. 1, West Teas field, Lea County, New Mexico (0.44 in. in the photo equals 0.17 mm). Some algal fenestral structure with coarser dolomite filling in the pore space. Larger moldic porosity also occurs. The fenestral structure is characteristic of the coastal tidal flat facies.

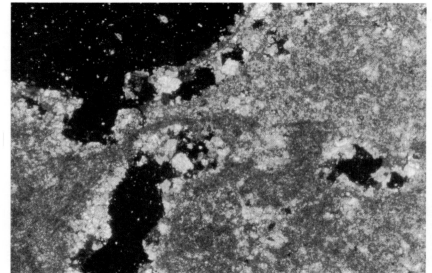

Figure 22.5. Yates Formation, 3258 ft (993.3 m), Cities Service State B. F. No. 1, West Teas field, Lea County, New Mexico (0.44 in. in the photo equals 0.17 mm). Recent work by Coniglio and Harrison (1983) suggests that the peloids and clotted texture observed are characteristic of a subaerial alteration zone. In this hypothesis, "micritic areas gradually grade into spherical or ellipsoidal peloids." These peloidal deposits are interpreted as diagenetic precipitates and are characteristic of caliche deposits on the Florida Keys.

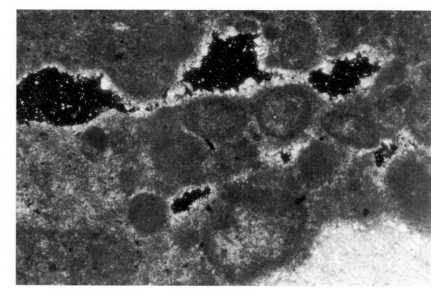

193

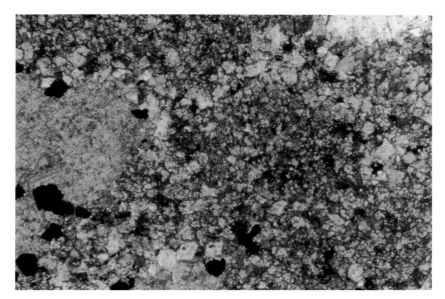

Figure 22.6. Yates Formation, 3260 ft (993.9 m), Cities Service State B. F. No. 1, West Teas field, Lea County, New Mexico (0.44 in. in the photo equals 0.17 mm). Sample was deposited where pyrite (opaque grains, lower left) could form. This suggests reducing conditions that were underwater most of the time. Coarser-grained dolomite appears to have filled in most of the original porosity between the bioclastic fragments.

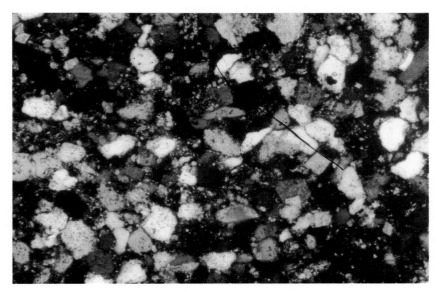

Figure 22.7. Yates Formation, 3262 ft (994.5 m), Cities Service State B. F. No. 1, West Teas field, Lea County, New Mexico (0.44 in. in the photo equals 0.17 mm). This sand is believed to be of terrigeous origin. These sands are transported to the site of deposition by both eolian and aqueous media. Most of the sand probably accumulated as dunes in playa lakes and coastal paralic environments. (Longacre 1980, p. 108.)

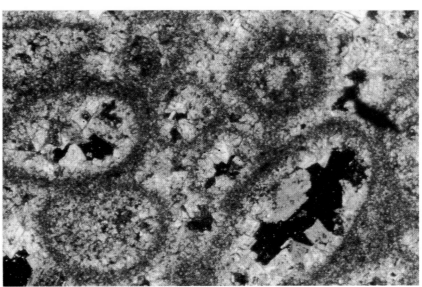

Figure 22.8. Yates Formation, 3398 ft (1036.0), Cities Service State B. F. No. 1, West Teas field, Lea County, New Mexico (0.44 in. in the photo equals 0.17 mm). Micrite coatings preserve outlines of grains that have subsequently leached and then infilled with sparry calcite that in some cases has been dolomitized. Again, this sediment reflects an alteration zone that has been subaerially exposed.

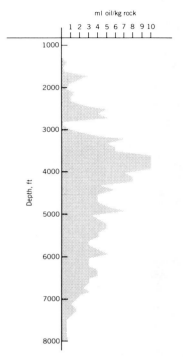

Figure 22.9. Source rock profile for a typical well in the deeper part of the Delaware basin.

mitized reef on the eastern edge of the Central basin platform; and lastly, the West Teas field illustrated the case where subaerial shelf deposits and shoreline facies meet. In all of these cases, the deeper parts of the closest basin seems to have provided the source materials, and fracture systems provided the conduits for the oil to reach the reservoir rock.

References

Coniglio, M. and R. S. Harrison, 1983, Holocene and Pleistocene Caliche from Big Pine Key, Florida, *Bulletin of Canadian Petroleum Geology*, Vol. 31, No. 1, pp. 3–13.

Longacre, S. A., 1980, Dolomite Reservoirs from Permian Biomicrites, in *Carbonate Reservoir Rocks, Notes for SEPM Core Workshop No. 1, Denver, Colorado 1980*, R. B. Halley and R. G. Loucks, Eds., Society of Economic Paleontologists and Mineralogists, pp. 105–117.

Stenzel, W. K., 1965, *Times of Migration and Accumulation of Petroleum in Abo Reef of Southeastern New Mexico—A Hypothesis*, American Association of Petroleum Geologists, Memoir 4, pp. 243–256.

23

Gulf Coast

Introduction

Meyerhoff (1980, p. 175) stated:

The Gulf of Mexico area is the fifth largest potential petroleum-producing region in the world, being exceeded only by the Arabian (Persian) Gulf Basin, the west Siberian Basin, the eastern Venezuela (Maturin) Basin, and the western Canada Basin. The last two are larger than the Gulf of Mexico Basin because of their enormous tar-sand deposits.

The fact that exploration is now concentrated in the deeper waters of the Gulf of Mexico does not mean that more oil will not be found onshore or that study of the older reservoirs is unprofitable. In such an area where there are literally thousands of fields, it is very nearly impossible to choose the so-called representative fields. The writer has elected to mention those fields with which he has some familiarity and where he has core materials available. Even from this small sample, some idea of the scope of possibilities can be obtained.

General Geologic Setting

Both north and west of the modern Mississippi River delta, carbonate sediments were deposited on a broad shallow shelf from the Jurassic through the Lower Cretaceous. Later in this chapter a representative from this group of sediments will be examined; however, the initial focus will be on the the clastic sediments that began to take over in Middle Cretaceous times and were dominant by Upper Cretaceous time. Most of these clastic sediments were provided by the ancestral Mississippi River. As illustrated in Figure 23.1, clastic wedges of younger sediments were laid down on the previous base, moving the shoreline seaward. Since Upper Cretaceous time the center of deposition where most of the clastic sediments are deposited has moved seaward 200 k. The generalized cross section of the Gulf of Mexico geologic province (Figure 23.2) shows that today's sedimentary profile was produced by the interaction between sediment supply, sea level changes, and isostatic subsidence in response to the ever-growing load of sediments.

The major tectonic elements, as will be observed in the south Texas example, are normal faults along which movement took place as the sediments were deposited (Figure 23.3). This concomitant fault movement and sedimentation commonly produces a thickening of the section as the fault is approached from the downthrown side. This sometimes gives a "reverse drag" appearance to the sands as they are compared in cross sections with electric logs.

Much of the major oil and gas production along the south Texas Gulf Coast is associated with linear trends as illustrated in Figure 23.4. This work will focus on fields and reservoir rocks located within these favorable trends. The first field to be discussed is the Romeo field in Jim Hogg County, Texas, which produces from the Eocene Yegua Formation (see Figure 23.5).

The second clastic sequence to be covered is the Oligocene Frio Formation (Figure 23.5). The May field in Kleberg County is the first example of a Frio sand reservoir to be studied. The second example of a Frio sand that is from a somewhat different environment involves the State Tract field in Corpus Christi Bay. Proceeding from one to the other of these fields, an attempt is made to mention the contrasting elements as well as the similarities.

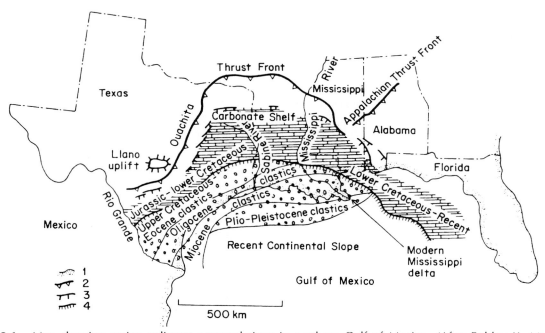

Figure 23.1. Map showing major sediment accumulations in northern Gulf of Mexico. (After Robley K. Matthews, Dynamic Stratigraphy: An Introduction to Sedimentation and Stratigraphy, (c), 1974, pp. 85, Reprinted by permission of Prentice-Hall, Inc., Englewood Cliffs, New Jersey.)

General Geologic Setting

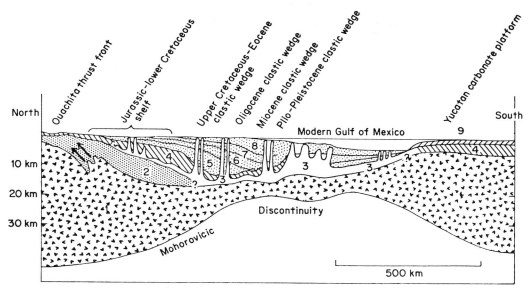

Figure 23.2. Generalized cross section of the Gulf of Mexico geological province. (1) Basement rocks, (2) Paleozoic sediments, Ouachita-Appalachian type, (3) Jurassic salt, (4) Jurassic & Lower Cretaceous shelf carbonates, (5–8) Upper Cretaceous-to-Pleistocene clastics. (After Robley K. Matthews, Dynamic Stratigraphy: An Introduction to Sedimentation and Stratigraphy, (c), 1974, pp. 86, Reprinted by permission of Prentice-Hall, Inc., Englewood Cliffs, New Jersey.)

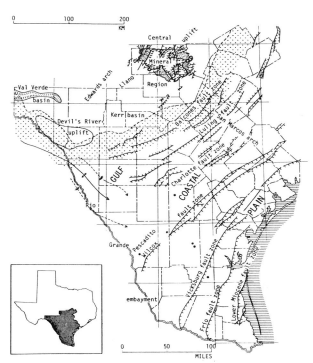

Figure 23.3. Tectonic map of south Texas showing major structural features. Large black dots are salt domes. Small dot pattern is subcrop of Paleozoic Quachita facies. (After Landes 1970, p. 265.)

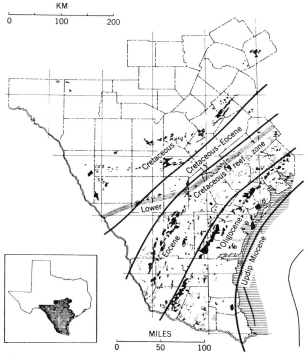

Figure 23.4. Generalized production trend map of south Texas showing the geologic age of the important fields beneath each zone. (After Landes 1970, p. 271.)

CENOZOIC
    Quaternary
        Recent
        Pleistocene
    Tertiary
        Pliocene
        Miocene
            Catahoula
        Miocene–Oligocene
            Anahuac
        Oliogecene
            Frio: leading reservoir in district ⟵
            Vicksburg
        Eocene
            Jackson
            Claiborne: includes Yegua ⟵
            Wilcox: includes Carrizo
            Midway: includes Poth

MESOZOIC
    Cretaceous
        Upper (Gulfian)
            Navarro: includes Olmos
            Taylor: includes San Miquel
            Austin
            Eagle Ford
            Woodbine
        Lower (Comanchean)
            Washita
            Fredericksburg
                Edwards
                Paluxy
            Trinity
                Glen Rose

    Jurassic
        Cotton Valley
        Smackover
        Louann Salt
        Werner
        Eagle Mills

PALEOZOIC
    Carboniferous
    Cambro-Ordovican
        Ellenburger

PRE-CAMBRIAN
        Granite, schist, and gneiss

Figure 23.5. Generalized Stratigraphic Chart of south Texas. (After Landes 1970, p. 269.)

Romeo Field, The Yegua Formation, Jim Hogg County, Texas

Geologic Background

The first important oil production from the Jackson trend was the Mirando Valley field in Zapata County back in 1921. As is the case in Pennsylvania, exploration reached its usual saturation point many years ago; however, new fields continue to be discovered. The reason for this apparent paradox is that there are unpredictable situations along the overall trend where stratigraphic traps occur that result from the pinchouts of some 50 separate sands over an approximate 1800-ft interval of the Jackson and Yegua sections. Complicating the problem are faults and structures. In addition, dry gas that is odorless is characteristic of some sections; therefore, some early wells missed producing zones.

In contrast to what will be observed for the Frio, the Jackson–Yegua strand lines appear to be unrelated to local structural conditions (West 1963, p. 70). Although a series of down-to-the-coast faults cut across the trend, no consistent changes in sand development have been observed between the high and low sides of these faults. The trapping mechanism, on the other hand, frequently involves a down-to-the-coast fault (down to the southeast) cutting across a strand line. Another condition that is necessary is that the reservoir not be faulted against another sand. This appears to cause a failure in the trapping of the oil (West 1963, p. 70).

As will be observed from the photomicrographs of the thin sections of the Yegua Sand in the Romeo field (Figure 23.8–23.12), the sands can have good permeability and porosity. West (1963, p. 75) indicates that sands in the Queen–Sabe Albercas fields have porosites in the 30–35% range, which is generally typical of good Jackson production. Furthermore, oil recovery per acre foot from Jackson sands is usually on the order of 500–1000 barrels per acre for the better fields.

Figure 23.6, which provides a west–east stratigraphic cross section of northwestern Jim Hogg County, shows the electric log characteristics of the

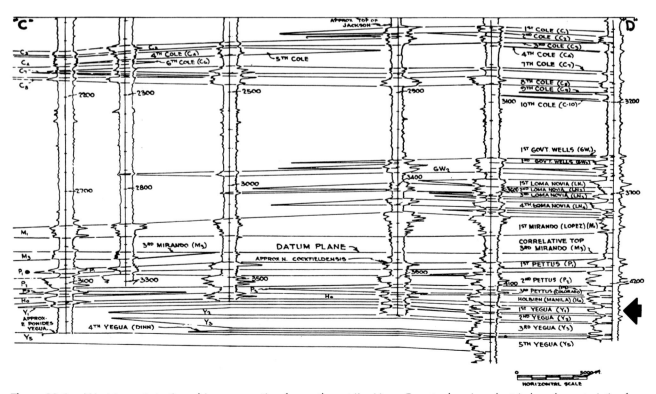

Figure 23.6. West-to-east stratigraphic cross section for northwest Jim Hogg County showing electric log characteristics for the Yegua sands. (After West 1963, p. 76.)

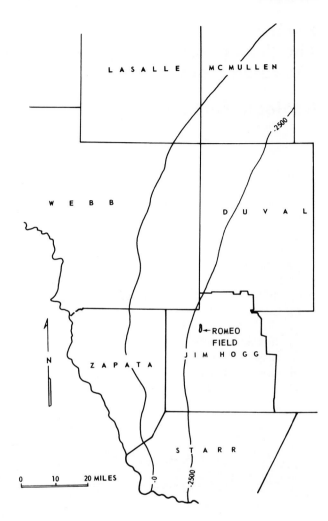

Figure 23.7. Shows the location of the Romeo field, south Texas, in relation to the Jackson trend, contoured on the top of the second member of the Pettus Sand. (Modified after West 1963, p. 68).

Yegua sands and the westward pinchouts of other sands higher in the section. Figure 23.7 shows the location of the Romeo field in relation to the Jackson trend. The Romeo field is listed in the literature as an anticlinal trap (Reiss et al. 1961). However, this author does not have the structure maps with which to illustrate this statement. Regardless of the structural details, the petrology of these reservoir sands appears to be rather uniform, shown in the thin-section photomicrographs (Figures 23.8–23.12).

Geologic Interpretation

West (1963, p. 70) has interpreted these sands as strand lines that oscillated back and forth as subsidence took place. Rainwater observes (1967, p. 182) that there was continuous subsidence of both the clastic and carbonate provinces throughout the middle Eocene. Sediments in the upper Eocene involving the Jacksonian Stage reveal a time of general transgression of the sea with no regressions. Small upper Eocene deltaic depocenters are located in the Texas Coastal Plain. The south Texas area provides a good example of this type of deposition. In the case of the Yegua sands, it appears that the sedimentation kept up with the transgressing sea and local subsidence, allowing the sequence of 50 separate sands to be deposited along the trend that paralleled the Eocene coastline.

It also seems clear that the volcanic ash contributed to the sediment that was being trasported to the coast. In Frio time, this continues to be an important contributor.

201

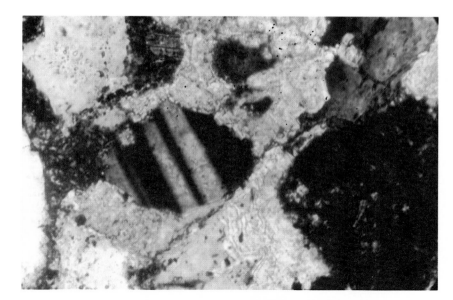

Figure 23.8. The Yegua Formation, 3554–3555 ft (1083.7 m) Cities Service No. 1 Gutierrez, Romeo field, Jim Hogg County, Texas (0.44 in. in the photo equals 0.08 mm). High magnification shows the carbonate cement replacing quartz, plagioclase feldspar, and a chert fragment. Some kaolinite is present in the pore space not filled with carbonate cement.

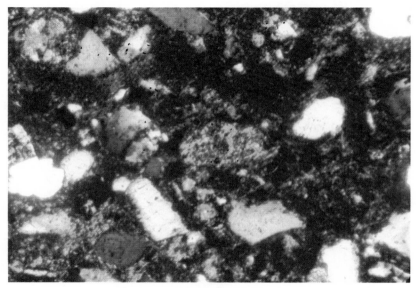

Figure 23.9. The Yegua Formation, 3556–3557 ft (1084.3 m), Cities Service No. 1 Gutierrez, Romeo field, Jim Hogg County, Texas (0.44 in. in the photo equals 0.08 mm). Clay matrix fills most of the pore space between the grains. In general, the grains are angular yet fairly well sorted. The greenish glauconite grain in the center is an indicator of a marine environment of deposition.

Figure 23.10. The Yegua Formation, 3571.5 ft (1088.9 m), Cities Service No. 1 Gutierrez, Romeo field, Jim Hogg County, Texas (0.44 in. in the photo equals 0.08 mm). The sand is loosely cemented with numerous weathered plagioclase feldspar grains and clay minerals (kaolinite) between the grains. Both the quartz grains and the feldspars are quite angular. Detrital carbonate fragments are abundant.

Figure 23.11. The Yegua Formation, 3574 ft (1089.6 m), Cities Service No. 1 Gutierrez, Romeo field, Jim Hogg County, Texas (0.44 in. in the photo equals 0.08 mm). Angular plagioclase feldspars and quartz grains are common. Detrital carbonate grains are also important. Volcanic rock fragments involving feldspar laths in a glassy matrix suggest that volcanic ash has contributed to the sediment.

Figure 23.12. The Yegua Formation, 3576 ft (1090.2 m), Cities Service No. 1 Gutierrez, Romeo field, Jim Hogg County, Texas (0.44 in. in the photo equals 0.08 mm). Detrital carbonate grains are more abundant. Plagioclase feldspars are also numerous along with angular quartz grains. Several grains of chert are observable along with kaolinite clays filling in some of the pore space.

References

Landes, K. K., 1970, *Petroleum Geology of the United States,* Wiley-Interscience, New York, 571 pp.

Matthews, R. K., 1974, *Dynamic Stratigraphy,* Prentice-Hall, Englewood Cliffs, NJ, 370 pp.

Meyerhoff, A. A., 1980, Future Petroleum Provinces of the Gulf of Mexico Region, *Transactions of the Gulf Coast Association of Geological Societies,* Vol. 30, p. 175 (abstract).

Rainwater, E. H., 1967, Resume of Jurassic to Recent Sedimentation History of the Gulf of Mexico Basin, *Transactions of the Gulf Coast Association of Geological Societies,* Vol. 17, pp. 179–186.

Reiss, B., J. Schulz, Jr., E. Sharp, and J. C. Wise, 1961, Developments in South Texas in 1960, *Bulletin of the American Association of Petroleum Geologists,* Vol. 45, No. 6, pp. 853–867.

West, T. S., Sr., 1963, Typical Stratigraphic Traps Jackson Trend of South Texas, *Transaction of the Gulf Coast Association of Geological Societies,* Vol. 13, pp. 67–78.

24

Frio Formation
The May Field,
Kleberg County, Texas

The Oligocene Frio Formation of the Texas Gulf Coast has produced almost 6 billion bbl of oil and 60×10^{12} ft^3 of gas for a total of more than 16 billion BOE (barrels of oil equivalent) in hydrocarbons (Galloway et al. 1982, p. 671). The prolific nature of this type of sediment makes it an attractive subject to analyze. Furthermore, the Gulf Coast of Texas is one of the most intensely explored petroleum provinces on earth, and it might be presumed that there is relatively little that can be added to what already exists in the literature. Nevertheless, a look at the petrologic detail can be rewarding in terms of what might occur in other frontier areas.

The May field is one of the numerous hydrocarbon accumulations associated with the ancient Norias Delta System. It is located close to the shores of Baffin Bay about 25 miles south of Corpus Christi (Figure 24.1).

The Frio sediments were deposited in an extracratonic basin that was characterized by rapid subsidence in areas where the sediment loading was greatest. The advancing wedge of sediments consists of an updip part composed of interbedded continental and marine sands and shales. These strata are underlain by several thousand feet of undercompacted marine mudstones deposited on the outer slope and in the basin. According to Galloway et al. (1982, p. 651), the instability caused by the rapid sediment loading on top of the plastic, water-saturated muds at the top of an inclined undersea slope resulted in large,

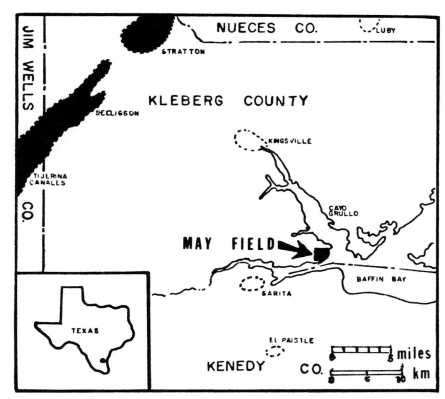

Figure 24.1. Shows location of the May field in relation to the Seeligson field.

down-to-the-coast faults that occurred as sediment deposition was taking place. It is this set of faults (see Figure 24.2) that is responsible for much of the trapped oil and gas in the Frio Formation.

Three major structural provinces are observable. The first is the Houston embayment of east Texas, which is characterized by salt diapirism and associated faulting plus the salt withdrawal subbasins.

The second major feature is the Rio Grande embayment of south Texas. In this region, large but more discontinuous trends of growth faults, deep-seated shale ridges, and massifs occur (Galloway et al.

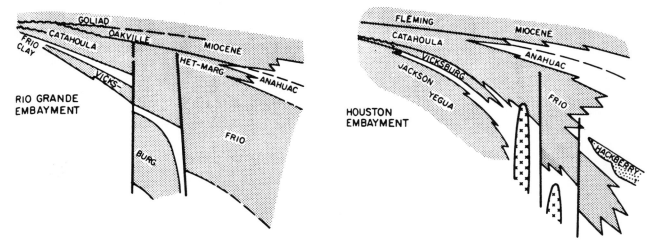

Figure 24.2. Schematic Frio stratigraphic sections for the Houston and Rio Grande embayments on the Texas Gulf Coast. (After Galloway et al. 1982, p. 653.)

Mineralogy

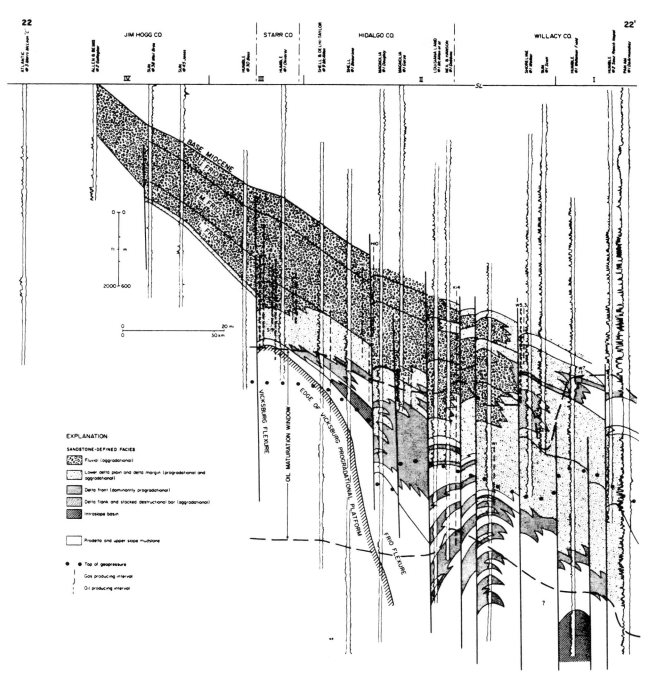

Figure 24.3. Stratigraphic section of the Frio Formation from Jim Hogg County to the Gulf of Mexico showing the Gueydan Fluvial System and the Norias Delta System. In addition, the approximate top and bottom of the hydrocarbon liquid window is marked as is the top of the geopressurized zone. (After Galloway et al. 1982, p. 662.)

1982, p. 651). In addition to these fault trends, oblique sets of deep, low-relief anticlines occur in this embayment.

The main depocenter in the south Texas Coastal Plain in Frio time involves the Norias Delta System (see cross-section in Figure 24.3). In this zone, the sand content ranges between 25 and 40% of the 12,000-ft-thick Frio section. The Norias delta was produced by the shifting of the drainage of axis of a single large river; therefore, the lateral extent of the system remained relatively fixed through time. The maximum width of the system is 125 miles (200 km),

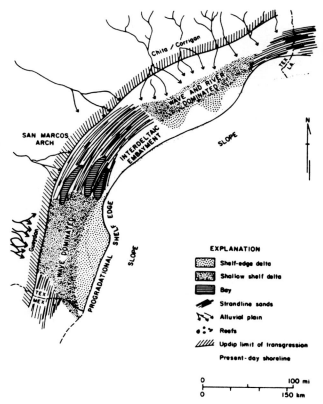

Figure 24.4. Map portraying the proposed paleography of the Texas Coastal Plain and shelf in Frio time. (After Galloway et al. 1982, p. 670.)

and seismic work indicates that sand-rich deltaic facies may extend offshore for 10 miles from Kenedy County (see Figure 24.4). Ultimately, downdip and at great depth, the sediments grade into deepwater slope deposits that are mostly shale.

Northward along the strike, the Norias System grades into what Galloway et al. (1982, p. 663) have called the Greta–Carancahua barrier–lagoon system. This latter area will be discussed in connection with the Corpus Christi Bay field and the associated reservoir rocks.

The May field is a marine part of the sands associated with the Norias delta. Directly updip and westward from the May field is the Seeligson field, part of which Nanz (1954, p. 104) interpreted as a distributionary and larger deltaic splay channel exhibiting bifurcating sands as it extends seaward (Figures 24.5 and 24.6).

Structurally, the May field is located on the downthrown side of one of the down-to-the-coast normal faults that are so common in the Frio sediments of the Gulf Coast.

Mineralogy

Initially, it was estimated that there would be a number of significant mineralogic changes in the Frio sands as the source area for the sediment became more active tectonically. The question, however, could not be resolved wihtout some assessment of the sampling interval required. The other problem that was related to the first involved the selection of those minerals that would provide the most amount of information. As the initial point counts were taken, it became clear that the differences between samples that would signal important tectonic events were not obvious. It was, therefore, decided to make thin sections of every available sample and to make percentage point count profiles of both the light and heavy minerals. Figures 24.7–24.12 show the profiles obtained for the light minerals for the samples available from the Arkansas Fuel Oil V. A. Hubert 1A well in the May field. Careful examination of these profiles reveals a rather striking uniformity over the whole interval from 7675 ft (2339.9 m) to 9370 ft (2856.7 m). A typical electric log through the main productive interval is shown in Figure 24.13.

The percentage of quartz changes slightly proceeding up the cored interval. From 8867 ft (2703.4 m) to 8227 ft (2508.2 m) and from 8803 ft (2683.8 m) to 8745 ft (2666.2 m), the quartz percent is notably higher than in the interval up to 7941 ft (2421.3 m). By contrast, the calcite cement is more common from 8570 ft (2612.8 m) to 7576 ft (2309.8 m). According to Loucks et al. (1979, p. B60), the timing of the emplacement of sparry calcite cement is later than the formation of quartz and feldspar overgrowths and the leaching phase.

Volcanic rock fragments (Figures 24.7 and 24.8) show a marked increase in the lower portion of the well (9370 ft (2856.7 m) to 8868 ft (2703.7 m). Although there is a considerable amount of this material throughout the whole sequence of cored Frio, the increase appears to reflect the time of the most intense volcanic activity.

As is shown later, straight-edged and pseudohexagonal biotite flakes are a more precise indicator of ash falls and can be used in this section as a good measure of volcanic activity. It should be emphasized, however, that thin-section point counts of biotite are not sensitive enough to produce a meaningful profile, however, counts of biotite in heavy-mineral amounts are useful.

The amount of detrital calcite present increases rather regularly from the bottom to the top of the

Mineralogy

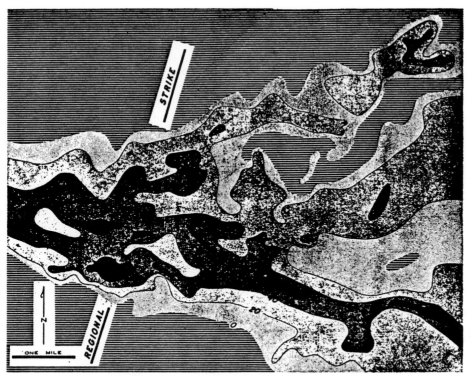

Figure 24.5. Isopach map of total sand thickness for the 19B Frio sand in the Seeligson field of Jim Wells and Kleberg counties, Texas. (After Nanz 1954, p. 104.)

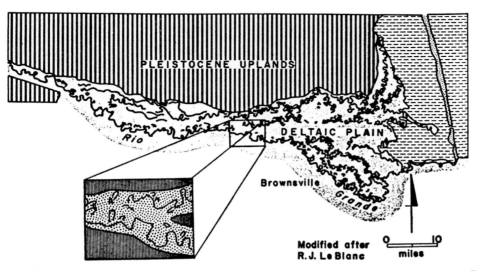

Figure 24.6. Distribution of Recent deltaic plain sediments of the Rio Grande River near Brownsville, Texas. (After Nanz 1954, p. 104.)

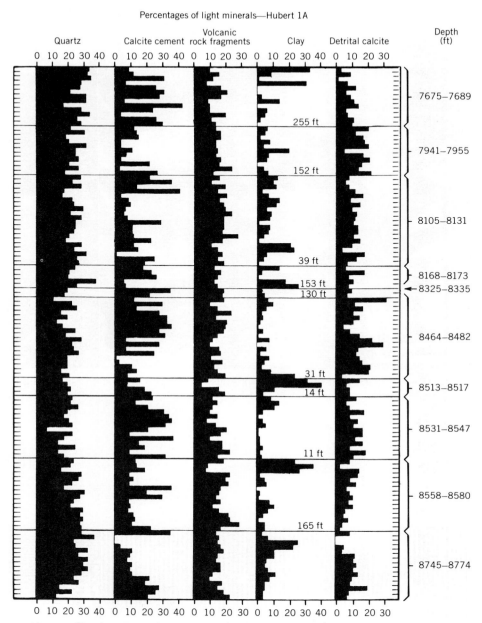

Figure 24.7. Light mineral composition profile changes with increasing depth, Frio sands in the May field.

Mineralogy

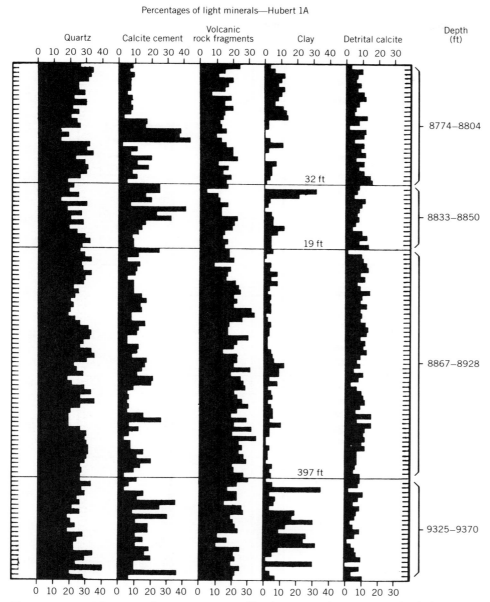

Figure 24.8. Light mineral composition profile changes with increasing depth for Frio sands in the May field, Kleberg County, Texas.

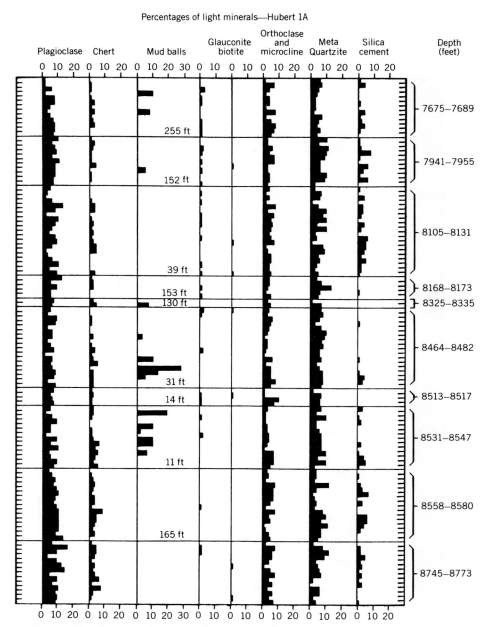

Figure 24.9. Light mineral composition profiles showing changes with increasing depth for Frio sands from the May field, Kleberg County, Texas.

Mineralogy

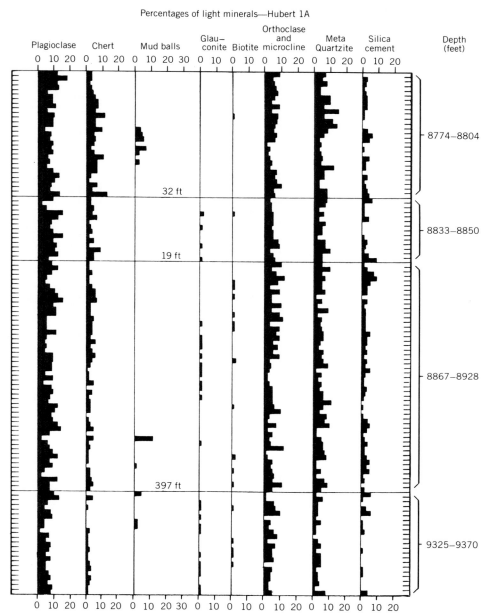

Figure 24.10. Light mineral composition profiles showing changes with increasing depth for Frio sands in the May field, Kleberg County, Texas.

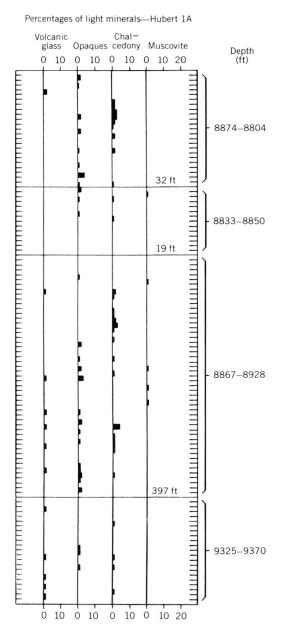

Figure 24.11. Light mineral composition profiles showing changes with increasing depth for Frio sands in the May field, Kleberg County, Texas.

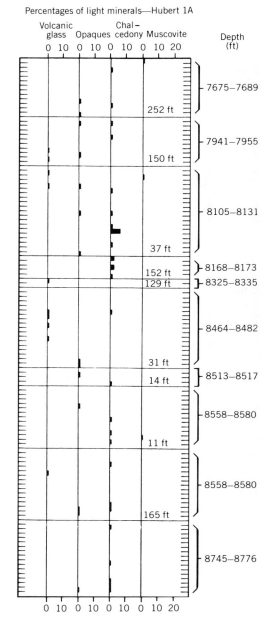

Figure 24.12. Light mineral composition profiles showing changes with increasing depth for Frio sands from the May field, Kleberg County, Texas.

cored interval (Figures 24.7 and 24.8). The detrital carbonate fraction includes shell fragments produced by shellfish growing in the area and limestone plus dolomite fragments eroded from interior outcrops. If the sediments were deposited on the outer shelf, many more foraminifera and particularly *globigerina* would be anticipated. Since these are not common, it is proper to suggest that the environment of deposition is close to the inner shelf. The fact that glauconite occurs rather uniformly throughout indicates continuous marine conditions (Figure 24.14). Therefore, the boundaries in terms of environmental indicators are

Mud Balls

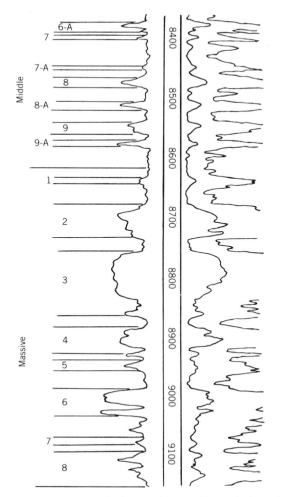

Figure 24.13. Typical electric log profile including the "middle" and "massive" Frio sand zones in the May field.

Figure 24.14. Profile of the total number of glauconite grains per thin section from the "middle sands" of the Frio reservoir of the V. A. Hubert 1A well, May field, Kleberg County, Texas.

reasonably clear and the inference is that the May field sediments were probably deposited nearshore on the inner shelf.

Plagioclase feldspar occurs in greater quantities in the lower half of the V. A. Hubert 1A well (Figures 24.9 and 24.10). This parallels the pattern observed for the volcanic rock fragments and gives additional support to the idea that there was increased volcanic activity in Lower Frio time.

Chert shows a slow increase in percentage (Figures 24.9 and 24.10) proceeding up from 9370 ft (2865.7 m) to a maximum in the interval from 8776 ft (2675.6 m) to 8797 ft (2682.0 m).

Studies of grainsize and the mineral composition of the coarsest fraction indicate that chert is commonly the coarsest material in the thin sections, even though its percentage of the whole is relatively small.

Mud Balls

One of the unusual features that shows up in the thin-section studies of the Frio sands in the May field is the occurrence of mud balls. As shown in Figure 24.9 and 24.10, these occur at rather irregular intervals with many feet of sediment in between. Mud balls have been observed along the shore at the mouth of the Rio Grande at Boca Chica. However, their existence beyond the banks that slumped into the river is usually very limited.

The rather large time intervals between occurrences as represented by many feet of sediment suggests that they might be associated with movement along the faults, which were contemporaneous with sedimentation.

Orthoclase feldspar and metaquartzite fragments tend to remain at a fairly constant percentage over most of the cored interval, denoting that the distant crystalline source areas contributed sediment without a major interruption (Figures 24.9 and 24.10).

Quartz cement occurs more heavily in the bottom half of the core, although the trend is not clearly defined (Figures 24.9 and 24.10). The chart compiled by Loucks et al. (1979, p. B62), as illustrated in Figure 24.15, suggests that this increase is to be expected in the 6000–8000-ft range (1829.3–2439.0 m).

Heavy Minerals

In attempting to piece together the history of the Frio sediments in the May field, it was also decided to study the heavy minerals. In this instance, the heavy mineral fraction includes those minerals from the disaggregated sample that sink in bromoform, that is, their specific gravity is greater than that of bromoform, which is 2.85.

The profiles of the percentages for heavy minerals with increasing depth in the V. A. Hubert 1A well are illustrated in Figures 24.16–24.21.

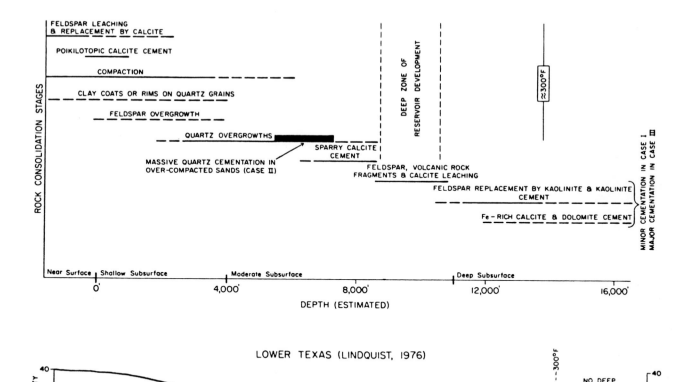

Figure 24.15. Shows rock consolidation stages with increasing depth in terms of porosity changes observed for the Frio Formation in "lower" Texas. (Modified after Loucks et al. 1979, p. B62.)

Heavy Minerals

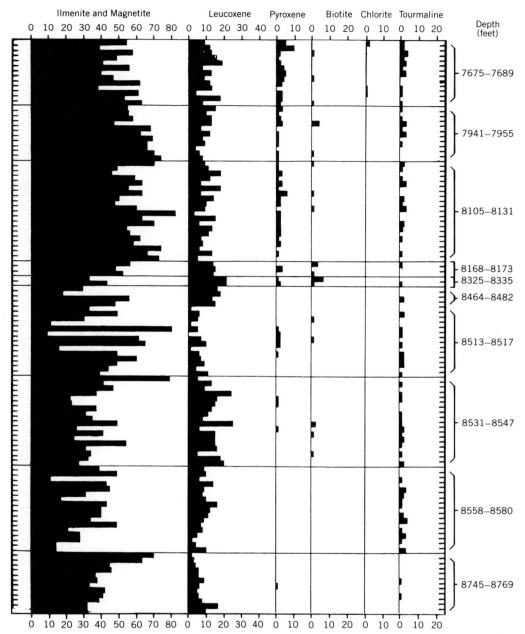

Figure 24.16. Heavy mineral percentage profiles showing changes with increasing depth in the V. A. Hubert 1A well, Mayfield, Kleberg County, Texas.

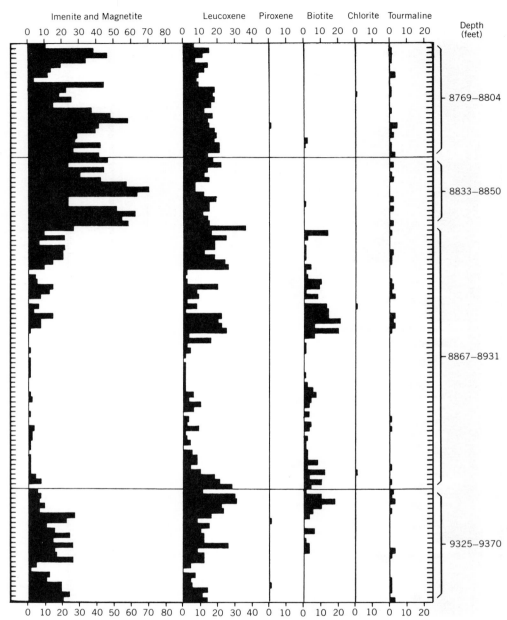

Figure 24.17. Heavy mineral percentage profiles showing changes with increasing depth in the V. A. Hubert 1A well, May field, Kleberg County, Texas.

Heavy Minerals

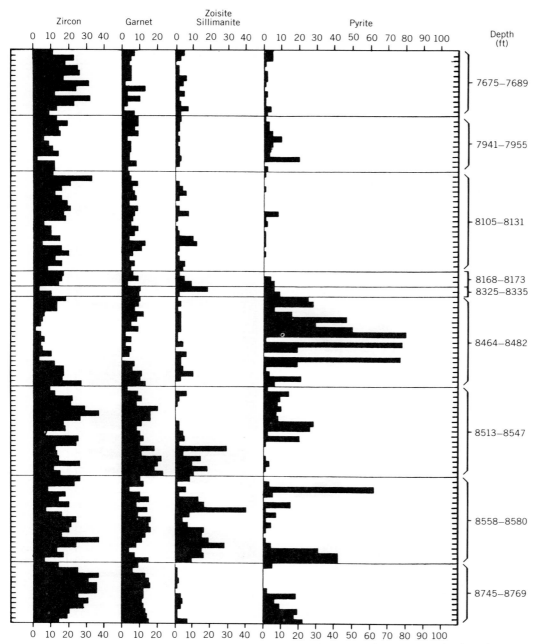

Figure 24.18. Heavy mineral percentage profiles showing changes with increasing depth in the V. A. Hubert 1A well, Mayfield, Kleberg County, Texas.

218   **Frio Formation, The May Field, Kleberg County, Texas**

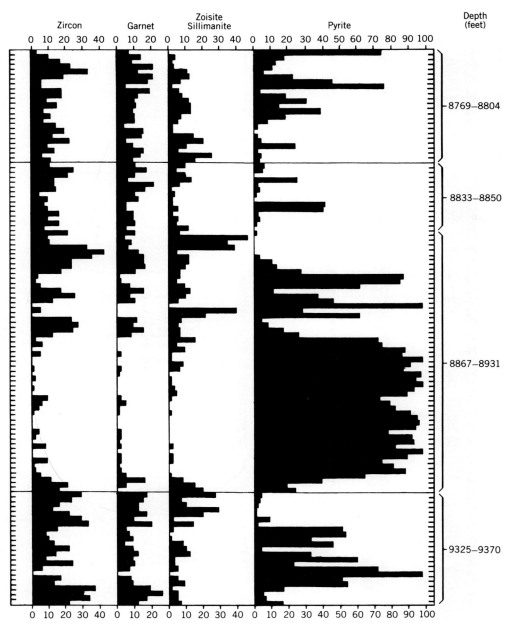

Figure 24.19. Heavy mineral percentage profiles showing changes with increasing depth in the V. A. Hubert 1A well, May field, Kleberg County, Texas.

Heavy Minerals

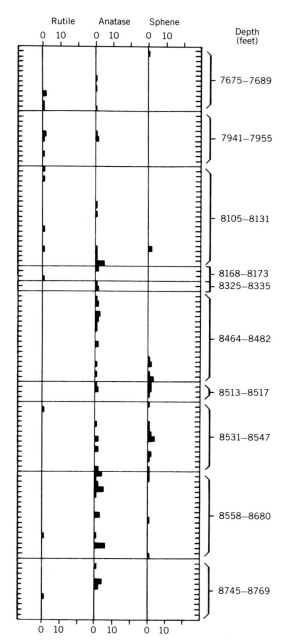

Figure 24.20. Heavy mineral percentage profiles showing changes with increasing depth in the V. A. Hubert 1A well, May field, Kleberg County, Texas.

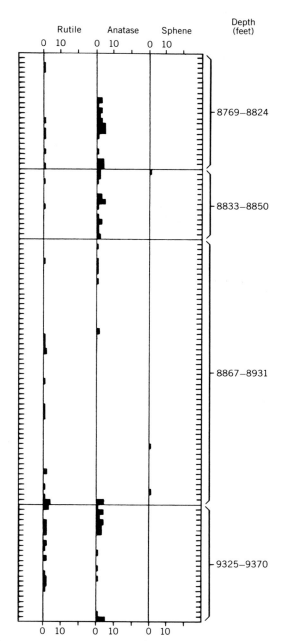

Figure 24.21. Heavy mineral percentage profiles showing changes with increasing depth in the V. A. Hubert 1A well, May field, Kleberg County, Texas.

Pyrite

Probably the most striking anomaly that appears in these profiles is the variation in the amount of pyrite (Figures 24.18 and 24.19). From 8930 ft (2722.6 m) to 8890 ft (2710.4 m), pyrite composes the large majority of the assemblage (Figure 24.19). When viewed in thin section, it becomes clear that the pyrite has grown around and included the quartz and feldspar grains. The unabraided edges of the pyrite crystals furthermore support the conclusion that the pyrite was formed in place after deposition of the sand. Pyrite is normally associated with reducing conditions; therefore, its occurrence in sizable quantities over an interval of 40 ft of sediment indicates that the sediment was not stirred up or exposed to oxygenation by wave action.

The grainsize data from the same samples vary gradually from foot to foot and show fairly good sorting, which confirms the relatively undisturbed nature of the depositional environment.

The major pay sands (8801–8881 ft, 2683.2–2707.6 m) are located directly above the zone that is pyritized; however, only small amounts of pyrite occur within the pay zone itself. It is also worth noting that the amount of pyrite generally decreases toward the top of the Frio section.

Metamorphic Heavy Minerals

Both garnet and sillimanite–zoisite group are considered to be diagnostic of a high-rank metamorphic source area. These minerals are largely absent in the 8892–8930 ft zone (2710.9–2722.6 m). However, this probably reflects the large amount of pyrite that dominates the assemblage at that point (Figure 24.19). From 8892 ft on up the core to 7675–7676 ft (2339.9 m) there is a gradual decrease in the amount of metamorphic-type heavy minerals. This suggests that the volume of sediment coming from the distant interior became relatively less important toward the top of the Frio.

Ilmenite and Magnetite

According to Pettijohn (1957, p. 513), the presence of quantities of ilmenite and magnetite indicates that at least part of the sediment is derived from basic igneous sources; therefore, the variation in these opaque minerals can be important. Proceeding from the bottom of the cored interval to the top there is a general increase in the amount of ilmenite and magnetite (Figures 24.16 and 24.17). Pyroxene begins to appear regularly at about the 8535-ft

Figure 24.22. Total number of biotite flakes per heavy mineral mount for Frio sands from the May field, Kleberg County, Texas. From A. F. O. V. A. Hubert 1A.

Ilmenite and Magnetite

(2602.1-m) level and increases in abundance up to the top of the cored interval. These indicators suggest that more volcanic rocks were being eroded, perhaps caused by new lava flows.

Another mineral that can be used as an indicator of volcanic activity in terms of ash falls is idiomorphic biotite. In order to obtain a better estimate of the amount of biotite, it was decided that the whole slide should be scanned and all the biotite flakes counted. Although this involved much more effort, the resulting profiles appear to be quite sensitive to the rapid change in biotite abundance. In Figures 24.22 and 24.23, it becomes clear that the zone from 9359 ft (2853.4 m) to 8869 ft (2704 m) contains numerous

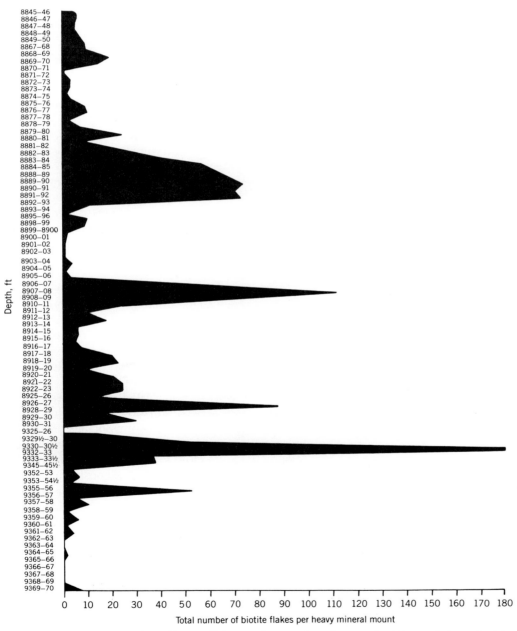

Figure 24.23. Total number of biotite flakes per heavy mineral mount for Frio sands from the May field, Kleberg County, Texas. From A. F. O. V. A. Hubert 1A.

sharp influxes of idiomorphic biotite. These influxes appear to reflect volcanic ash falls. Apparently, volcanic activity near the base of the Frio contributed mostly ash to the sediment. The simultaneous occurrence of pyroxene and minor amounts of biotite in the Upper Frio suggests that the source of the volcanic material may not have been the same as that for the Lower Frio. In other words, inasmuch as such minerals as pyroxene are unstable (Pettijohn 1957, p. 516), their occurrence indicates that the source of this material was nearer to the basin of deposition. Callender and Folk (1959) state that extensive volcanism began in the middle Eocene and continued in ever-increasing amounts into the Oligocene. The present evidence suggests that the area of volcanic activity moved northward in Oligocene time, which would correspond to the increase observed by Callender and Folk in Bastrop County, Texas.

It is interesting to note that in the major pay zone, 8724–8870 ft (2659.8–2704.3 m), the volcanic activity as reflected by biotite decreases to almost zero. The reducing conditions as evidenced by the pyrite and the ash falls that occurred at the same time set the stage for the preservation of organic matter.

The remaining heavy minerals, namely, tourmaline, anatase, rutile, chlorite, and zircon, do not have particularly informative profiles.

Thin Sections

The photomicrographs in Figures 24.24–24.29 illustrate some of the features discussed by Loucks et al. (1979, p. B60). These include poikilotopic calcite, feldspar overgrowths, and quartz overgrowths, all of which are typical of Frio sands (see Figure 24.30).

Grainsize

The grainsize distributions for the major producing zone of the V. A. Hubert 1A well are provided in Figure 24.31. In contrast to earlier distributions illustrated in this book, these are based on sieve analyses.

What emerges from examination of these distributions is that the most common modal value is at 0.125 mm. The overall variation over the major productive interval is not large. Loucks (1979, p. B61) plotted the relationship between the percentage of quartz to the average grainsize for the Frio sands of the lower, middle, and upper Texas Gulf Coast and found that the lower Texas samples were distinctive in that the quartz content was significantly lower than for the other two zones (Figure 24.32). The Frio sands of the May field appear to be in accord with this observation, with quartz content averaging between 20 and 30% and the modal grainsize occurring at between 0.177 and 0.125 mm.

Source Rocks

In discussing the source rocks for the Frio oils, Galloway et al. (1982, p. 679) state:

Gulf Coast Tertiary source rocks are known to be generally lean, and the Frio is typical in that regard. In all, TOC (total organic carbon) data collected from various industry and government sources for 140 Frio samples averaged slightly below 0.3 wt.% organic carbon. Few samples exceeded the 0.4% lower limit for significant Gulf Coast petroleum source rocks suggested by Dow (1978). No obvious patterns in regional source richness are apparent (Galloway et al. 1982). Data did not, however, permit detailed correlation with depositional facies.

Plots of the correlation indexes for crude oils from Kleberg County suggest that at least two source rock varieties are involved. Figure 24.33 shows that crude oils from three of the fields, namely, Seeligson, Borregos, and Stratton, form one type and the crude oil from Alazan North forms another. Specifically, the Alazan oil is significantly more paraffinic.

From the work of Nanz (1954), it is known that the reservoir sands at Seeligson closely resemble those of the nonmarine deltaic plain associated with the present Rio Grande River. Both the Seeligson and Stratton fields are close to each other and are parallel to the coast. Both crude oils are considerably more naphthenic. Moving from the Seeligson area to that of the May field, a distance of about 26 miles (43 k) seaward, we find that glauconite is present in every sand sample. From this, it is logical to conclude that the May field was deposited in a marine environment, and we would expect crude oils from fields of this type to be more paraffinic, reflecting marine source materials.

Unfortunately, distillation data for the May field are not available from the literature. Therefore, another approach is required. It is possible to move

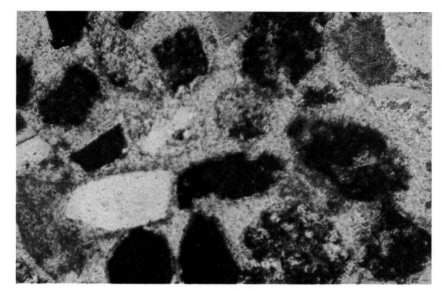

Figure 24.24. Frio Formation, depth unknown, Arkansas Fuel Oil, V. A. Hubert 1A, May field, Kleberg County, Texas (0.44 in. in the photo equals 0.45 mm). Example of poikilotopic calcite cement. According to Loucks et al. (1979, p. B60), this type of cementation indicates very early "diagenetic freezing" where porosity was totally occluded before any other cementation stage occurred. This type of cementation takes place in the very shallow subsurface. Note how the calcite is replacing the quartz grain (upper right). Other grains also show evidence of replacement by calcite.

Figure 24.25. Frio Formation, 8478.5 ft (2584.9 m), Arkansas Fuel Oil, V. A. Hubert 1A, May field, Kleberg County, Texas (0.44 in. in the photo equals 0.45 mm). Quartz grain in the center shows the typical type of quartz overgrowth observed in the Frio sands. Detrital calcite and sparry calcite (light brown) are also common features. Quartz overgrowths in the Frio start to occur between 2000 and 4000 ft in depth of burial (Loucks et al. 1979, p. B60). Other grains include weathered feldspar (right center), a grain of chalcedonic quartz (center left), and a volcanic rock fragment (top center).

Figure 24.26. Frio Formation, 8804 ft (2684.2 m), Arkansas Fuel Oil, V. A. Hubert 1A, May field, Kleberg County, Texas (0.44 in. in the photo equals 0.45 mm). Large plagioclase feldspar (striped grain, left center) shows overgrowths. One corner has also pressed into a detrital carbonate grain (light brown). Volcanic rock fragments (lower left) and chert (upper left) are common.

Figure 24.27. Frio Formation, 8804 ft (2684.2 m), Arkansas Fuel Oil, V. A. Hubert 1A, May field, Kleberg County, Texas (0.44 in. in the photo equals 0.45 mm). Typical glauconite grain (green, center) occurs amid plagioclase feldspar, chert, quartz, sparry calcite cement, and detrital carbonate fragments.

Figure 24.28. Frio Formation, 8879.5 ft (2707.2 m), Arkansas Fuel Oil, V. A. Hubert 1A, May field, Kleberg County, Texas (0.44 in. in the photo equals 0.45 mm). The overall texture is characterized by a loosely cemented framework. Some sparry calcite cement is supporting a very open structure involving plagioclase feldspar, detrital calcite and angular quartz grains.

Figure 24.29. Frio Formation, 8877.5 ft (2706.5 m), Arkansas Fuel Oil, V. A. Hubert 1A, May field, Kleberg County, Texas (0.44 in. in the photo equals 0.45 mm). Again angular quartz, chert, detrital calcite, and volcanic rock fragments compose most of the sediment. Some sparry calcite helps hold the framework open. Note the yellowish brown biotite fragment in the upper right center.

Source Rocks

Diagenetic Feature	Basis for Paragenetic Sequence (Relative Timing)	Basis for Depth Estimation
Feldspar leaching and replacement by calcite	Common feature in paleosoils of the Catahoula Formation (Frio equivalent) (Galloway 1977). Feldspar overgrowths in leached voids of feldspars reveal early leaching.	Soil feature at or near surface.
Poikilotopic calcite cement	Lack of any other cements indicates very early "diagenetic freezing"; porosity was totally occluded before any other cementation stage took place. Immediately adjacent to poikilotopic calcite cement are sandstones that contain several cement types and the grains are normally compacted.	Evidence of early cementation and loose packing places this cement as a soil or very shallow subsurface feature.
Clay coats	Common feature in paleosoils of the Catahoula Formation (Galloway 1977). Also overlain by quartz overgrowths.	Soil feature at or near surface.
Clay rims	Quartz overgrowths are absent where chlorite clay rims are thick; similar situation as described by Tillman and Almon (1977).	Occur before quartz. Galloway (1974) documented their formation between 1000 and 5000 ft in sandstones of similar composition.
Feldspar overgrowths	Euhedral against quartz overgrowths, which indicates they formed before quartz.	Occur before quartz.
Quartz overgrowths	Overlie clay coats, absent around clay rims, anhedral against feldspar overgrowths, and absent where poikilotopic calcite is present. Euhedral against sparry calcite, kaolinite, and iron-rich carbonate cements. Also leached-grain embayments into the quartz overgrowth indicates leaching occurred after quartz cementation.	Galloway (1975), for sandstones of similar composition, found quartz overgrowths to start between 2000 and 4000 ft. Selected samples of Frio sandstones indicated no overgrowths at 2500 ft and well-developed overgrowths at 5300 ft.
Sparry calcite cement	Anhedral against quartz and feldspar overgrowths. Underwent dissolution during moderate subsurface leaching stage.	Timing relative to quartz and feldspar overgrowths and to leaching puts this cement later. Lindquist (1976) noted sparry calcite to be well developed at depths of 8000 ft.
Leaching of feldspars, volcanic rock fragments, and calcite	Leaching of calcite cement that replaced feldspars in soil zone and also leaching of sparry calcite cement. Leached feldspars, volcanic rock fragments, and leached-grain embayments in quartz overgrowths filled by kaolinite and iron-rich carbonate cements.	Lindquist (1976) noted leaching starting around 8500 feet. Also, if fluids released by the transformation of montmorillonite to illite caused the leaching, this approximates the leaching stage at around 8000–9000 ft.
Kaolinite cement and feldspar replacement by kaolinite	Replaces feldspars and fills leached porosity in feldspar, primary pore space, and resurrected primary pore space (previously filled with sparry calcite). Anhedral against quartz and fills leached-grain embayments in quartz overgrowths.	Occurs after leaching. In south Texas, kaolinite cementation begins around 10,000 ft.
Iron-rich calcite and dolomite cements	Filled leached porosity. Also, pore fluids that leached the earlier calcites had to be acidic whereas pore fluids that deposited late iron-rich carbonate cements had to be basic; therefore, a period of time had to elapse while pore fluid chemistry changed.	Occurs after leaching and after kaolinite cementation because, according to Lindquist (1970), kaolinite is stable at a lower pH than carbonates.

Figure 24.30. Factors controlling reservoir rock quality as related to mineralogy, timing, and depth of burial for Frio sandstones of the Texas Gulf Coast. (After Loucks et al. 1979, p. B60.)

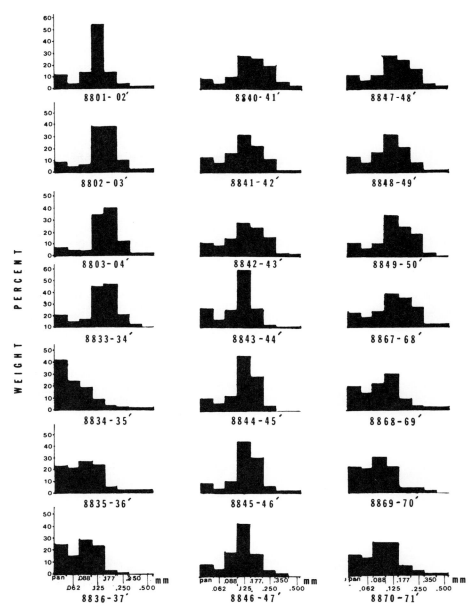

Figure 24.31. Grainsize distributions (sieved) for the major producing zone of the Arkansas Fuel Oil Company, V. A. Hubert 1A, May field, Kleberg County, Texas.

north along the strike and to locate Frio fields that have both the strong nonmarine component and those with a totally marine setting. In Figure 24.34, the correlation index curves for Frio age crude oils from Plymouth, Portilla, and Mustang Island. Both Plymouth and Portilla are located in San Patricio County, which is somewhat inland and updip from Corpus Christi (see Figure 24.35, top); whereas Mustang Island is on the seaward side of the trend. The general cross section of the Corpus Christi area and the Frio sands is shown in Figure 24.35. In this instance, it is clear from the correlation index curves (Figure 24.34) that the crude oil from the Frio sands in the Mustang Island field is much more paraffinic

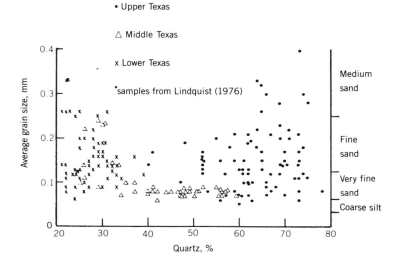

Figure 24.32. Relationship of percentage of quartz to average grainsize between lower, middle, and upper Texas Gulf Coast. (After Loucks et al. 1979, p. B61.)

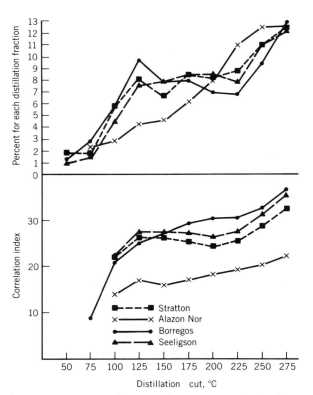

Figure 24.33. Correlation index curves and distillation fractions for crude oils from Frio sand fields in Kleberg County, Texas. (Data from McKinney et al. 1966.)

than either of the crude oils from the Frio sand in the Plymouth or Portilla fields. It seems quite likely that the increased paraffinicity of the Mustang Island crude oil reflects more marine source materials. Galloway et al. (1982, p. 680) state the following with regard to source materials for the Frio:

In summary, although Frio mudstones contain low percentages of organic carbon and are dominated by gas-prone woody and herbaceous organic matter types, the volumes of potential source rock lying within or below the oil maturation window are immense. In the more deeply buried down dip plays, the possibility of internal source must be considered viable. In shallow plays, a deeper, extrinsic hydrocarbon source is required.

Hunt (1984, p.1266) has shown that for shale samples taken from a well drilled off the shore of South Padre Island, Texas, the threshold of intense oil generation occurs at 7874 feet (2.4 k). Since this well was drilled a relatively short distance from the May Field, it seem likely that the reservoir at a depth of 8801–8081 ft (2683.2 to 2707.6m) is well within the oil-generating envelope for this particular sedimentary province.

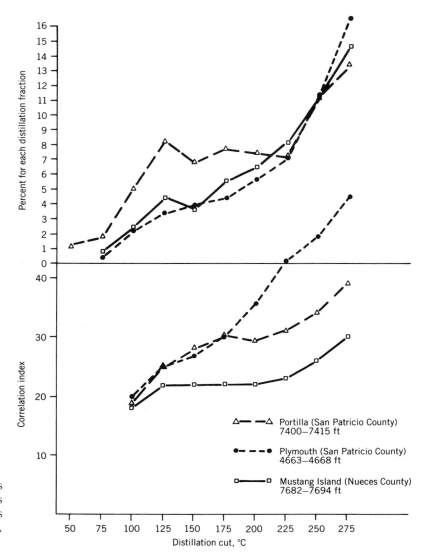

Figure 24.34. Correlation index curves and distillation fractions for Frio crude oils from fields in San Patricio and Nueces counties. (Data from McKinney et al. 1966, and McKinney and Garton, 1957.)

The deeper extrinsic source appears to be related to regional growth faulting (down-to-the-coast faults) and such features as shale diapirs, which presumably create channels for the vertical migration of hydrocarbons generated at depth.

The author is therefore, not convinced that the "lean" sediments that surround the sand could not produce significant amounts of petroleum. What we see, in the final analysis, seems to be a mixture of perhaps several sources.

Comparison with Samples from Greta–Carancahua Barrier Strand–Plain System at Corpus Christi

As mentioned previously in this section, the sediments from the producing Frio sands in the May field are somewhat different from those in the stacked barrier island reservoirs directly to the north. In order to

Comparison with Samples from Greta–Carancahua Barrier Strand–Plain System at Corpus Christi

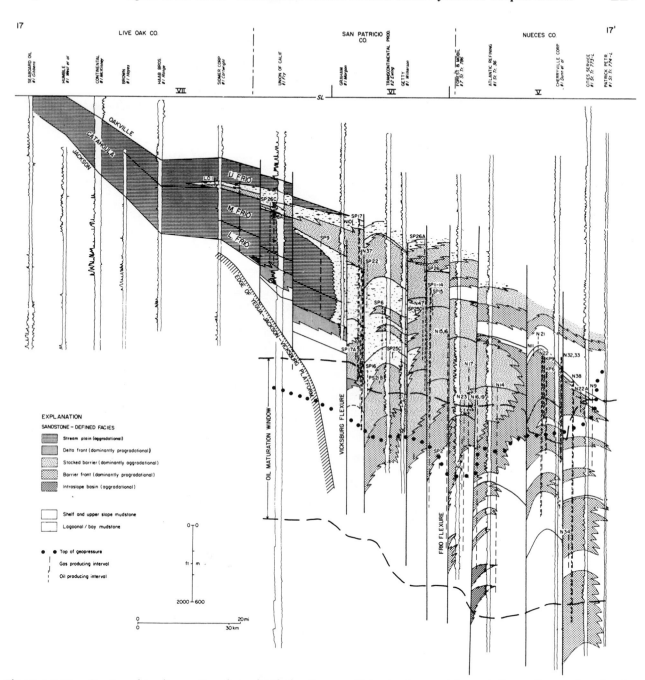

Figure 24.35. Stratigraphic dip section through Choke Canyon–Flatonia stream-plain and Greta–Carancahua barrier strand-plain system. Stacked barrier and updip lagoonal facies of Frio are well developed. (After Galloway et al. 1982, p. 667.)

Figure 24.36. Frio Formation, 7006.5 ft (2126.1 m), Cities Service State Tract No. 9, Corpus Christi Bay, Nueces County, Texas (0.44 in. in the photo equals 0.17 mm). Frio sand with kaolinite clay surrounding the individual grains. Note that most of the grains are quartz with minor amounts of chert and plagioclase feldspar (lower left). Individual sand grains appear to have coatings of clay that show as zones of lighter colored matrix.

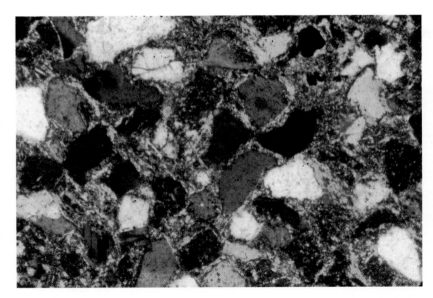

Figure 24.37. Frio Formation, 7427.5 ft (2264.5 m), Cities Service State Tract No. 10, Corpus Christi Bay, Nueces County, Texas (0.44 in. in the photo equals 0.17 mm). The sand is highly angular and not well sorted. More plagioclase feldspar appears to be present along with muscovite flakes. It should be noted that muscovite is not present in the samples from the May field, suggesting that a source different from those that contributed to the Norias Delta System to the south is providing.

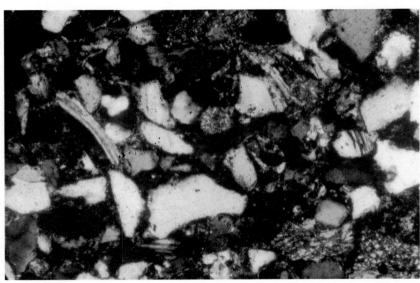

Figure 24.38. Frio Formation, 7972.5 ft (2430.6 m), Cities Service State Tract No. 16, Corpus Christi Bay, Nueces County, Texas (0.44 in. in the photo equals 0.17 mm). An example of poikilotopic calcite cement filling in the porosity. Note the heavy mineral grains' (lower right) brighter color and the dark grains along the top. Chert and quartz grains are the major constituents of the sand.

Recent Equivalents

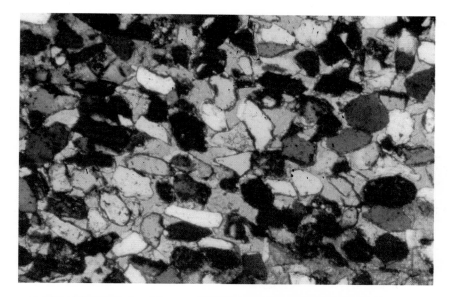

Figure 24.39. Frio Formation, 8377.5 ft (2554.1 m), Cities Service State Tract No. 16, Corpus Christi Bay, Nueces County, Texas (0.44 in. in the photo equals 0.17 mm). More poikilotopic calcite cement. In some cases the calcite cement is replacing the quartz grains. Elongate quartz grains appear to be oriented in layers (diagonally across the photomicrograph.)

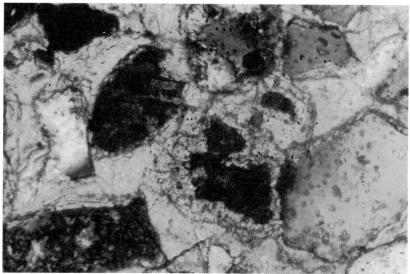

Figure 24.40. Frio Formation, 8377.5 ft (2554.1 m), Cities Service State Tract No. 16, Corpus Christi Bay, Nueces County, Texas (0.44 in. in the photo equals 0.08 mm). Higher-magnification view of calcite cement replacing both quartz and feldspar.

obtain a better feeling for the differences, photomicrographs were taken of several Frio sands as they occur in the Corpus Christi Bay field (Figures 24.36–24.40). The approximate location of the Cities Service Oil Company State Tract wells in Corpus Christi Bay is shown in Figure 24.35. Certain features that were mentioned by Galloway (1982) as being characteristic of the Frio reservoir sands show up clearly in the photomicrographs.

Recent Equivalents

The whole present-day coastline along Texas and Louisiana is closely analagous to the Tertiary shorelines, which are productive (see Figure 24.41). There is vast literature dealing with the Recent sediments of the Gulf Coast, and to review them would require a full-length text in itself. Nevertheless, it can be stated

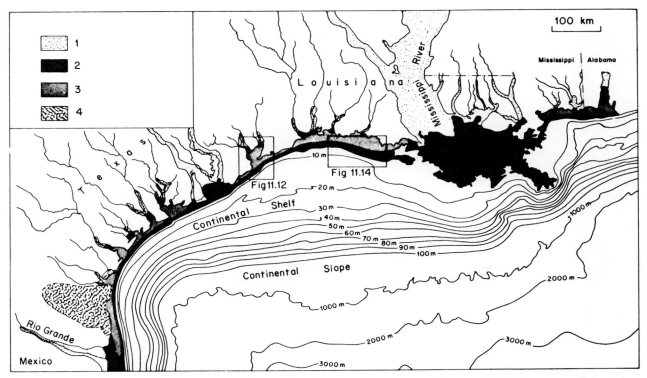

Figure 24.41. Generalized features of the Recent coastal sedimentation for the northwest Gulf of Mexico. (1) Alluvial plain deposits; (2) subaerial coastal sediments, deltas, prograding barrier bars and cheniers; (3) lagoons and marshes; (4) eolian sand blown inland from the barrier beach. (After Robley K. Matthews, Dynamic Stratigraphy: An Introduction to Sedimentation and Stratigraphy, (c), 1974, pp. 176, Reprinted by permission of Prentice-Hall, Inc., Englewood Cliffs, New Jersey.)

that all this effort has helped our understanding of the many complex phenomena that occur in these sediments.

Summary

The May field provides an example of an oil and gas accumulation related to the Norias Delta System (Rio Grande embayment). Its location on the northern flank of this system places it in a transitional zone between the typical deltaic deposits and the barrier island strand line facies of the Greta–Carancahua System. Evidence from the glauconite content suggests that the producing zone was of totally marine origin. It is likely, therefore, that the growth faulting associated with the May field was generally located a short way seaward of the barrier island–lagoonal complex.

If we accept the hypothesis proposed by Galloway et al. (1982, p. 686), that is, that the hydrocarbons found within Frio reservoirs are not indigenous to the Frio Formation but were derived from underlying units, then the growth faults provide both the trapping mechanism and the channel for hydrocarbon emplacement.

References

Callender and R. L. Folk, 1958, Idiomorphic Zircon, Key to Volcanism in the Lower Tertiary Sands of Central Texas, *American Journal of Science*, Vol. 256, pp. 257–269.

Galloway, W. E., 1974, Deposition and Diagenetic Alteration of Sandstone in Northwest Pacific Arc-Related Basins, Implications for Graywacke Genesis: *Geological Society of America Bulletin*, Vol. 85, pp. 379–390.

Galloway, W. E., 1977, Catahoula Formation of the Texas Coastal Plain—Depositional Systems, Mineralogy, Structural Development, Ground-Water Flow History, and Uranium Distribution, University of Texas,

References

Austin, *Bureau of Economic Geology Report of Investigations* Vol. 87, 59 pp.

Galloway, W. E., D. K. Hobday, and K. Magara, 1982, Frio Formation of Texas Gulf Coastal Plain: Depositional Systems, Structural Framework, and Hydrocarbon Distribution, *American Association of Petroleum Geologists Bulletin,* Vol. 66, No. 6, pp. 649–688.

Hunt, J. M., 1984, Generation and Migration of Light Hydrocarbons, *Science,* Vol. 226, No. 4680, pp. 1265–1270.

Lindquist, S. J., 1976, Sandstone Diagenesis and Reservoir Quality, Frio Formation (Oligocene), South Texas: Unpublished Masters Thesis, University of Texas, Austin, Texas, 147 pp.

Loucks, R. G., D. G. Bebout, and W. E. Galloway, 1979, Relationship of Porosity Formation and Preservation to Sandstone Consolidation History—Gulf Coast Lower Tertiary Frio Formation, in *AAPG Continuing Education Short Course Note Series #11,* D. Bebout, G. Davies, C. H. Moore, P. S. Schoelle, and N. C. Wardlow, Eds. American Association of Petroleum Geologists, Tulsa, Oklahoma, pp. 55–66

Matthews, R. K., 1974, *Dynamic Stratigraphy,* Prentice-Hall, Englewood Cliffs, NJ, 370 pp.

McKinney, C. M. and E. L. Garton, 1957, *Analyses of Crude Oils from 470 Important Oilfields in the United States,* Bureau of Mines Report of Investigations 5376.

McKinney, C. M., F. P. Ferrero, and W. J. Wenger, 1966, *Analyses of Crude Oils from 546 Important Oilfields in the United States,* Bureau of Mines Report of Investigations 6819, 345 pp.

Nanz, R. H., Jr., 1954, Genesis of Oligocene Sandstone Reservoir, Seeligson Field, Jim Wells and Kleberg Counties, Texas, *Bulletin of the American Association of Petroleum Geologists,* Vol. 38, No. 1, pp. 96–117.

Pettijohn, F. P., 1957, *Sedimentary Rocks,* 2nd ed., Harper & Brothers, New York, 718 pp.

Tillman, R. W. amd W. R. Almon, 1977, Diagenesis of Frontier Formation Offshore Bar Sandstones, Spearhead Ranch Field, Wyoming (Abstract), *Proceedings of the Rocky Mountain Section American Association of Petroleum Geologists and Society of Economic Paleontologists and Mineralogists, 26th Annual Meeting,* April 2–6, 1977, Denver, Colarado, p. 53.

# 25

The Cadeville Sandstone, Schuler Formation, Calhoun Field, Jackson Parish, Louisiana

Geologic Background

The Calhoun field is located 80 miles (128.7 km) east of Shreveport, Louisiana, in Jackson, Lincoln, and Quachita parishes (Figure 25.1) and lies within the Gulf Coastal Plain province. The Upper Jurassic Cotton Valley Group, which contains the Cadeville Sandstone (Figure 25.2), is locally split into the Bossier Formation and the Schuler Formation. Both of these were deposited as wedges of sediment that thicken southward into the North Louisiana basin. The Schuler Formation contains the Cadeville Sand and consists mainly of dark gray and black marine shales that thin northward and interfinger with nonmarine sandstones and conglomerates in southern Arkansas (Figure 25.3).

Included within the Schuler Formation is the "B lime," which is about 110 ft thick. The Cadeville Sand occurs as a lens within the B lime with a sand thickness ranging to 38 ft. (See electric log in Figure 25.4.)

The regional structural setting involves a south dip of about 200 ft/mile that is locally modified by the Monroe and Sabine uplifts. These broad features are

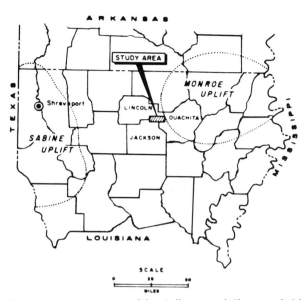

Figure 25.1. Location of the Calhoun and Cheniere fields. (After Exum 1973, p. 302.)

Figure 25.2. Stratigraphic chart of the Mesozoic rocks of northern Louisiana showing the relative positions of the Cadeville and Hosston sands. (After Exum 1973, p. 303.)

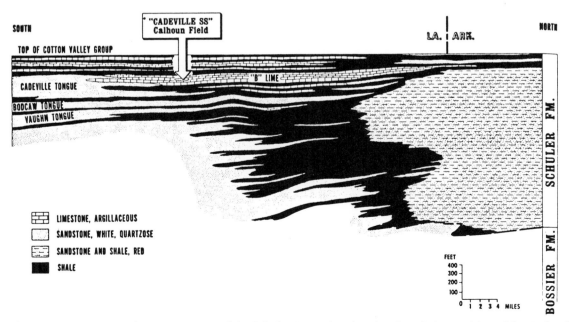

Figure 25.3. Stratigraphic cross section of the Schuler Formation showing the relative position of the Cadeville Sand. (After Exum 1973, p. 304.)

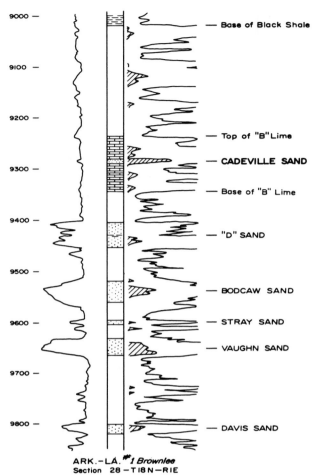

Figure 25.4. Typical electric log through the upper part of the Schuler Formation showing the Cadeville producing sand in the Calhoun field, Louisiana. (After Exum 1973, p. 305.)

interrupted by local faults and folds. In the Calhoun field, the structure is an east–west trending anticline about 2 miles wide and 4 miles long, with approximately 150 ft of structural closure.

According to Exum (1973, p. 302), the entrapment mechanism in the Cadeville Sand is clearly stratigraphic with production limited to the top and west flank of the Calhoun structure. The Calhoun field itself (independent of the structure) is 11.5 miles long in an east–west direction and averages about 2 miles wide, occupying an area of approximately 14,000 acres.

An interesting fact from the reservoir engineering point of view is that the original reservoir pressure was 8201 psi, which is almost double the 4275 psi pressure predicted by a usual gradient of 0.45 psi/ft.

The average permeability is 213 millidarcies, and the average porosity is 14.8%.

Petrology

The Cadeville Sand is predominantly of subrounded quartz that is mostly fine to very fine sand. Some rock fragments, chert, and plagioclase feldspar are also present. Frequently, the quartz sand is mixed with pelecypod and gastropod fragments that according to Exum (1973, p. 311) account for at least 90% of the fauna.

Authigenic minerals are common and include quartz, calcite, pyrite, dolomite, and dickite (Exum 1973, p. 308).

One of the most unusual features of the Cadeville is the occurrence of a zone of serpulid worm tubes. This latter feature is photographed in Figures 25.5–25.8. It occurs in several wells located on 320-acre spacing; hence, its occurrence is not of the single sample variety.

According to Exum (1973, p. 311), three types of porosity are observed in the Cadeville, intergranular, intragranular, and moldic. The intergranular porosity occurs in the quartz sand and is the major variety of porosity. The intragranular porosity occurs inside gastropod shells and worm tubes. This type of porosity was quite important prior to the formation of the in-filling sparry calcite. Unfortunately, most occurrences of the intragranular type have been filled in.

The last variety is moldic porosity, which is not observed to any great degree in the thin sections available to this author. The sandier zones, as illustrated in Figures 25.9–25.11, show that the primary intergranular porosity has been reduced by the growth of quartz cement on the quartz grains.

Environment of Deposition

Looking at the isopach map of the gross Cadeville Sand as it occurs in the Calhoun field (Figure 25.12), one gets the impression that we are dealing with a sandbar or barrier beach. Why such a deposit should occur in the midst of the carbonate mudrocks of the B lime is not obvious. The abundance of marine fossils indicates that the deposit is marine in origin and the

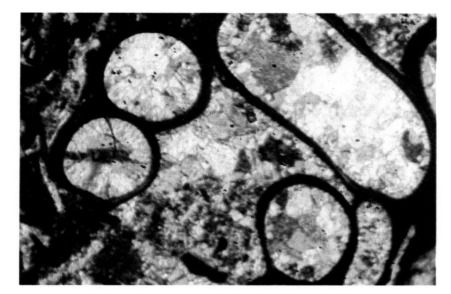

Figure 25.5. Cadeville Sandstone, Schuler Formation, 9877 ft (3011.3 m), Cities Service Tremont K-1, Calhoun field, Jackson Parish, Louisiana (0.44 in. in the photo equals 0.17 mm). Serpulid worm tubes filled in with sparry calcite. These structures are built of concentric laminae of calcite. In this instance, the tubes are adjoining and the walls are unbroken.

Figure 25.6. Cadeville Sandstone, Schuler Formation, 10,072 ft (3070.7 m), Monsanto, Hodde No. 1, Calhoun field, Jackson Parish, Louisiana (0.44 in. in the photo equals 0.17 mm). Fossil hash composed of calcite-filled serpulid worm tubes and thin-shelled pelecypod fragments. Some carbonate mud fills in the spaces between fossil fragments.

Figure 25.7. Cadeville Sandstone, Schuler Formation, 10,072 ft (3070.7 m), Monsanto, Hodde No. 1, Calhoun field, Jackson Parish, Louisiana (0.44 in. in the photo equals 0.17 mm). Serpulid worm tube fragments crushed and infilled with sparry calcite. Some pyrite (black) included in a zone across the photomicrograph. The walls of the worm tubes are thicker than in Figures 25.5 and 25.6 but not as thick as those illustrated by Scholle (1978, p. 41.)

Figure 25.8. Cadeville Sandstone, Schuler Formation, 10,072 ft (3070.7 m), Monsanto, Hodde No. 1, Calhoun field, Jackson Parish, Louisiana (0.44 in. in the photo equals 0.17 mm). More crushed and fragmented serpulid worm tubes with sparry calcite infilling. Initially the porosity was very high; however, the sparry calcite appears to have filled in almost all the void space.

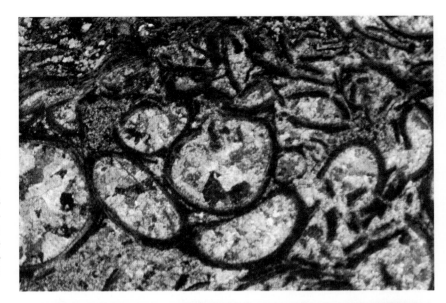

Figure 25.9. Cadeville Sandstone, Schuler Formation, 9870 ft (3009.2 m), Cities Service Tremont K-1, Calhoun field, Jackson Parish, Louisiana (0.44 in. in the photo equals 0.17 mm). Clean quartz sand with numerous quartz overgrowths cementing the grains and cutting down on the overall porosity. Some clay matrix material is present.

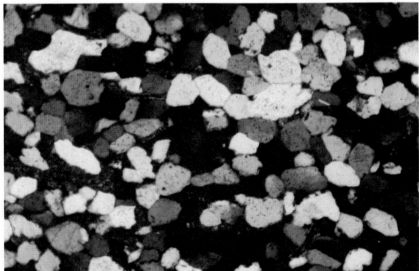

Figure 25.10. Cadeville Sandstone, Schuler Formation, 9871 ft (3009.5 m), Cities Service Tremont K-1, Calhoun field, Jackson Parish, Louisiana (0.44 in. in the photo equals 0.17 mm). More angular quartz sand exhibiting both calcite and quartz cementing materials. Small grains of calcite appear to be more or less equally dispersed around the boundaries of the quartz.

Source Rocks

Figure 25.11. Cadeville Sandstone, Schuler Formation, 9874 ft (3010.4 m), Cities Service Tremont K-1, Calhoun field, Jackson Parish, Louisiana (0.44 in. in the photo equals 0.17 mm). Clean quartz sand that is highly cemented with quartz overgrowths. Small amounts of carbonate are also present along with some pyrite.

low-angle cross-stratification suggest a shoal-water environment. Exum's (1973, p. 313) interpretation is that of a sandbar in a shallow, wave-agitated sea, and this appears to be the most likely hypothesis.

Recent Sediment Equivalents

Sandbars completely surrounded within mudstone are not common features; however, study of Recent lime mud environments at Flamingo near the tip of the Florida peninsula suggests that were a source of sand available, such a deposit would easily form.

Source Rocks

The fact that Cadeville Sand is completely enclosed in the envelope of the B lime strongly suggests that the large amount of organic matter associated with the sandbar and its surroundings provided the source materials. The pyrite that is frequently found along

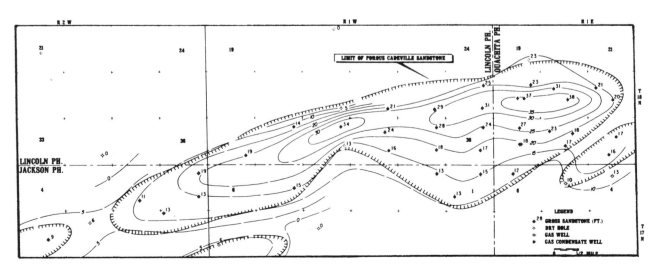

Figure 25.12. Isopach map of the gross Cadeville Sand in the Calhoun field, Louisiana. (After Exum 1973, p. 307.)

with the fossil shell materials certainly indicates that the necessary reducing conditions did play a part. Furthermore, the zone with the abundant serpulid worm tube fragments also supports the fact that marine life was present in sufficient quantity to produce the hydrocarbons.

References

Exum, F. A., 1973, Lithologic Gradients in Marine Bar, Cadeville Sand, Calhoun Field, Louisiana, *American Association of Petroleum Geologists Bulletin,* Vol. 57, No. 2, pp. 301–320.

Scholle, P. A., 1978, *A Color Illustrated Guide to Carbonate Rock Constituents, Textures, Cements and Porosites,* American Association of Petroleum Geologists, Memoir 27, Tulsa, OK, 241 pp.

26

The James Limestone, Pearsall Formation, Fairway Field, Anderson and Henderson Counties, Texas

Geologic Background

The Fairway field is located approximately 60 miles (100 km) southeast of Dallas, Texas, near the eastern end of the Anderson–Henderson County line (Figure 26.1). The major producing zone is from the James Limestone Member of the Pearsall Formation, which is Lower Cretaceous in age (Figure 26.2). As Terriere (1976, p. 157) points out, the reservoir rock is an unusually good example of a subsurface reef complex and therefore deserves some additional study. The productive limits of the field include 23,000 acres with 152 wells on 160 acre spacing. The original reservoir pressure was 5226 psia, and the average pay thickness is approximately 70 ft (Calhoun and Hurford 1970, p. 1217). By the end of 1974, the Fairway field had produced 123,703,000 bbl of 48° AP gravity oil and had estimated reserves of 76,234,000 bbl (Terriere 1976, p. 157). The trap in the Fairway field is partly structural and partly stratigraphic. The location of the major reef appears to have been controlled and enhanced by a growing structure. The stratigraphic units in the vicinity of

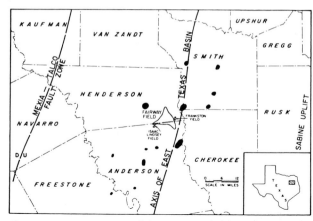

Figure 26.1. Location of the Fairway field. Salt domes are shown as black spots. (After Terriere 1976, p. 158.)

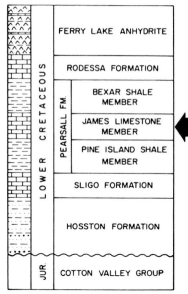

Figure 26.2. Partial columnar section for the Fairway field area. (After Terriere 1976, p. 159.)

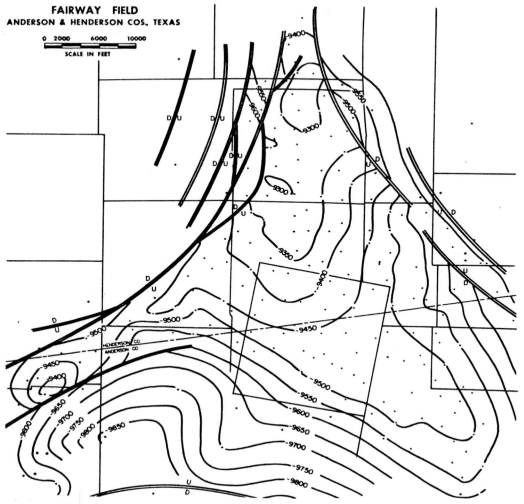

Figure 26.3. Structure of the top of the James Limestone Fairway field. (After Terriere 1976, p. 160.)

Geologic Background

the Fairway field include the Hosston Formation which overlies the Jurassic–Cretaceous boundary, and three overlying limestone units—the Sligo, James, and Rodessa, which are separated by units of dark gray calcareous shale. Regionally, where the James is not a reef, it occurs as a thin argillaceous limestone.

Overlying the Rodessa Formation is the Ferry Lake Anhydrite, which, as the name implies, contains anhydrite with thin beds of shale and limestone.

The relationship between the James reef and the geologic structure is illustrated in Figure 26.3. The shape of the reef is that of a southeast plunging nose cut off by a group of faults on the northeast. The thickness of the reservoir reef parallels the structure (Figure 26.4). As pointed out by Terriere (1976, p. 158), the parallelism of the isopach lines with the corner lines of the top of the reef suggests that the reef growth was related to the growing structure. Whether or not the Fairway structure is somehow related to a salt dome or salt flowage at depth is still a point of debate.

Representative electric logs of the James limestone reef as it occurs in the Fairway field is shown in Figure 26.5.

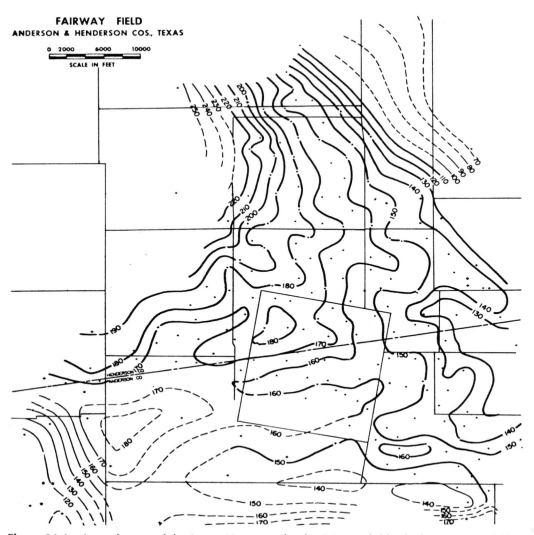

Figure 26.4. Isopach map of the James Limestone for the Fairway field. Thicknesses are in feet. (After Terriere 1976, p. 161.)

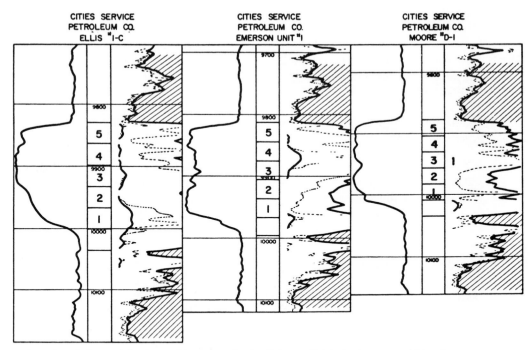

Figure 26.5. Representative electric logs of the James Limestone in the Fairway field, Anderson and Henderson counties, Texas. (After Terriere 1976, p. 170.)

Thin Sections

In the case of the Fairway field, the author is again faced with a shortage of samples with which to discuss the various lithologic varieties present. Terriere (1976, p. 159) found that there were five rock types that could be recognized from megascopic description of the slabbed cores. In the materials available to this author, only two types are present; nevertheless, the photomicrographs of these two varieties (see Figures 26.6–26.11) provide some idea of the sediment types.

Crude Oil Characteristics

Fortunately, the Bureau of Mines has published data concerning the correlation index for crude oil from the Fairway field. Examination of Figure 26.12 reveals that the correlation index is fairly paraffinic, which is in keeping with the marine reef environment of deposition. It should also be noted that there is little or no evidence of biodegradation.

One of the unusual features of the cores from the Fairway reef is the occurrence of a black organic material that is frequently powdery. In some instances, it is present in shiny masses with conchoidal fracture and has been called "gilsonite" in some core descriptions (Terriere 1976, p. 120). The material occurs within secondary pores and therefore has entered the rock after its original deposition. Terriere (1976, p. 170) suggests that the maturation process has produced both the high gravity oil and the black sootlike residue.

Geologic Interpretation

The reservoir rock of the Fairway field is one of the better examples of a reef complex exhibiting numerous fossil remains, many of which are framework-building organisms characteristic of the reef core. These include corals, stromatoporids, algae, rudistids, and bryozoans. In some cases, large bivalve shells (large clams) are found upright in the lime rock, indicating that no reworking has taken place.

245

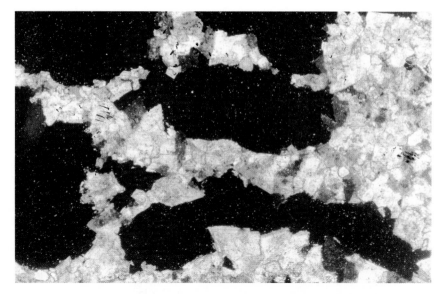

Figure 26.6. James Limestone Member, Pearsall Formation, 9933 ft (3028.4 m), Cities Service Moore No. 1, Fairway field, Henderson County, Texas (0.44 in. in the photo equals 0.17 mm). Open moldic porosity with fairly coarse grained calcite forming the framework. (Black is open pore space.)

Figure 26.7. James Limestone Member, Pearsall Formation, 9933 ft (3028.4 m), Cities Service Moore No. 1, Fairway field, Henderson County, Texas (0.44 in. in the photo equals 0.17 mm). Moldic pore space. (Black is open pore space.) Smaller grain-size in clusters suggest other pellet-type structures. Coarse calcite appears to have filled in some of the open pores. One small quartz grain probably of wind-blown origin appears in the upper left corner.

Figure 26.8. James Limestone Member, Pearsall Formation, 9933 ft (3028.4 m), Cities Service Moore No. 1, Fairway field, Henderson County, Texas (0.44 in. in the photo equals 0.17 mm). Coarse-grained calcite dominates and in some cases has filled in the pore space; however, considerable moldic porosity remains.

Figure 26.9. James Limestone Member, Pearsall Formation, 9933 ft (3028.4 m), Cities Service Moore No. 1, Fairway field, Henderson County, Texas (0.44 in. in the photo equals 0.17 mm). Radiating porosity pattern reveals the micritized structure of a colonial coral. Some fine-grained quartz grains (white) have been included in the structure and are probably wind-blown sand.

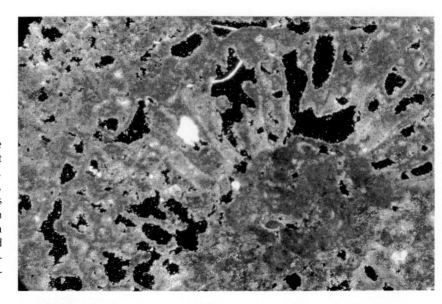

Figure 26.10. James Limestone Member, Pearsall Formation, 9933 ft (3028.4 m), Cities Service Moore No. 1, Fairway field, Henderson County, Texas (0.44 in. in the photo equals 0.17 mm). Open porosity held by a framework of calcite-cemented micritized miliolids and other unidentified foraminifera.

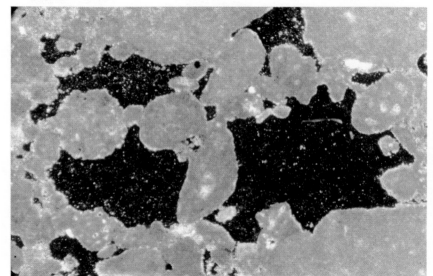

Figure 26.11. James Limestone Member, Pearsall Formation, 9933 ft (3028.4 m), Cities Service Moore No. 1, Fairway field, Henderson County, Texas (0.44 in. in the photo equals 0.17 mm). Fossil hash composed of mostly mililoids and textularids, which have all been micritized. Considerable open pore space exists, although some calcite cement appears to be surrounding the inner surfaces of several of the pores.

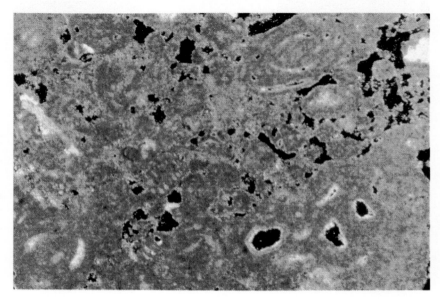

Engineering Studies

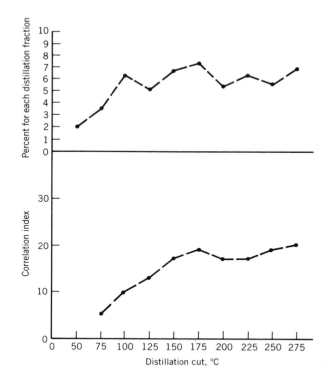

Figure 26.12. Correlation index curve and distillation fractions for crude oil from the James Lime, 9880–9890 ft (3012–3015 m), Fairway field, Anderson County, Texas. (Data from McKiney et al. 1966.)

Engineering Studies

Work on the use of radioactive isotopes to trace the configuration of gas displacement fronts in the alternate gas–water miscible recovery project at Fairway was carried out by Calhoun and Hurford (1970). This study deserves a brief review because of the interesting flow patterns that emerged.

Phase I of the tracer program began in August 1966 when 10 Ci of tritiated hydrogen was injected in one well and 10 Ci of krypton 85 was injected into another well at some distance from the first well. In December of the same year, a second set of injections was initiated, and three more wells were injected with 10 Ci of tritiated hydrogen, 10 Ci of krypton 85, and 10 Ci of tritiated methane. (Each well was injected with a diffrent tracer.)

Some of the flow paths that were produced after approximately 3 years of operation are illustrated in Figure 26.13. A few of the conclusions that resulted from this study are worth noting:

1. Tracers indicated the need for controlled injections and withdrawals to even out sweep configurations.

2. Frontal configurations developed from tracers and cumulative injection volumes indicate that the initial gas slugs fingered irregularly and did not advance as a uniform displacement band.

3. Gas sweep configurations or the direction of gas fingering can change markedly in a short time interval as the injection rate is varied.

4. Reversals in direction of flow can occur where oil can miscibly displace gas in flow channels previously swept by gas, offsetting the advantage of miscible displacement of oil by gas.

5. In alternate gas–water injection projects, the travel time from the injector to the producing well of both injected fluids can be measured with radioactive tracers. When breakthrough occurs for two injected fluids, the relative velocities of these fluids behind the displacing fronts can be estimated from the travel times and the injected volumes of each fluid.

6. The amount of mixing that occurs with two injected fluids and the flow paths that develop when two displacement mechanisms are operating at the same time can be estimated from the tracer measurements.

248   The James Limestone, Pearsall Formation, Fairway Field, Anderson and Henderson Counties, Texas

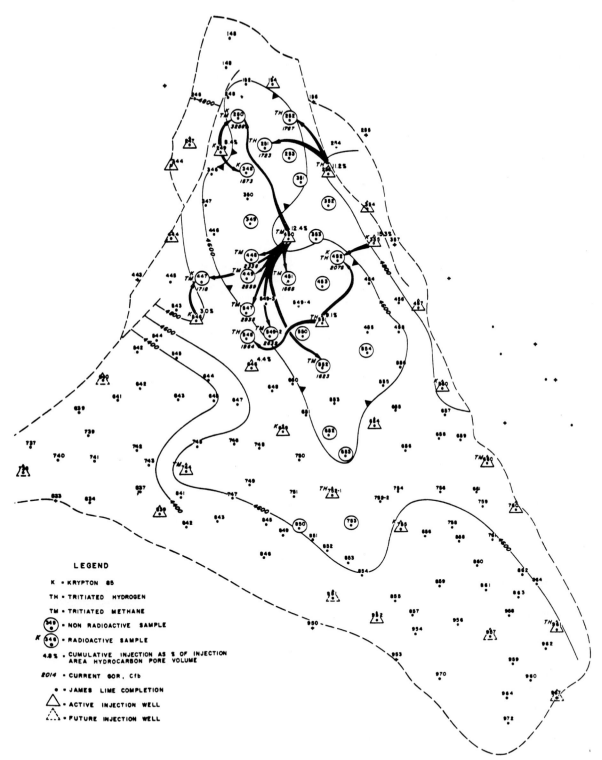

Figure 26.13. Isobaric map of zone A of the Fairway field showing flow paths of radioactive isotopes 3 years after the tracers were injected. (After Calhoun and Hurford 1970, p. 1221.)

Summary

The Fairway field is a case where faulting appears to have aided in the formation of an environment where a reef could flourish over a considerable period of time. The fact that several zones can be readily distinguished shows that changes in environment did occur; however, they were not so drastic as to terminate the basic reef character over a 70-ft interval. The changes in average porosity and permeability for each zone are shown in Table 26.1. Even though only one relatively porous zone was chosen for the radioactive tracer study, the relationship of the channels between wells was quite complicated.

Additional reefs of this type may be present in this area, and the challenge appears to involve the correct evaluation of reef-related structural detail.

Table 26.2 outlines the reserves and productive capacity of the Fairway field.

TABLE 26.1 Average Porosity and Permeability Values, Fairway Field (Terriere, 1976, p. 176)

Limestone Type	Number of Values	Average Porosity (%)	Average Permeability (millidarcies)
Ia	276	10.8	37.2
Ib	322	9.4	14.0
II	185	7.2	8.0
IIIa	87	9.2	28.6
IIIb	84	10.8	43.7
IV	106	8.2	12.4

TABLE 26.2 Summary Report of Reserves and Productive Capacity, Natural Gas, December 31, 1974, Fairway Field[a,b]

	Crude Oil (MMbbl)	Lease Condensate (MMbbl)	Associative (BCF)	Nonassociative (BCF)	Liquids (MMbbl)
	Wet Basis				
Hydrocarbons originally in place	438.1	NA	629.0	NA	—
Proved ultimate recovery 2[c]	207.4	NA	377.4	NA	—
Cumulative production 2[c]	120.1	NA	68.1	NA	—
Proved reserves	87.3	NA	309.3	NA	—
	Dry Basis				
Proved reserves	—	—	263.0	NA	26.0
Reserves in shut-in reservoirs	0	NA	0	NA	0
Indicated secondary and tertiary reserves	0	—	0	—	—
Production					
1973 (total)	17.0	NA	10.2	NA	3.5
1974 (total)	13.6	NA	9.0	NA	3.8

(Continued)

TABLE 26.2 (Continued)

	Crude Oil (MMbbl)	Lease Condensate (MMbbl)	Associative (BCF)	Nonassociative (BCF)	Liquids (MMbbl)
Long-term projection of production (annual total)					
1975	11.1	NA	9.4	NA	—
1976	9.4	NA	10.2	NA	—
1977	7.8	NA	10.5	NA	—
1978	6.5	NA	10.7	NA	—
1979	5.3	NA	10.7	NA	—
1980	4.4	NA	10.6	NA	—
1981	3.6	NA	10.3	NA	—
1982	3.0	NA	9.9	NA	—
1983	2.5	NA	9.3	NA	—
1984	2.0	NA	8.2	NA	—
	(Mbbl)	(MMbl)	(MMCF)	(MMCF)	(Mbbl)
Daily Averages					
December 1974 production	34.2	NA	25	NA	10.2
Short-term productive capacity (60-day basis)	34.2	NA	25	NA	—

[a] Includes the James Lime reservoir and minor reserves in the Rodessa, Pettit, and Massive anhydrite.
[b] From Office of Policy and Analysis and Office of Energy Resource Development, 1975, Final Report, Vol. II, *Oil and Gas Resources and Productive Capacities,* Federal Energy Administration, Washington, D.C., 151 pp.
[c] James Lime Formation only.

References

Calhoun, T. G. and G. T. Hurford, 1970, Case History of Radioactive Tracers and Techniques in Fairway Field, *Journal of Petroleum Technology,* Vol. 22, pp. 2217–2224.

McKinney, C. M., E. P. Ferrero, and W. J. Wenger, 1966, *Analyses of Crude Oils from 546 Important Oilfields in the United States,* Bureau of Mines Report on Investigations 6819, 345 pp.

Terriere, R. T., 1976, *Geology of Fairway Field, East Texas, North American Oil and Gas Fields,* American Association of Petroleum Geologists, Memoir 24, Tulsa, OK, pp. 157–176.

27

Rocky Mountains and Northern Great Plains

General Geologic Setting

The Rocky Mountain chain traverses much of the North American continent as it extends from northern Mexico to Alaska. For the purpose of this work, the area considered reaches from the northeastern corner of Arizona to the middle of Alberta, Canada. Figure 27.1 shows the major sedimentary basins and the structural framework of the Rocky Mountain region. Providing examples from the hundreds of oil and gas-productive reservoirs in this broad area is an enormous task and is beyond the scope of this book. On the other hand, it is possible to discuss a number of the reservoir rocks and to give some idea of the overall diversity.

The first reservoir type to be discussed is from southwestern Nebraska and, strickly speaking, this example is outside the province as normally defined; however, the reservoir provides a transitional case in moving from the production in Texas and Kansas to that of the northern Great Plains and the numerous basins that lie between the mountain ranges of Wyoming. The general connection can be appreciated in terms of the overall regional tectonics wherein western Nebraska is crossed from northwest to southeast by the Chadron–Cambridge arch. This broad granite–cored uplift connects the Black Hills uplift in western South Dakota (near the Wyoming line) to the central Kansas uplift.

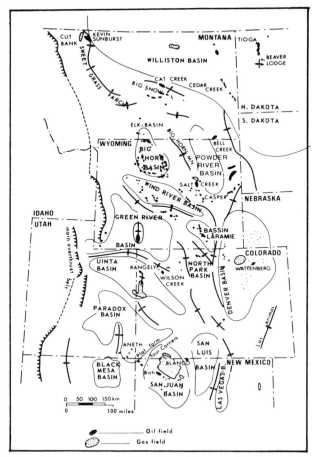

Figure 27.1. Map showing the sedimentary basins and the main structural framework of the Rocky Mountain province. (After Perrodon 1983, p. 285.)

and Mesozoic eras, a number of downwarps of the crust occurred, producing the Rocky Mountain geosyncline, which with several submergences and the deposition of thick continental sequences during times of emergence provided a thick sequence of sediments. The aggregate thickness of this sedimentary column amounts to more than 10 miles (Landes 1970, p. 355).

Beginning in the late Mesozoic time, the Laramide Revolution pushed the basement rocks upward along the length of the geosyncline. The accompanying folding, faulting, and, in some areas, igneous activity produced a varied terrain that again was worn down before the end of the Tertiary period and then re-elevated. Erosion during this last period of uplift provided the present-day panorama of the Rocky Mountains.

One of the leading oil-producing states in this province is Wyoming, and much of the discussion that follows is based on Wyoming outcrops and oil fields.

Wyoming will provide the focus of the first portion of the U.S. section of the Rocky Mountains discussion with its many basins, both topographic and sedimentary, surrounded by majestic peaks. Thin sections from reservoir rocks within specific fields in the Rocky Mountain area were not available; however, the outcrop samples of the reservoir horizons were obtained and are quite instructive.

The Rocky Mountains are formed from igneous, metamorphic, and sedimentary rocks of all kinds. The sedimentary rocks range in age from Precambrian to Recent, but the most widespread and thickest deposits are of Cretaceous age.

The most important periods of tectonic activity, which produced metamorphism, overthrusting, folding, and extrusions and intrusions, occurred in both the Precambrian and late Mesozoic to early Cenozoic (Laramide Revolution). During the Paleozoic

The "Basal Sand" Des Moines Series (Pennsylvanian), Sleepy Hollow Field, Red Willow County, Nebraska

Geologic Background

The structure of the Sleepy Hollow field can best be described as a spur or offshoot from the Cambridge arch. The size and scope of the Cambridge arch is illustrated in Figures 27.2 and 27.3, which show both its horizontal extent and its vertical influence through time. The more specific relation of the Sleepy Hollow field to the structural irregularities on top of the arch are shown in Figure 27.4 along with that of other nearby fields.

Considerable debate has occurred over the age of the productive "basal sand" at Sleepy Hollow. Larson (1962, p. 2088) favored the Early Pennsylvanian rather than the Cambrian age; and this conclusion appears to be correct. Figure 27.5 provides a west-to-east cross section for southern Nebraska, which shows the stratigraphic setting at the beginning of Lansing–Kansas City deposition. From this illustration, it is easy to see why there could be considerable

Geologic Background

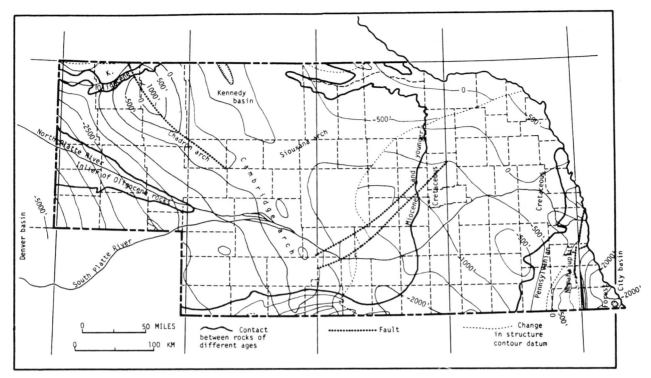

Figure 27.2. Tectonic and simplified geologic map of Nebraska. Structure contours outlining the Cambridge arch are based on the Precambrian floor. (After Cohee et al. 1962.)

disagreement over the age of the sediments directly above the granite.

As Larson pointed out (1962, p. 2088), the area in which the Sleeply Hollow field is located has undergone two periods of major uplift and erosion and one period of diastrophism. After the deposition of the Cambrian Reagan Sand and the Cambro-Ordovician Arbuckle, these same sediments were removed by uplift and erosion along the arch. More uplift and erosion took place at the end of the Devonian, and finally regional diastrophism occurred at the close of the Mississippian. In this latter stage, the uplift resulted in all the pre-Pennsylvanian sediments being stripped off down to and including the Precambrian itself.

In the case of the Sleepy Hollow field, some of the Cambrian sand appears to have been eroded and redeposited as a reasonably clean quartz sand. The buildup of the basal sand is illustrated in the cross section of the field (Figure 27.6). The thin-section photomicrographs for the basal sand as defined in the Sleepy Hollow field are examples of the clean sand that appear to have been derived from erosion of the Cambrian Reagan Sandstone (Figures 27.7 and 27.8).

It is obvious that the basal sand from the Skelly No. 4 Kennedy has two populations of sand-size material involved. The precise significance of this bimodal distribution is not clear; however, there is the possibility that some of the Precambrian is also contributing to the sediment.

An example of known Reagan Sand is also provided in the photomicrographs taken from the Ray field in Norton County, Kansas; thin-section photomicrographs of the Cities Service No. 4 Finnegan from the Reagan Sand are shown in Figure 27.9 and 27.10 for comparison purposes. The major difference between the basal sand in the Sleepy Hollow field and the Reagan Sand in the Ray field appears to be the occurrence of large rounded quartz grains in the samples from Sleepy Hollow.

As illustrated in Figure 27.11, the distance between the Ray field and the Sleepy Hollow field is approximately 70 miles. Therefore, the observed difference would not be surprising if the two sands were derived from the Cambrian. In this case, it appears more logical to assume that the basal sand at Sleepy Hollow was produced by erosion of the Reagan Sandstone.

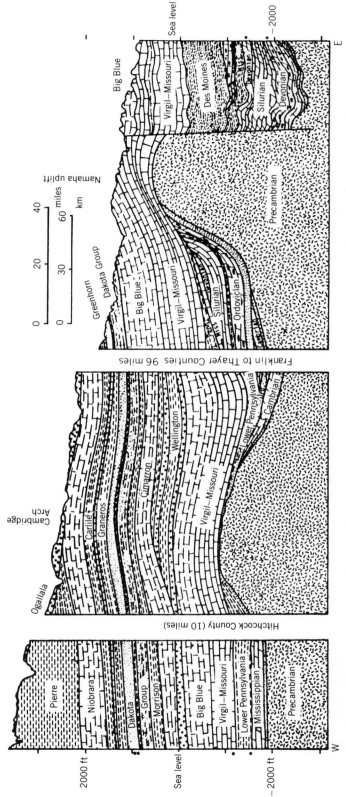

Figure 27.3. West-to-east block and gap cross section along the southern border of Nebraska showing the structure over the Cambridge arch. (After Landes 1970, p. 102.)

Geologic Background

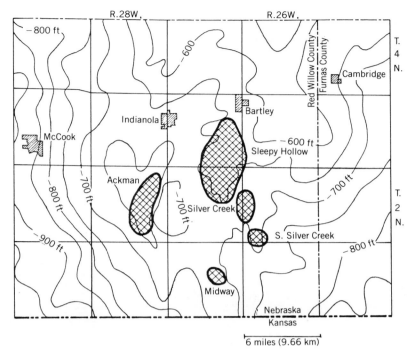

Figure 27.4. Oil fields in relation to regional structure along the crest of the Cambridge arch. Structure is based on the top of the Lansing–Kansas City groups. (After Landes 1970, p. 105.)

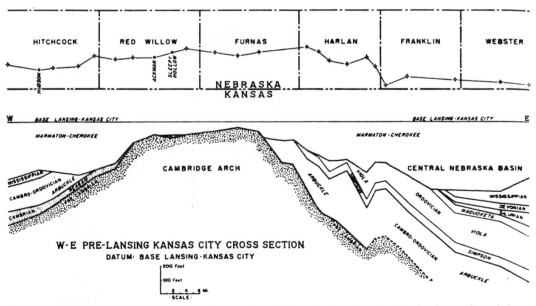

Figure 27.5. West–east cross section for southern Nebraska showing the Cambridge arch and the stratigraphic relationships at the beginning of deposition of the Lansing–Kansas City Formation. (After Larson 1962, p. 2082.)

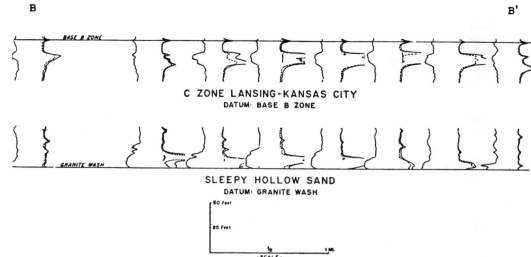

Figure 27.6. West–east cross section through the Sleepy Hollow field. The lower section portrays the buildup of the Sleepy Hollow basal sand. (After Larson 1962, p. 2087.)

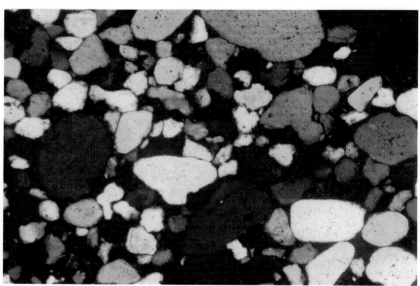

Figure 27.7. Basal sand, Des Moines Series, 3450 ft (1051.8 m), Skelly No. 4 Kennedy, Sleepy Hollow field, Red Willow County, Nebraska (0.44 in. in the photo equals 0.17 mm). Quartz sand that is bimodal in terms of grainsize. Most of the quartz grains are well rounded and in a number of instances are well cemented with quartz overgrowths.

Figure 27.8. Basal sand, Des Moines Series, 3450 ft (1051.8 m), Skelly No. 4 Kennedy, Sleepy Hollow field, Red Willow County, Nebraska (0.44 in. in the photo equals 0.17 mm). Quartz sand with large well-rounded grains indicate two sources of supply. Some smaller grains are feldspars, suggesting that at least some of the smaller grains are related to the erosion of the central granite core.

Figure 27.9. Reagan Sand, Cambrian, 3665.72 ft (1117.4–1119.5 m), Cities Service No. 4 Finnegan, Ray field, Norton County, Kansas (0.44 in. in the photo equals 0.17 mm). Relatively pure quartz sand is in some cases very well rounded and in others it is not. Sorting appears to be good. Half of the photomicrograph shows poikilotopic calcite cementing. The remaining half appears to be cemented with quartz.

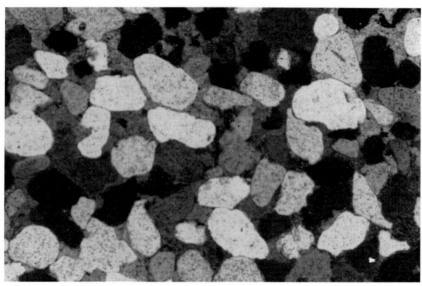

Figure 27.10. Reagan Sand, Cambrian, 3665.72 ft (1117.4–1119.5 m), Cities Service No. 4 Finnegan, Ray field, Norton County, Kansas (0.44 in. in the photo equals 0.17 mm). Well-rounded quartz sand with poikilotopic calcite cement filling most of the pore space. Some quartz cementing suggests that the quartz cementation took place first.

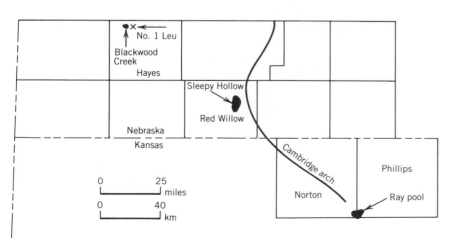

Figure 27.11. Map showing approximate locations for the Sleepy Hollow, Blackwood Creek, and Ray fields.

The next section deals with the Blackwood Creek field and provides a case where a coarse-grained arkose is the major reservoir rock. The origin of this sediment is clearly the core granite material from the rising Cambridge arch. The photomicrographs from the Cities Service No. 1 Leu, (Figures 27.14–27.16), a wildcat well to the east of the Blackwood Creek field, shows the same type of sediment.

Basal Sand, Pennsylvania, Blackwood Creek Field, Hayes County, Nebraska

Geologic Background

Although the Blackwood Creek field is located to the west of the crest of the Cambridge arch, the sediment character appears to be a granite wash (Figures 27.12 and 27.13). In this section, the reservoir rock from the basal sand as it occurs in the British American N-1B Sullivan in the Blackwood Creek field is compared to the core from a wildcat well a short distance to the east (Cities Service Oil Company No. 1 Leu). The latter well is interesting because the oil saturation occurs only in the top section of the core, and although the lower part has considerable porosity and permeability, no oil is observed (see Table 27.1). It should be noted that the zone at the top is cemented with both calcite and quartz cement, (Figure 27.14).

As can be seen from the photomicrographs of the Leu well (Figure 27.15 and 27.16), the granite rock fragments in some cases are cemented with anhydrite. This suggests that the environmental conditions were such that evaporates were being deposited in the nearshore saline areas. In other words, the granite wash materials were being dumped into a marine sea. At least at these levels, we are not dealing with a nonmarine arkose. This observation is supported by the recovery of crinoid columnals and fusilinids from one of the shales (Figures 27.17 and 27.18).

The overall thickness of these arkosic sediments and their distribution over several miles suggests that other traps may exist along the Cambridge arch and the Chaldron arch to the northwest.

Geologic Interpretation

A look at the core analyses in Table 27.1 for the Cities Service No. 1 Leu indicates that even though there is evidence of considerable anhydrite cementation, the permeability in a number of zones in the top portion of the core appears to be remarkably good. The oil saturation also occurs in this top zone. However, the values are relatively low.

The fact that fusulinids (Figure 27.17), brachiopods, bryozoans (Figure 27.18), ostracodes, and crinoid columnals (Figure 27.18) were recovered from the red and gray shale zones at 4806.5 ft (1465.4 m) and 4818.5 ft (1469.1 m) suggests that the shallow Pennsylvanian seas were rich with organic matter.

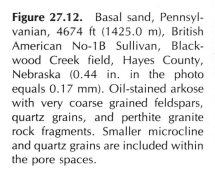

Figure 27.12. Basal sand, Pennsylvanian, 4674 ft (1425.0 m), British American No-1B Sullivan, Blackwood Creek field, Hayes County, Nebraska (0.44 in. in the photo equals 0.17 mm). Oil-stained arkose with very coarse grained feldspars, quartz grains, and perthite granite rock fragments. Smaller microcline and quartz grains are included within the pore spaces.

Figure 27.13. Basal sand, Pennsylvanian, 4674 ft (1425.0 m), British American No-1B Sullivan, Blackwood Creek field, Hayes County, Nebraska (0.44 in. in the photo equals 0.17 mm). One major granite rock fragment composed of feldspar and quartz. Quartz crystals have filled in the fractures, which appear to have occurred in two directions, one along the boundary between the feldspar and quartz and the other at right angles to it. The fracturing associated with vertical uplift apparently was healed with quartz cement prior to erosion.

Figure 27.14. Basal sand, Pennsylvanian, 4785.5 ft (1459.0 m), Cities Service No. 1 Leu, Section 17-8N-34W (Wildcat), Hayes County, Nebraska (0.44 in. in the photo equals 0.17 mm). Coarse arkosic sandstone showing areas with anhydrite cement. The quartz appears to have grown originally in fraction zones and subsequently eroded, redeposited, and cemented again.

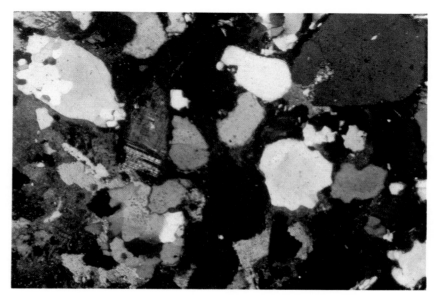

Figure 27.15. Basal sand, Pennsylvanian, 4778.5 ft (1456.9 m), Cities Service No. 1 Leu, Section 17-8N-34W (Wildcat), Hayes County, Nebraska (0.44 in. in the photo equals 0.17 mm). Poorly sorted arkosic sand with both silica and calcite cement blocking the pores. In the left center, a piece of unweathered plagioclase feldspar occurs.

TABLE 27.1 Core Analysis Results from the Basal Sand, Cities Service Oil Co. No. 1 Leu, Wildcat, Hayes County, Nebraska

	Depth (ft)	Permeability (Millidarcies)		Porosity (%)	Oil (% Pore)	Total Water (% Pore)
		Horizontal	Vertical			
1	4777–4778	1702		22.6	18.6	46.5
2	4778–4779	2105		21.6	19.0	48.2
3	4779–4780	34		16.5	4.8	78.9
4	4780–4781	2.9		10.2	10.1	58.9
5	4781–4782	935		21.2	5.7	68.8
6	4782–4783	1930		24.4	9.0	67.6
7	4783–4784	374		24.0	5.0	52.9
8	4784–4785	1.4		14.9	8.7	61.0
9	4785–4786	26		11.3	29.1	27.4
10	4790–4791	3.5		22.7	3.5	63.8
11	4791–4792	615		15.0	25.3	62.7
12	4792–4793	101		19.0	24.7	60.5
13	4823–4824	135		18.8	0.0	70.7
14	4824–4825	27		20.1	1.0	78.1
15	4825–4826	162		17.3	0.0	73.9
16	4826–4827	196		20.7	1.0	81.2
17	4827–4828	665		19.0	0.0	85.3
18	4828–4829	834		19.4	0.0	88.0
19	4829–4830	413		19.3	0.0	78.1
20	4830–4831	154		23.7	2.5	82.1
21	4831–4832	243		27.3	0.0	76.2
22	4832–4833	399		23.4	0.0	87.3
23	4833–4834	370		24.1	0.0	88.7
24	4834–4835	334		22.9	0.0	79.0
25	4839–4840	1.1		8.6	0.0	74.4
26	4844–4845	64		20.7	0.0	76.7
27	4845–4846	14		20.9	0.0	79.8
28	4846–4847	6.4		24.1	0.0	82.2
29	4847–4848	23		22.4	0.0	82.5
30	4848–4849	9		21.5	0.0	78.6

Figure 27.16. Basal sand, Pennsylvanian, 4782.5 ft (1458.1 m), Cities Service No. 1 Leu, Section 17-8N-34W (Wildcat), Hayes County, Nebraska (0.44 in. in the photo equals 0.17 mm). Coarse-grained arkose with anhydrite cement. Orthoclase feldspar shows varying degrees of weathering. The anhydrite has plugged most of the pores at this particular location. Quartz grain at the top shows evidence of having one side exposed to a fracture zone where quartz crystals began to grow again.

Figure 27.17. Fusulinid (center) *Wedekindellina* euthysepta recovered from shales at 4806.5 ft (1465.4 m) in core materials from the Cities Service No. 1 Leu, Hayes County, Nebraska. This particular fusulinid is found in lower Des Moines sediments. Magnification 12×.

Figure 27.18. Crinoid columnal (upper left) and a bryozoan (center, diagonal) recovered from shales at 4818.5 ft (1469.1 m) in core materials from the Cities Service No. 1 Leu, Hayes County, Nebraska. Magnification 12×.

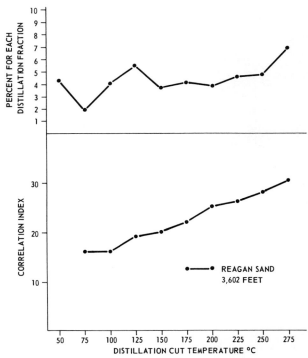

Figure 27.19. Correlation index curve and distillation fractions for crude oil from the Ray field, Norton and Phillips counties, Kansas. (Data from McKinney et al. 1966, p. 93.)

Storms and wind tides probably carried the crinoids and other fossil debris into the red-bed zones where they were preserved by rapid sediment deposition.

The correlation index plot for oil from the Reagan sand in Figure 27.19 is typical of older crude oils where there is no indication of biodegradation.

References

Cohee, G. V., 1962, Tectonic Map of the United States, U.S. Geological Survey and American Association of Petroleum Geologists.

Landes, K. K., 1970, *Petroleum Geology of the United States,* Wiley-Interscience, New York, 571 pp.

Larson, W. S., 1962, Ackman Field and Environs, Southwestern Nebraska, *Bulletin of the American Association of Petroleum Geologists,* Vol. 45, No. 11, pp. 2079–2089.

McKinney, C. M., E. P. Ferrero, and W. J. Wenger, 1966, *Analysis of Crude Oils from 546 Important Oilfields in the United States,* Bureau of Mines Report of Investigations 6819, 345 pp.

Perrodon, A., 1983, *Dynamics of Oil and Gas Accumulations,* Elf Aquitaine, Pau, France, 368 pp.

28

Wyoming

In terms of an example of the Rocky Mountain province, the state of Wyoming is a good one. It is the leading oil-producing state in the Rocky Mountain group, and its production extends from the Cambrian to the Eocene age (see Figure 28.1). One of the fortunate aspects as far as the geologist is concerned is the occurrence of numerous outcrops covering all of the major rock types and reservoir varieties.

In this chapter, outcrop samples are discussed that have underground equivalents in major oil reservoirs. Specifically, in the case of the Cretaceous Frontier Sand, a brief summary of the Salt Creek field will be provided. A similar example will be discussed for the Tensleep Sandstone. However, in the case of the Phosphoria Formation, a broad discussion of the source rock is included. It is hoped that the reader will be able to obtain a reasonable idea of what these rocks are like in terms of mineralogy and texture and thereby be able to compare these samples with others from the Rocky Mountain Belt that follow in Chapters 31 and 32.

ERA	PERIOD	GROUP OR FORMATION	RESERVOIRS
Cenozoic	Pliocene	North Park and/or Brown's Park	
	Miocene		
	Oligocene	White River	
	Eocene	Bridger / Green River / Wind River	"Wasatch" sands
	Paleocene	Fort Union	Fort Union sands
Mesozoic	Cretaceous	Lance	
		Pierre — Lewis	
		Pierre — Mesaverde	Various sands, including Almond at top
		Pierre — Steele	Sussex / Shannon — Cody Shale, with reservoir sandstone near base
		Niobrara	
		Frontier	Frontier and Wall Creek sands ←
		Mowry	
		Muddy	Muddy sands; Newcastle Sand; Dynneson Sand
		Thermopolis	
		Inyan Kara	Fall River and "Dakota" sands / Lakota sands
	Jurassic	Morrison	
		Sundance	Sundance sands
	Triassic	Nugget	Nugget sand
		Red Peak or Chugwater	Crow Mountain sand; Curtis sand
		Dinwoody	
Paleozoic	Permian	Phosphoria or Goose Egg	Phosphoria (Embar) Limestone ←
	Pennsylvanian	Wells or Tensleep or Casper	Tensleep sand; Minnelusa Formation ←
		Amsden	Darwin sand
	Mississippian	Brazer / Madison	Madison Limestone
	Devonian	Darby	
	Ordovician	Big Horn	Big Horn Dolomite
	Cambrian	Gallatin	Deadwood sand
		Gros Ventre	
		Flathead	Flatsand sand

Figure 28.1. Composite stratigraphic column for Wyoming oil and gas reservoirs; not to scale. (After Landes 1970, p. 402.)

Salt Creek Field

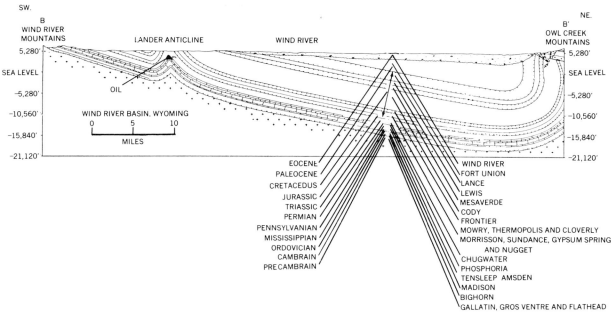

Figure 28.2. Cross section of the Wind River Basin, Wyoming. (After Wyoming Geological Association, P.O. Box 545, Casper, Wyoming 82604.)

Geologic Background

The vertical structural movements in Wyoming have been so large that erosion has exposed the pre-Cambrian crystalline rocks, and these have become significant mountain ranges. Located between these somewhat linear pre-Cambrian cores are synclinal basins that contain thick sedimentary sequences. The basins are both structural and topographic features; and it is in these basins where most of the oil and gas fields have been found.

Outcrops of the marine Paleozoic and Mesozoic sediments occur around the edges of the deeper basins, providing the geologist with an opportunity to sample some of the classical reservoir type rocks. The Wind River basin is particularly noteworthy because of the accessibility of the outcrops. Figure 28.2 shows a cross section of the Wind River basin and illustrates the outcrop areas that occur on both sides of the basin; namely, in the Wind River Mountains and the Owl Creek Mountains. The samples discussed in this text were taken from the outcrops along the Granite Mountains on the south and the outcrops along the Wind River Canyon where it cuts through the Owl Creek Mountains on the north.

The Frontier Formation

The first reservoir rock to be considered comes from the Frontier Formation. This particular rock sequence is of interest because of the excellent production from the Frontier (Second Wall Creek Sand) in the giant Salt Creek field. The location of the Salt Creek with respect to the total footage of Frontier Sandstone is shown in Figure 28.3. In terms of overall thickness in proceeding from west to east, the Frontier sands thin rather rapidly, as illustrated in Figure 28.4. The same statement holds true in proceeding from the northwest to the southeast into Nebraska, as is illustrated in Figure 28.5.

Salt Creek Field

The giant Salt Creek field extends for 9 miles (14.5 km) in a north–south direction and 4 miles (6.4 km) from east to west. More than 2500 producing wells have been drilled within its boundaries. The major oil-productive zone is the Second Frontier Sand, which ex-

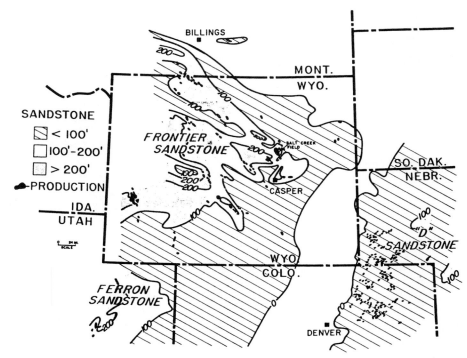

Figure 28.3. Isopach map of the total sandstone footage within the Frontier and Correlative formations showing the location of the Salt Creek field. (After Barlow and Haun 1970, p. 151.)

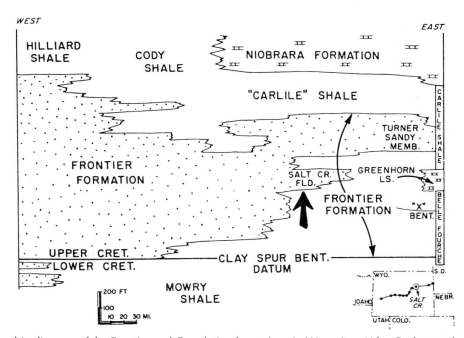

Figure 28.4. West-to-east stratigraphic diagram of the Frontier and Correlative formations in Wyoming. (After Barlow and Haun 1970, p. 150.)

Salt Creek Field

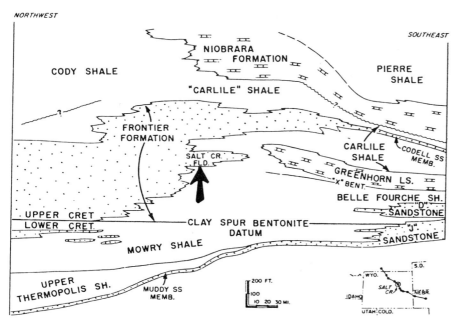

Figure 28.5. Northwest-to-southwest stratigraphic diagram of the Frontier and Correlative formations in Wyoming. (After Barlow and Haun 1970, p. 150.)

tends over an area greater than 22,000 acres. Barlow and Haun (1970, p. 147) indicated that the cumulative production from this field is in excess of 420 million barrels, which places it in the "giant" class.

The generalized structure map contoured on the top of the Dakota Formation (Figure 28.6) shows the location of the Salt Creek field in relation to the distribution of the Second Frontier Sandstone (Kf_2) and the southwest flank of the Powder River basin. Work on the history of the Cretaceous in Wyoming suggests that the Salt Creek anticline, which is shown in Figure 28.7, had very little, if any, structural development before to the latest Cretaceous and early Laramide orogony. Barlow and Haun (1970, p. 147) state that the Salt Creek oil must have accumulated initially in stratigraphic traps and that secondary migration resulted in the structural accumulation that has proved to be highly productive.

The Salt Creek field is close to the eastern edge of the thick Frontier Sandstone area where littoral and offshore environments occur as part of the most seaward portion of a deltaic sequence. A complex interbedding of shale and sandstone plus sandstone lenses and disconformities make correlation of subdivisions within the Frontier from one basin to another very difficult.

Locally, the west-to-east electric log cross section through the Salt Creek field (Figure 28.8) shows the offshore barlike appearance of the Second Frontier Sand.

According to Barlow and Haun (1970, p. 153), the average median diameter of sand grain in 31 samples of sandstone and sandy siltstone from the Kaycee and Buffalo areas (Figure 28.9) is 0.13 mm with a range 0.007–0.26 mm.

The pebble bed at the top of the Second Frontier contains some interesting constituents, as shown in Table 28.1. Apparently the chert pebbles contain cri-

TABLE 28.1 Constituent Pebbles in Top of Second Frontier Sandstone

Mineral or Rock Type	Percent
Black-coated chert	40
Uncoated chert	27
Andesite porphyry	21
White quartz	5
Pink quartz	3
Quartzite	3
Clear quartz	1
Total	100

(After Barlow and Haun, 1970.)

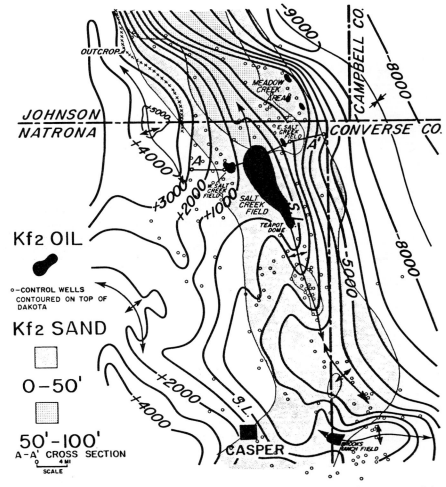

Figure 28.6. Generalized structure map contoured on the top of the Dakota Formation showing the southwest part of the Powder River basin and the distribution of the Second Frontier Sandstone (Kf$_2$) in relation to the Salt Creek field. (After Barlow and Haun 1970, p. 152.)

noid columnals, bryozoans, fusilinids, and so on, which suggest a Pennsylvanian age. The pebble zones also contain fresh shark teeth, which are interpreted to indicate short-distance transport as well as periods of slow deposition with few strong currents. The most logical source for the porphyry pebbles mentioned in Table 28.1 is the Beartooth–Yellowstone Park area in northwestern Wyoming and southwestern Montana.

Source Beds

The oil source rocks for the Frontier production appear to be the associated Frontier shales. The early work of Hunt and Jamieson (1958) indicated that within an 800-square-mile area that included the Salt Creek field and part of the southwest Powder River basin, there are 3 billion bbl of oil within the Frontier shales. This, in itself, would suggest that there is still considerable exploration potential within the Frontier.

As mentioned previously, the best production has come from the Second Frontier (or Wall Creek) Sandstone. Other important reservoirs in the same structure include the First Frontier (or Wall Creek) Sandstone, the Lakota, the Third Sundance (Jurassic), and the Pennsylvanian Tensleep.

It should be emphasized that the photomicrographs of the Frontier contained in this chapter were obtained from outcrop samples and are not true reser-

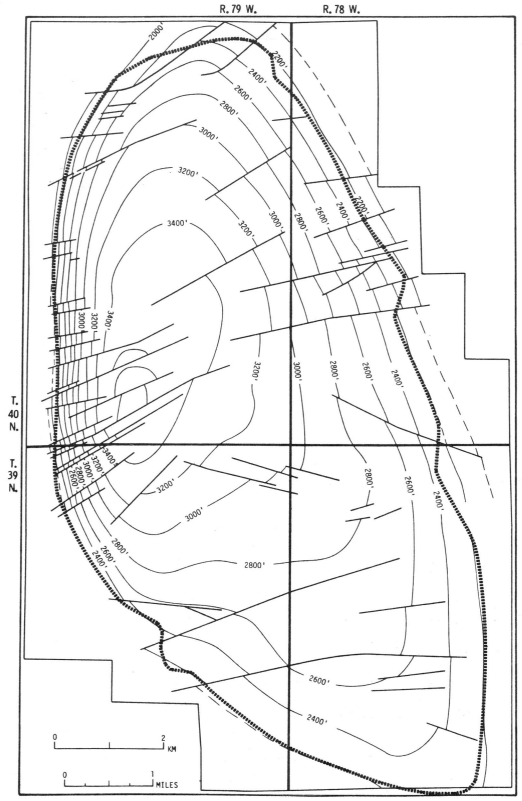

Figure 28.7. Structure contour map of the Salt Creek oil field, Natrona County, Wyoming. Contours are based on the top of the Second Wall Creek Sand. The heavy broken line shows original field boundary. (After Landes 1970, p. 404.)

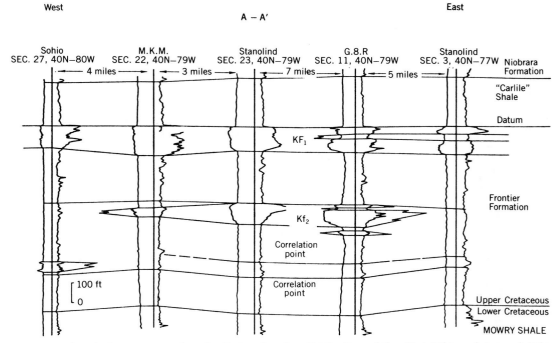

Figure 28.8. West-to-east electric log cross section A–A' showing the distribution of the First (Kf_1) and Second (Kf_2) Frontier sandstones across the Salt Creek field, Natrona County, Wyoming. (After Barlow and Haun 1970, p. 153.)

voir rocks. Nevertheless, it is this author's opinion that the similarities are sufficient to provide the reader with a good feeling for the overall mineralogy.

Geologic Environment

In order to obtain a more detailed picture of what the environment of deposition was like for the Frontier sandstones of the Salt Creek field, it is appropriate to study the outcrops of the Frontier in the Kaycee–Tinsdale Mountain area. Merewether, Cobban, and Spencer (1976) have carried out detailed outcrop studies plus correlations and have provided interpretations of the environments. The location map of their study area and what is referred to as the Baily Flats well are provided in Figure 28.10.

Goodell (1962) describes the Frontier sedimentation and gives the broad view of the geologic events, which are summarized as follows: The source area for much of the Frontier sediment was in central Idaho where sedimentary, volcanic, and low-rank metamorphic rocks were eroded. Environments of deposition for the Frontier Formation in Wyoming vary from nonmarine in the west to marine in the east and they migrate laterally, reflecting marine transgressions and regressions. In the outcrop, the Frontier shows a cyclic depositional history where the better developed sandstone reveal a vertical gradation from paludal at the base, to fluvial in their middle, to marine at the top.

In terms of the overall source areas for those sediments, Hawn et al. (1972) indicate that the main sources for the Frontier clastics were the areas uplifted during the emplacement of the Idaho batholith. A look at the lithologic log in Figure 28.11 shows that there are several bentonite zones, indicating that the contribution from volcanic sources were certainly a significant part of the Frontier sediment.

When the individual zones shown on the lithologic log are considered in terms of their environment of deposition, an oscillating set of shoreline conditions emerges as illustrated by Merewhether et al. (1976, p. 40) in Figure 28.12.

Geologic Interpretation

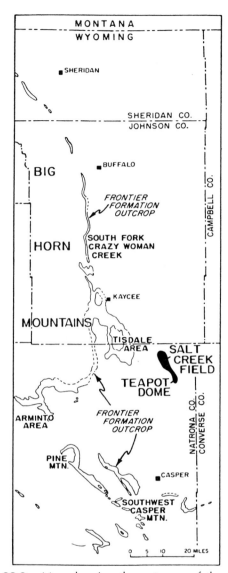

Figure 28.9. Map showing the outcrops of the Frontier Formation and localities mentioned in the text. (After Barlow and Haun 1970, p. 154.)

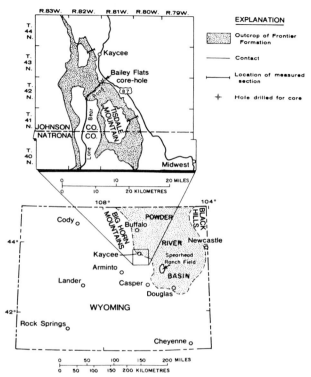

Figure 28.10. Map of the Kaycee–Tisdale Mountain area, showing outcrop area of the Frontier Formation and the Baily Flats core-hole midwest is the approximate location of the Salt Creek field. (After Merewether, E. A., W. A. Cobban and C. W. Spencer—Upper Cretaceous Frontier fm. in Kaycee–Tisdale Mountain Area, Johnson Co., Wyo. 1976.)

Mineralogy

Whole-rock core analyses by x-ray diffraction carried out by Merewhether et al. (1976, p. 38) indicate that quartz, potassium feldspar, mica, illite, chlorite, mixed-layer clays, and possibly kaolinite are present in the lower Frontier shales. As can readily be seen from the photomicrographs in Figures 28.13 and 28.14, the quartz is present in many different forms.

Geological Interpretation

The outcrops of the Wall Creek Sandstone Member of the Frontier in the Tisdale–Kaycee area form a prominent cliff that permits an observer to examine the changes that occur over a considerable horizontal distance. The light gray to light brown color of the sandstone appears to be fairly uniform and contains shell fragments, fine coaly fragments, a few pyrite

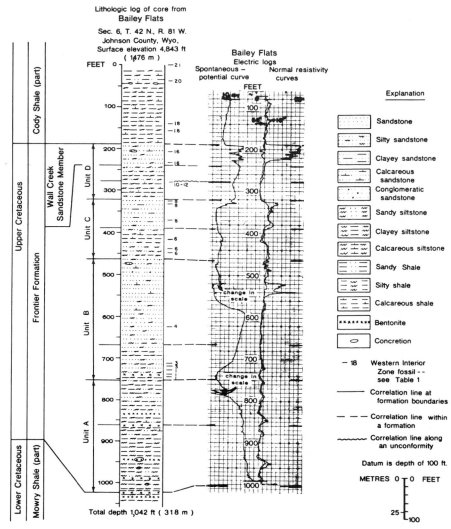

Figure 28.11. Lithologic log and electric log of the Frontier Formation at Baily Flats, showing characteristic marker bed at the contact of the Wall Creek Member with the Cody shale. (After Merewether, E. A., W. A. Cobban and C. W. Spencer—Upper Cretaceous Frontier fm. in Kaycee–Tisdale Mountain Area, Johnson Co., Wyo. 1976.)

Geologic Interpretation

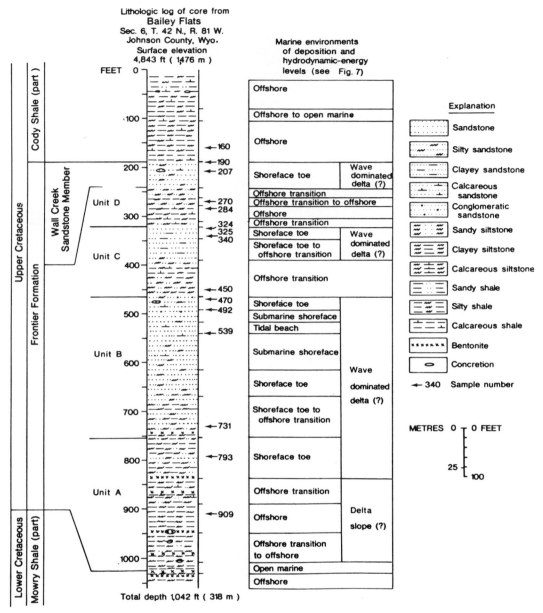

Figure 28.12. Lithologic log and an interpretation of the depositional environments of the Frontier Formation and Bailey Flats, Wyoming. (After Merewether, E. A., W. A. Cobban and C. W. Spencer—Upper Cretaceous Frontier Fm. in Kaycee–Tisdale Mountain Area, Johnson Co., Wyo. 1976.)

Figure 28.13. Frontier Formation, outcrop sample from Rattlesnake Ridge, Natrona County, Wyoming (0.44 in. in the photo equals 0.08 mm). The assemblage is dominated by quartz grains and detrital chert. The grain at the lower right is a variety of megaquartz, which is frequently associated with chert (see Scholle 1979, p. 27). The grain in the upper right center has structures of fibrous microquartz, suggestive of a cavity or fracture lining.

Figure 28.14. Frontier Formation, outcrop sample from Rattlesnake Ridge, Natrona County, Wyoming (0.44 in. in the photo equals 0.08 mm). Poorly sorted quartz and chert grains dominate. In the right center, note the megaquartz rock fragment that appears to have formed in a fracture where the growth occurred initially as a fibrous quartz growing out from the fracture walls. Small amounts of quartz cement occur at the grain-to-grain contacts.

grains, and occasional biotite flakes. The upper part of the Wall Creek is cross-bedded in the outcrops.

Merewether et al. (1976, p. 42) used the scanning electron microscope and found subangular to subrounded sand grains surrounded with kaolinite crystals. The kaolinite was probably a postdepositional feature and does not have any bearing on the environment of deposition.

According to Merewether et al. (1976, p. 43), the rate of sedimentation is between 141 ft (43 m) and 170 ft (52 m) per million years. This is a fairly low rate of sedimentation compared with the seashore marine rates observed for Late Cretaceous rocks in other areas of Wyoming, which range from 600 to 665 ft (183–203 m) per million years.

Examination of the correlation indexes for three Frontier crude oils (Figure 28.15) suggests that the oil from the Elk Basin field was subjected to conditions that were different from those that prevailed at the Salt Creek and Big Muddy fields.

The Wall Creek section of the Frontier was probably deposited in a shoreline environment that may have been closely related to a wave-dominated delta. Merewether et al. (1976, p. 43) suggest that the

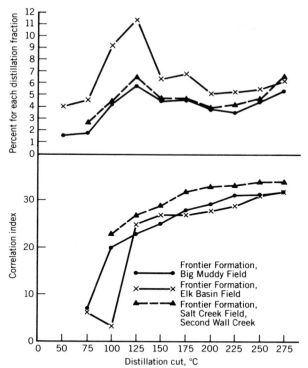

Figure 28.15. Correlation index curves and distillation fractions for crude oils from the Frontier Formation in three different Wyoming fields. (Data from McKineny et al. 1966, pp. 320, 324, and 339.)

sandstone units of the Frontier may "represent rapid eastward shifts in the position of the shoreline."

In terms of exploration potential, the Frontier probably still has considerable oil to be found. However, the search will not be easy.

References

Barlow, J. A., Jr., and J. D. Haun, 1970, Regional Stratigraphy of Frontier Formation and Relation to Salt Creek Field, Wyoming, in *Geology of Giant Petroleum Fields,* American Association of Petroleum Geology, Memoir 14, pp. 147–157.

Goodell, H. G., 1962, The Stratigraphy and Petrology on the Frontier Formation, in *Symposium on Early Cretaceous Rocks of Wyoming and Adjacent Areas,* Wyoming Geology Association, 17th Annual Field Conference Guidebook, pp. 173–210.

Haun, J. D., L. A. Hale, H.G. Goodell, D. G. McCublin, R. J. Weimer, and G. R. Wulf, 1972, Cretaceous System, in *Geologic Atlas of the Rocky Mountain Region, United States of America,* Rocky Mountain Association of Denver, CO pp. 190–228.

Hunt, J. M., and G. W. Jamieson, 1958, Oil and Organic Matter in Source Rocks of Petroleum, in *Habitat of Oil,* Weeks, L. G., Ed., American Association Petroleum Geologists, pp. 735–746.

Landes, K. K., 1970, *Petroleum Geology of the United States,* Wiley-Interscience, New York, 571 pp.

McKinney, C. M., E. P. Ferrero, and W. J. Wenger, 1966, *Analyses of Crude Oils from 546 Important Oilfields in the United States,* Bureau of Mines Report of Investigations 6819, 345 pp.

Merewether, E. A., W. A. Cobban, and C. W. Spencer, 1976, The Upper Cretaceous Frontier Formation in the Kaycee–Tisdale Mountain Area, Johnson County, Wyoming, in *Geology and Energy Resources of the Powder River Basin,* R. Laudon, Ed., Wyoming Geological Association Guidebook, 28th Annual Field Conference, pp. 33–44.

Rocky Mountain Association of Geologitst, 1951, Rocky Mountain Region, *American Association of Petroleum Geologists Bulletin,* Vol. 35, No. 2, pp. 274–315.

Scholle, P. A., 1979, *A Color Illustrated Guide to Constituents, Textures, Cements, and Porosites of Sandstones and Associated Rocks,* Memoir 28, American Association of Petroleum Geologists, Tulsa, OK 207 pp.

29

Tensleep Formation, Central Wyoming

The Tensleep Formation as it occurs over much of the western half of Wyoming resembles in many ways the classical orthoquartzite of Krynine. In this case, however, it is composed of alternating orthoquartzite and dolomite or dolomitic limestone. From the petroleum engineer's point of view, the fact that most of the dolomite and much of the sandstone are horizontally laminated is of importance in optimizing secondary recovery. The sandstones are commonly cross-bedded where they are thickest, dipping in the range 30°–40°, which adds to the fluid flow problems.

The work of Morgan, Cordiner, and Livingston (1978), describing the details of the Tensleep reservoir as it occurs in the Oregon Basin field (see Figure 29.1), will be summarized because it is particularly germane to the theme of this book and deals with what both geologists and engineers need to consider when attempting to squeeze oil out of the Tensleep.

Fractures

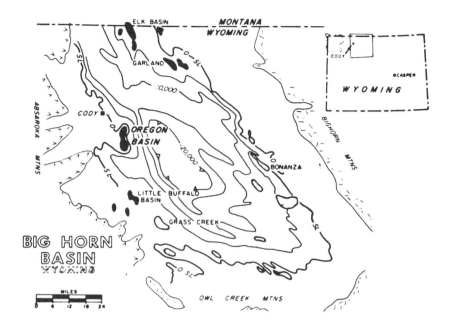

Figure 29.1. Map of the Big Horn basin, Wyoming. Contours are based on the top of the Tensleep Sandstone with an interval of 5000 ft (1500 m). Also shown in the location of the Oregon basin field. (After Morgan et al. 1978, p. 610.)

General Mineralogy

The sands are highly quartzose and generally well sorted, although outcrop samples from the Atlantic City, Wyoming, area exhibit some relatively poorly sorted zones. The quartz grains are normally subrounded to rounded and fine to medium sand size. Within the sand zones there appear to be no fossils. The dolomite, on the other hand, commonly contains fine-grained fossil debris. In summary, the main reservoirs of the Tensleep are orthoquartzites with minor amounts of dolomite incorporated in the sand.

Geologic Interpretation

Houlik (1973, p. 507) believes that the Tensleep Formation was produced by sedimentation in nearshore environments where many changes in facies occurred. These include changes between coastal sabkhas or supratidal flats (laminated dolomite), continental sabkhas (laminated sandstone), and dune fields (cross-bedded sandstone). The oscillating environments explain local variations in the Amsden–Tensleep contact.

Over the more than 50 years that oil has been produced from the Tensleep at the Oregon Basin field, the engineers have considered the reservoir to be a relatively uniform, thick sandstone. Morgan et al. (1978, p. 610) show that not only are the Embar and Tensleep separate reservoirs but also that a combination of original depositional layering and postdepositional cementation causes the Tensleep reservoir to be highly stratified.

More specifically, nonreservoir dolomite layers combined with anhydrite-cemented layers separate the Tensleep into zones that act as discrete reservoirs. This layering causes both the injected and produced fluids to flow horizontally, usually in a single zone, and these zones are separated so that little or no cross-flow occurs.

Sedimentary structures ranging from well-formed cross-stratification to massive beds deserve consideration. According to Morgan et al. (1978, p. 617), the cross-stratification causes some low-magnitude directional permeability in the Tensleep where maximum horizontal permeability parallels the dip of the cross-stratification in the direction of the sediment flow and minimum horizontal permeability is normal to the dip of the cross-stratification. If the sand is very fine or has other cements that inhibit permeability, then directional permeability can be important.

Fractures

Recognizing whether or not fractures permit fluid flow between zones or whether they have a dominant effect is critical. In the Oregon Basin field faults and fractures are common (Figure 29.2). In particular, vertical fractures are scattered through the Tensleep, being more abundant on the crestal areas of the structures. According to Morgan et al. (1978, p. 618), most fractures are closed, short, and vertical. It is important to note that in this case closed fractures impede horizontal flow but do not form barriers. In other words, there is lower permeability normal to the fractures, which in turn disrupts radial flow and causes some directional permeability.

Morgan et al. (1978, p. 618) also indicate that, at least at the Oregon Basin, the fluid transmissibility has been greatly increased in some wells by hydraulic fracturing. It is hypothesized that this hydraulic fracturing caused the opening of closed natural fractures.

Porosity and Permeability

Porosity variations in the Tensleep cores from the Oregon basin field are caused by the amounts of dolomite, anhydrite, and dolomite cement present. In the photomicrographs that are included in this chapter, the dolomite is the only cementing agent other than quartz. It occurs in the form of very small rhombs that are sprinkled throughout the sand. The samples photographed (Figures 29.3 and 29.4) are from the Wind River Canyon where it cuts through the Owl Creek Mountains roughly 70 miles southeast of the Oregon basin field and, therefore, some differences are to be expected.

The close relationship between porosity and the amount of nonreservoir material (i.e., dolomite and anhydrite) is illustrated in Figure 29.5. Morgan et al. (1978, p. 618) point out that so much of the variation is accounted for by these two variables that such factors as quartz cement, texture differences, clay content, solution, and so on, are of relatively minor significance.

The overall rock and fluid properties for the Tensleep in the South dome part of the Oregon basin field are shown in Table 29.1. Although the average porosity is 13.8% and the average permeability is 68

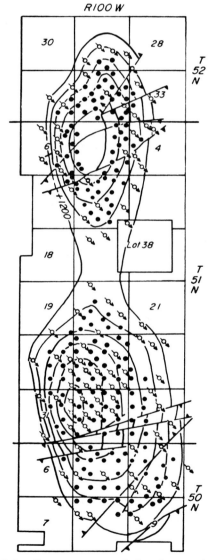

Figure 29.2. Structure map of the Oregon basin field based on the top of the Tensleep Sandstone contour interval, 200 ft (60 m). (After Morgan et al. 1978, p. 611.)

TABLE 29.1   Oregon Basin South Dome Tensleep Rock and Fluid Properties

Average depth to Tensleep, ft	3840
Formation thickness, ft	60–120
Average porosity, %	13.8
Average air permeability, millidarcies	68
Productive acres	5007
Initial oil saturation, % PV	88.8
Initial oil in place STB/acre-ft	823
Initial reservoir pressure, psi	1565
Reservoir temperature, °F	112
Solution gas at BP, ft^3/bbl	254
Crude viscosity at 112°F, cp	6.5
Crude gravity, °API	21

(After Morgan et al., 1978.)

Porosity and Permeability

Figure 29.3. Tensleep Formation, outcrop sample from Wind River Canyon, mile marker 69.7, south of Thermopolis, Wyoming (0.44 in. in the photo equals 0.08 mm). Angular, reasonably well sorted mostly quartz sand with small dolomite grains scattered in between. In several places, quartz cement is present where the quartz grains are in contact.

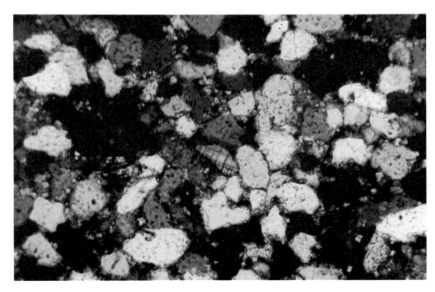

Figure 29.4. Tensleep Formation, outcrop sample from Wind River Canyon, mile marker 69.7, south of Thermopolis, Wyoming (0.44 in. in the photo equals 0.08 mm). Angular quartz sand that is somewhat less well sorted than in Figure 29.3. Note the microcline grain (center), which indicates a granitic source area.

millidarcies, there is a wide range of values that need to be appreciated. Figure 29.6 shows a comparison of two porosity–permeability plots from the same well and illustrate the effects of sampling. One set of data involved permeability measurements on samples taken adjacent to the porosity sample. The other set of data shows the variation when both permeability and porosity are measured on the same sample. As Morgan et al. (1978, p. 620) indicate, if all core analysis samples had the same orientation with respect to the sedimentary stuctures, a closer relationship between porosity and permeability would be likely. The linear trend for higher porosity–permeability values is much less apparent, which implies that predicting these values in higher ranges is likely to be inaccurate. At the lower end of the linear trend, it is apparent that 7% porosity correlates with about 1 millidarcy. According to Morgan et al. (1978, p. 620), below 7% porosity, little or no oil staining occurs, and the pores are so small that they contain only water. Relative permeabiltiy data indicate that connate water saturation (irreducible water saturation) is around 10%. This low value is believed to be the result of large open pores and a low clay mineral

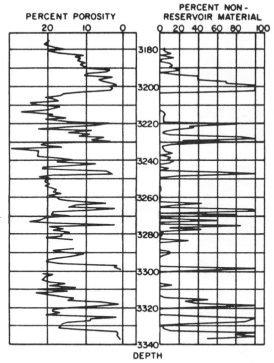

Figure 29.5. Profiles through the Tensleep Sandstone in the Oregon basin field showing the relation between porosity and nonreservoir rock. Measurements made on the slabbed surface of the core. Depth shown in feet. (After Morgan et al. 1978, p. 618.)

content in the Tensleep. The relative permeability data indicate that the Tensleep has good reservoir properties that allow oil to be mobile over a water saturation range of from 10% to almost 70% (Figure 29.7).

Discussion

The major points stressed by Morgan et al. (1978, p. 629–630) follow:

1. Major reservoir variations in the Tensleep are the result of erosion and the emplacement of nonreservoir dolomite or dolomite and anhydrite cement.
2. The layers of nonreservoir material separate the Tensleep Sandstone into separate reservoirs with little or no cross-flow.
3. Within these restricted layers, the flow of injected or produced fluids is horizontal.
4. The reservoir continuity is further disrupted by post-Tensleep erosion at the unconformity. Extra cementation is also associated with stream valleys that cut across the exposed Tensleep surface.

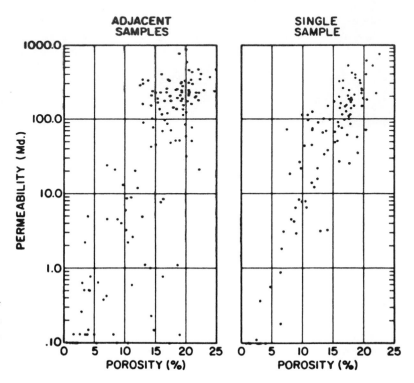

Figure 29.6. Comparison of two sets of porosity—permeability data from the same core of the Tensleep Sandstone in the Oregon basin field. (After Morgan et al. 1978, p. 619.)

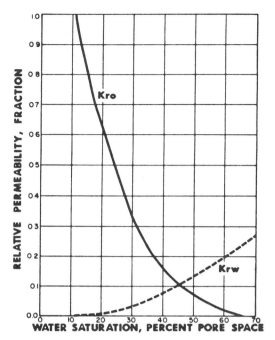

Figure 29.7. Relative permeability curves for the Tensleep reservoir rock as it occurs in the Oregon basin field. (After Morgan et al. 1978, p. 628.)

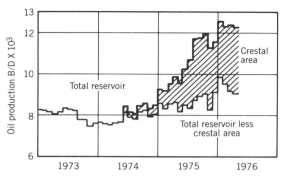

Figure 29.8. Total oil production from the Tensleep Sandstone in the Oregon basin field compared to oil production from the South Dome crestal area where combined geologic and engineering studies applied to water-injection development have enhanced oil recovery. (After Morgan et al. 1978, p. 632.)

5. Recognition of these distinct and discontinuous reservoir segments helps achieve better designed recovery programs. In other words, considerable oil "would be bypassed" if the Tensleep were considered a uniform reservoir sand.

The final point that Morgan et al. (1978, p. 632) make is that the close cooperation of geologists and engineers helps greatly in acheiving an optimum design for an oil recovery project (Figure 29.8). This latter theme is going to become more important as recovery projects become more expensive and more difficult to bring about.

Source Rocks

The Tensleep Formation lies immediately below the Phosphoria, which is the subject of the next chapter. The Phosphoria has the characteristic of being both a source rock and a reservoir rock. At the same time, it is believed to have been the source rock for a considerable amount of the Tensleep production; therefore, the next section will focus on the source rock aspects of the Phosphoria.

References

Houlik, C. W., Jr., 1973, Interpretation of Carbonate Detrital Silicate Transitions in Carboniferous of Western Wyoming, *Bulletin of the American Association of Petroleum Geologists,* Vol. 57, No. 3, pp. 498–509.

Morgan, J. T., F. S. Cordiner, and A. R. Livingston, 1978, Tensleep Reservoir, Oregon Basin Field, Wyoming, *Bulletin of the American Association of Petroleum Geologists,* Vol. 62, No. 4, pp. 609–632.

30

Phosphoria Formation, Central Wyoming

According to Claypool et al. (1978, p. 98), the Phosphoria Formation is believed to be the source of much of the oil in the upper Paleozoic rocks of the northern and central Rocky Mountain region. The fact that the Phosphoria functions as both a source rock and a reservoir rock makes the study of this rock units particularly interesting.

Geologic Setting

The Phosphoria Formation was deposited along the shelf in a large open bay along the Cordilleran seaway. The sedimentary sequence involves two transgressive–regressive cycles that appear to reflect the large movements of the Sonoran orogeny to the west.

The Meade Peak and Retort Phosphatic Shale members of the Phosphoria Formation will be discussed in some detail in terms of their source potential. The stratigraphic positions of these members and other parts of the Phosphoria Formation are shown in cross section A–A (Figure 30.1). The Meade Peak and Retort members are composed of dark gray shale, phosphorite, and marine chert. Proceeding eastward from the Idaho–Wyoming line, these beds intertongue with the carbonate members of the Park City Formation in central Wyoming. The black carbonaceous shales in the Meade Peak and Retort members express the maximum transgression phase in each cycle.

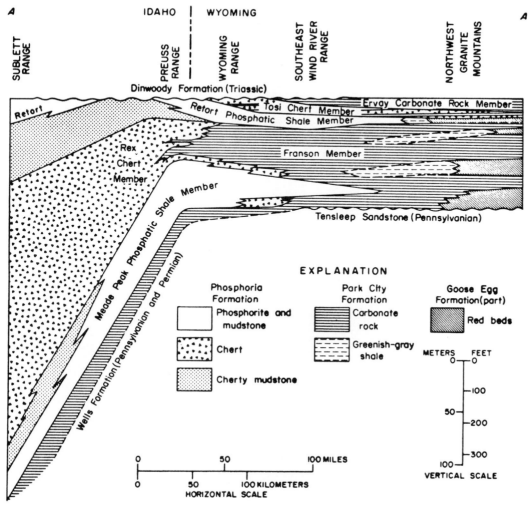

Figure 30.1. Cross section showing stratigraphic relations of Permian rocks from 50 miles southwest of Pocatello, Idaho, to 30 miles due east of Lander, Wyoming. (After Claypool et al. 1978, p. 100.)

The entire region was tilted northward between the deposition of the Meade Peak and that of the Retort resulting in a northward shift in the center of maximum organic carbon deposition. Overthrusting in the western part of the area took place in Jurassic time and resulted in at least 60 km of crustal shortening (Claypool et al. 1978, p. 101).

From the organic geochemistry point of view, the depth of burial of the Permian is important in relating the maximum temperatures and pressures to the degree of maturity exhibited by the source rocks. The maximum depth of burial for most of the Phosphoria appears to have been during the Late Cretaceous period when thick sediments were deposited rapidly.

Photomicrographs

The first set of samples from which the thin sections and hence the photomicrographs of the Phosphoria were made (Figures 30.2–30.4) were collected at a location called Wedding of the Waters, where the Wind River becomes the Big Horn River (Figure 30.5). This particular location is on the Big Horn basin side of the Owl Creek Mountains where the river has cut down through the whole sedimentary sequence, providing good exposures of both the Tensleep and Phosphoria formations.

It cannot be claimed that the Phosphoria at this

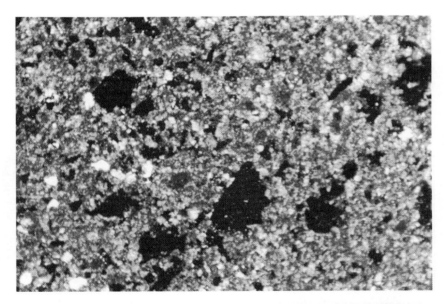

Figure 30.2. Phosphoria Formation (top), outcrop sample from near Wedding of the Waters, Wind River Canyon, south of Thermopolis, Wyoming (0.44 in. in the photo equals 0.17 mm). Porous carbonate with numerous small quartz grains, pellets, and fossil ghosts. In some cases, the fossils have been replaced by dolomite.

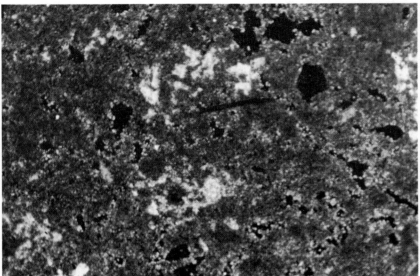

Figure 30.3. Phosphoria Formation (top), outcrop sample from near Wedding of the Waters, Wind River Canyon, south of Thermopolis, Wyoming (0.44 in. in the photo equals 0.17 mm). Micrite with pellets. Other fossil remains appear to have been dolomitized beyond recognition. Small amounts of anhydrite fill in some of the pore space. In general, the porosity is excellent. Solvent extracts of several samples from this outcrop were high in hydrocarbons, suggesting that this zone could have been a truncated reservoir.

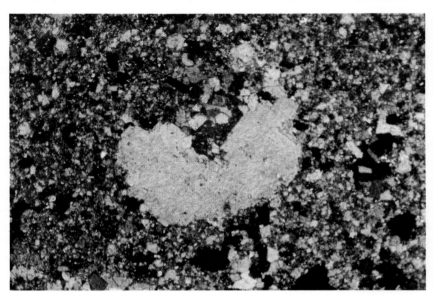

Figure 30.4. Phosphoria Formation, outcrop sample taken from the edge of the Wind River basin on the northern flank of the Rattlesnake Mountains (0.44 in. in the photo equals 0.17 mm). The carbonate has been stained to reveal calcite (pink). It is clear that the crinoid columnal in the center has not been dolomitized although most of the surrounding carbonate sediment has been transformed into dolomite. Some pyrite, indicating a reducing environment, is also visible.

Photomicrographs

point is a reservoir rock; however, it should be noted that the porosity is good. This may be the result of leaching caused by surface waters; nevertheless, both samples are porous enough to be part of a regular Phosphoria reservoir.

Solvent extracts indicated that the outcrop does, in fact, contain a fairly high amount of extractable hydrocarbons; therefore, to include these samples in a text about petroleum reservoirs is not too far from the mark.

The second sampling location for the Phosphoria (Figure 30.5) along the flanks of the Rattlesnake

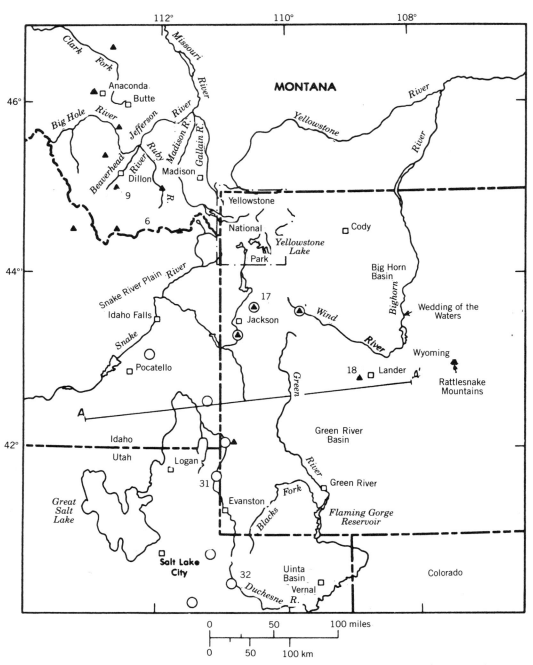

Figure 30.5. Map showing approximate locations of Wedding of the Waters and Rattlesnake Mountain outcrops of the Phosphoria Formation. Sample locations with numbers have gas chromatographic analyses illustrated in Figure 30.7. (After Claypool et al. 1978, p. 99).

Mountains does not exhibit good porosity nor does it contain extractable hydrocarbons in abundance. Clearly, the first location is that which appears to come closest to resembling the actual subsurface reservoir.

The Phosphoria As a Source Rock

Claypool et al. (1978, pp. 104–111) studied the hydrocarbon extracts from the Meade Peak and Retort Shale members to determine the relationship between those samples that had undergone metamorphism and those that were classified as mature. As the plot of extracted hydrocarbons versus organic carbon reveals (Figure 30.6), those samples in the metamorphic field are clearly low on hydrocarbons.

The chromatograms of the saturated hydrocarbons extracted from the two black shale members of the Phosphoria show the characteristics of the immature, mature, and metamorphosed in Figure 30.7. The locations of the samples used in this figure are shown in Figure 30.5, where the sample numbers on the map correspond to those listed with the chromatograms.

The samples illustrated as thermally immature in Figure 30.7 are classified on the basis of their kerogen and extractable organic composition. More specifically, they have saturated hydrocarbon mixtures in which normal alkanes are less important than the presumed isoprenoids pristane and phytane. In this immature chromatogram type, there are unresolved, complex, branched cyclic saturated hydrocarbons that show up as the large hump in samples A and B of Figure 30.7.

Samples that are thermally mature have saturated hydrocarbon chromatograms where the normal alkanes are comparable in peak height and in importance to the pristane and phytane peaks. The "mature" examples recorded in Figure 30.7 as C and D are shown as samples 18 and 32 on the map (Figure 30.5). Sample 18 is relatively close to both of the outcrop sampling areas from which the photomicro-

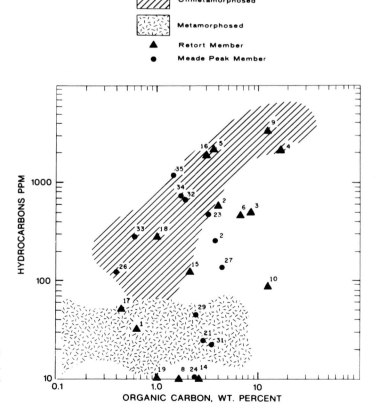

Figure 30.6. Hydrocarbon content versus organic carbon content in metamorphosed and unmetamorphosed rocks from the Phosphoria Formation. (After Claypool et al. 1978, p. 106.)

Organic Geochemistry of Phosphoria Black Shale

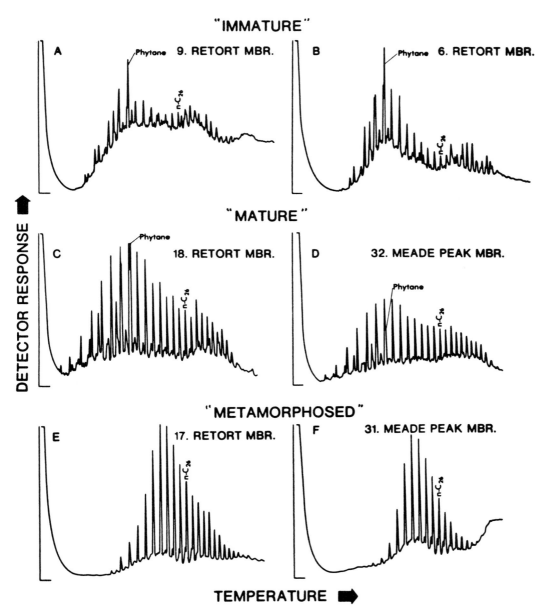

Figure 30.7. Gas chromatographic analyses of saturated hydrocarbons from the Meade Peak and Retort members of the Phosphoria Formation. "Mature" sample from the Retort Member taken near Lander, Wyoming. (After Claypool et al. 1978, p. 109.)

graphs in this chapter were obtained and is probably representative of the hydrocarbons they contain.

The incipiently metamorphosed samples have chromatograms of their saturates that show a predominance of normal alkanes as illustrated by samples E and F (17 and 31) in Figure 30.7. In terms of crude oil production, it appears that in the general region of the Idaho–Wyoming thrust belt the Phosphoria sediments have been "overcooked." By contrast, the Retort samples from southwestern Montana never have been deeply buried with the attendant heat and pressure and are still in the immature stage.

According to Claypool et al. (1978, p. 111), the degree of maturity reflected in the analyses of outcrop samples is mainly the result of the rock temperature at the maximum depth of burial. This means that the Phosphoria within the overthrust belt has experienced sufficient temperature to cause the loss of some of the previously existing hydrocarbons, whereas the Phosphoria equivalents deposited on the shelf to the east in central Wyoming have undergone temperatures sufficient to cause a maximum conversion of organic matter to liquid hydrocarbons. Claypool et al. (1978, p. 112) indicate that Phosphoria-derived oils have been buried to depths in excess of 2.1 km but less than 4.9 km at the end of the Cretaceous. Claypool et al. (1978, p. 112) go on to point out that the Pennsylvanian age Tensleep Sandstone that immediately underlies the Phosphoria contains most of the Phosphoria-derived oil in central Wyoming. It is worth noting that the Tensleep is barren except in areas inferred to have been buried under between 2 and 5 km of sediment at the end of the Cretaceous.

Claypool et al. (1978, pp. 113–117) have carried out a detailed discussion concerning the quantitative evaluation of the oil potential for the Phosphoria that should be helpful to explorationists in Wyoming who are concerned with both Phosphoria and Tensleep reservoirs.

References

Claypool, G. E., A. H. Love, and E. K. Maughan, 1978, Organic Geochemistry, Incipient Metamorphism, and Oil Generation in Black Shale Members of Phosphoria Formation, Western Interior United States, *Bulletin of the Association of Petroleum Geologists,* Vol. 62, No. 1, pp. 98–120.

31

Paddle River Gas Field, Nordegg Member, Alberta, Canada

The overall reserves of gas found in the Paddle River field are not as large as some of the fields found in the same geologic setting; however, there is reason to believe that the field is somewhat unique. Table 31.1 indicates gas reserves of 20×10^9 ft^3 spread over 10,000 acres; therefore, the field deserves some attention concerning how it originated.

Geologic Background

From the foothills of southwestern Alberta to its eastern outcrop edge on the Precambrian shield, the Alberta Plains Paleozolic sediment forms a wedge that thins from 5000 ft at the western edge to a few feet at the eastern edge. On top of this wedge, a number of unconformities occur, and the subcropping sediments have been exposed and eroded prior to the next episode of deposition. The overall configuration of these subcrops for Alberta is shown in Figure 31.1. It will be observed that the Paddle River field is located at the western edge of the Pekisko Formation subcrop in a position similar to that of the Mihmehik–Buck Lake field about 60 miles to the south. Figure 31.2 shows the general stratigraphic section for the latter field, and this is about the same for the Paddle River field. The legend for Figure 31.2 indicates that the Pekisko is a

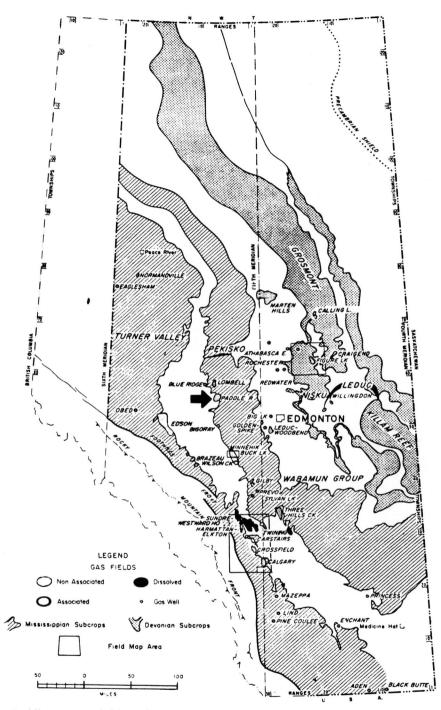

Figure 31.1. Shows the location of the Paddle River gas field in relation to the Devonian and Mississippian subcrops. (After Prather and McCourt 1968, p. 1272.)

Geologic Background

TABLE 31.1 Gas Reserves and Areas of Pekisko Unconformity Fields (Prather and McCourt 1968, p. 1278)

Field Name	Gas Reserves[a] 1×10^9 ft^3 Associated	Gas Reserves[a] 1×10^9 ft^3 Nonassociated	Area (acres)
Aden	—	10	3,000
Black Butte	—	9	1,300
Gilby	—	320	10,000
Lombell	—	40	4,000
Minnehik–Buck Lake	—	420	45,000
Paddle River	—	20	10,000
Prevo		34	6,500
Sylvan Lake	15	75	12,000
Three Hills Creek	—	140	25,000
Twining	Spare	39	6,000
Twining North	38	—	5,000
Wilson Creek	—	20	4,500

[a] Reserves are considered probable from well control.

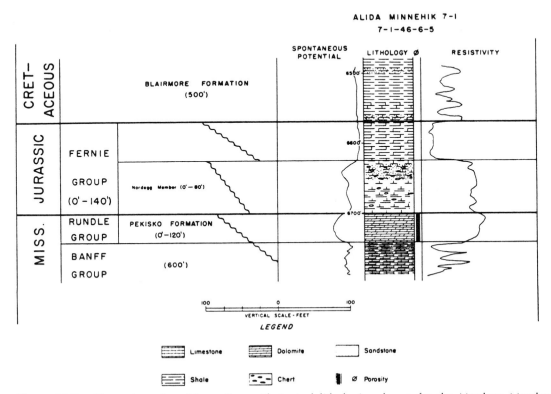

Figure 31.2. General stratigraphic section with typical lithologies shown for the Nordegg Member and the Pekisko Formation. (After Prather and McCourt 1968, p. 1278.)

dolomite and that the cherty part of the section is in the Nordegg Member of the Fernie Group. A brief glance at the mineral composition of the cored materials from the Paddle River field and it becomes obvious that the core samples are indeed from the Nordegg. In addition, it becomes clear from examination of the thin-section photomicrographs (Figures 31.3–31.12) that in the case of the Paddle River field there is a considerable amount of fracturing in the Nordegg.

Mineral Composition and Grainsize

The most striking characteristics of the mineral composition of the Nordegg Member are the rapid changes in the quartz, chert, and carbonate cement (Table 31.2 and Figure 31.13). In the case of the sample from 4836.5 ft (1474.5 m), we are dealing with a carbonate-cemented sand. Two feet deeper and the core is highly chertified. It should be noted that the chert contains pores that, although filled in with quartz, nevertheless have space for gas.

The grainsize information (Figure 31.14) for the long axes of quartz grains must be considered with reference to the mineral composition. Specifically, the initial sandstone (Figure 31.3) is merely highly cemented. However, in the remaining thin sections, (Figures 31.4–31.12) the detrital quartz grains are not the major constituent of the rock and are seen floating in a mass of chert. In view of this, the fact that the initial sample is coarser than the others is not surprising.

Source Materials

Deere and Bayliss (1969, p. 143) describe the chert from the Nordegg Member as "dark-brown to black, microcrystalline, calcareous and bituminous with an average composition of 75 percent calcite, 5 percent dolomite, and 10 percent bituminous material." This last figure is quite interesting because it suggests that the Nordegg could provide the source material for any oil or gas reservoirs associated with it. The fact that a number of reservoirs occur along the unconformity (Table 31.1) tends to support this concept.

The problem concerning how the Nordegg became silicified is worth consideration. Bovell (1979, pp. 109, 110) suggests spicules and possibly other unidentified spherical fossils that are now filled with calcite were at one time siliceous and composed of biogenous opal. The fact that biogenous opal is undersaturated within the interstitial fluids in marine sediments indicates that any biogenous opal is likely to have dissolved soon after deposition. Thus, in the Nordegg Member, the concentration of silica in the interstitial fluids probably increased until it became greater than the solubility of disordered cristobalite and quartz. According to Bovell (1979, p. 110), the disordered cristobalite replaced some of the organic shells, filled the molds, and later inverted to chalcedony. All of these changes apparently took place before a significant amount of burial occurred (Bovell, 1979, p. 112).

Geologic Interpretation

In the Paddle River field, the Lower Jurassic Nordegg Member lies on a subcrop composed of the westerly dipping Banff and Pekisko formations of Mississippian age. When the Nordegg was being deposited, Central Alberta was part of a continental shelf adjacent to a relatively deep marine basin to the west. The source of sediment coming onto the shelf was probably from the east (Bovell 1979, p. 145). Although the Nordegg can be subdivided into a half dozen recognizable zones, its important petrologic constituents are derived mainly from fossil debris that in many cases has been replaced by carbonate and silica. The collophane peloids are unusual features that occur in several zones and are characteristic of the Nordegg at Paddle River.

The lower Nordegg appears to have been deposited on or close to the outer shelf where anoxic bottom conditions prevailed and organic matter was abundant. By contrast, the upper Nordegg shows evidence of a high-energy, oxygenated environment characteristic of the inner shelf. As the shoreline advanced seaward, the inner shelf sediments were deposited on top of the high-organic outer shelf strata.

According to Bovell (1979, p. 147), the significant diagenetic processes include phosphatization, silicification, and compaction. The phosphatization occurred during the depositional phase, whereas the solution and replacement of silica and carbonate from the various fossil zones occurred after burial. The compaction stage where sutured grain boundaries and stylolites formed generated carbonate-rich

Figure 31.3. Nordegg Member, 4836.5 ft (1474.5 m), Honolulu et al. Paddle River 6-34-56-8-W5M, Paddle River field, Alberta, Canada (0.44 in. in the photo equals 0.08 mm). Example of poikilotopic calcite cementing of fine quartz sand. Several grains of chert are present; however, most of the sediment is quartz. This particular view is not typical of the whole sand as the point count data clearly indicate (Table 31.2).

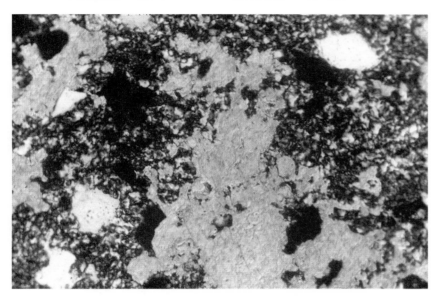

Figure 31.4. Nordegg Member, 4838.7 ft (1475.2 m), Honolulu et al. Paddle River 6-34-56-8-W5M, Paddle River field, Alberta, Canada (0.44 in. in the photo equals 0.08 mm). The initial sediment, which contained widely scattered quartz grains, has been chertified. After this, filling of the pore space with calcite occurred. Calcite (stained pink) invaded the chert reservoir rock enclosing the open channels and pore space.

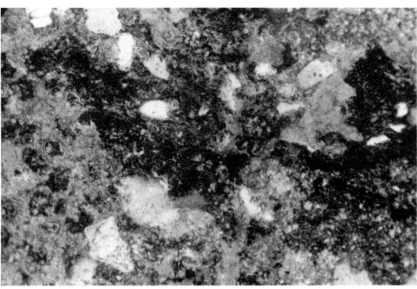

Figure 31.5. Nordegg Member, 4847.5 ft (1477.9 m), Honolulu et al. Paddle River 6-34-56-8-W5M, Paddle River field, Alberta, Canada (0.44 in. in the photo equals 0.08 mm). Highly disturbed texture with concentration of clay minerals along a stylolite that has subsequently been fractured. The fracture has been filled with calcite. The fracture crosses the chert zone; therefore, the calcite filling took place at a later date. Spherulitic structures composed of chalcedony are probably the remains of sponge spicules.

294

Figure 31.6. Nordegg Member, 4866.4 ft (1483.7 m), Honolulu et al. Paddle River 6-34-56-8-W5M, Paddle River field, Alberta, Canada (0.44 in. in the photo equals 0.17 mm). Porous chert zone showing ghosts of fossil structures that have been replaced by chert. Some zones show evidence of megaquartz filling of the pore space.

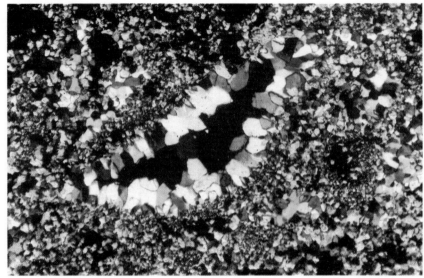

Figure 31.7. Nordegg Member, 4868.5 ft (1484.3 m), Honolulu et al. Paddle River 6-34-56-8-W5M, Paddle River field, Alberta, Canada (0.44 in. in the photo equals 0.17 mm). Moldic porosity within chert is being filled in with megaquartz. The pore is rimmed with bladed to equant quartz crystals. Other areas show zones where pores have been completely filled with megaquartz. Megaquartz is defined as having crystals larger than 20 μm (Scholle 1979, p. 117).

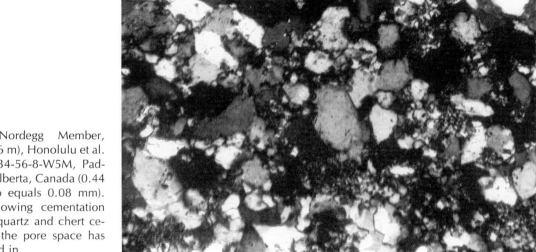

Figure 31.8. Nordegg Member, 4872.7 ft (1485.6 m), Honolulu et al. Paddle River 6-34-56-8-W5M, Paddle River field, Alberta, Canada (0.44 in. in the photo equals 0.08 mm). Quartz sand showing cementation with both megaquartz and chert cement. Much of the pore space has been totally filled in.

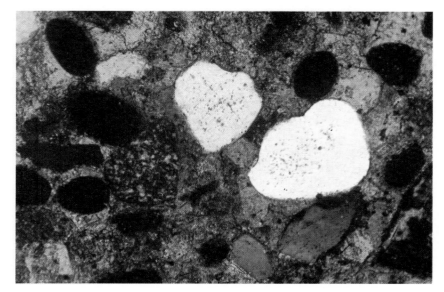

Figure 31.9. Nordegg Member, 5111.5 ft (1558.4 m), Honolulu et al. Paddle River 6-1-57-9W5M, Paddle River field, Alberta, Canada (0.44 in. in the photo equals 0.08 mm). Another example of poikilotopic calcite (stained pink) surrounding relatively large rounded quartz grains along with a large chert grain. Smaller dark isotropic rounded grains called peloids are a form of collophane. X-ray diffraction studies by Bovell (1979, p. 93) indicate that the phosphate formed in these peloids is carbonate apatite or fluorapatite.

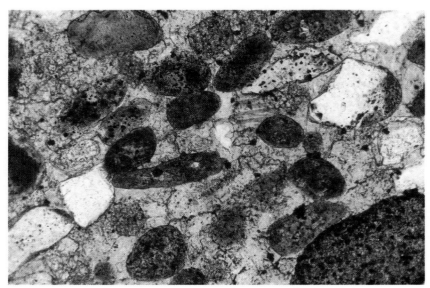

Figure 31.10. Nordegg Member, 5111.5 ft (1558.4 m), Honolulu et al. Paddle River 6-1-57-9W5M, Paddle River field, Alberta, Canada (0.44 in. in the photo equals 0.08 mm). View shows phosphate peloids in transmitted light cemented with calcite (stained pink). Note quartz grain at the upper right appears to have a phosphate coating.

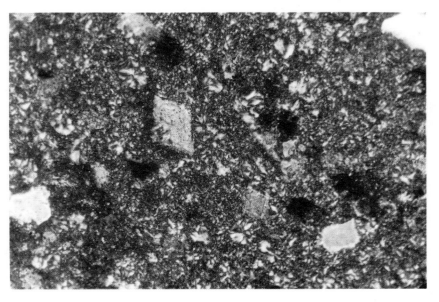

Figure 31.11. Nordegg Member, 5118 ft (1560.4 m), Honolulu et al. Paddle River 6-1-57-9-W5M, Paddle River field, Alberta, Canada (0.44 in. in the photo equals 0.08 mm). Chert forms most of the specimen. Numerous round areas filled in with chalcedonic quartz (chalcedony occurs as radiating bundles of fibers) are spread throughout. Several zoned dolomite rhombs also appear to have grown in place. Chalcedonic quartz frequently fills in cavities, suggesting that the chert was initially quite porous.

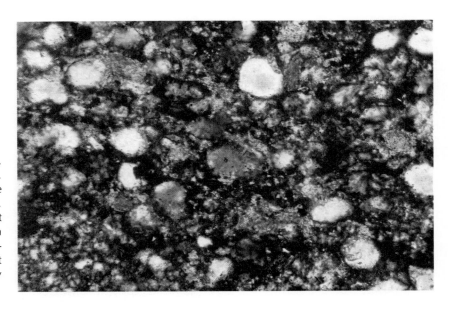

Figure 31.12. Nordegg Member, 5137 ft (1566.2 m), Honolulu et al. Paddle River 6-1-57-9W5M, Paddle River field, Alberta, Canada (0.44 in. in the photo equals 0.04 mm). Chert occurs as the major constituent with quartz grains and rounded chalcedonic areas, again indicating that these spots probably were initially open pores that have been filled in.

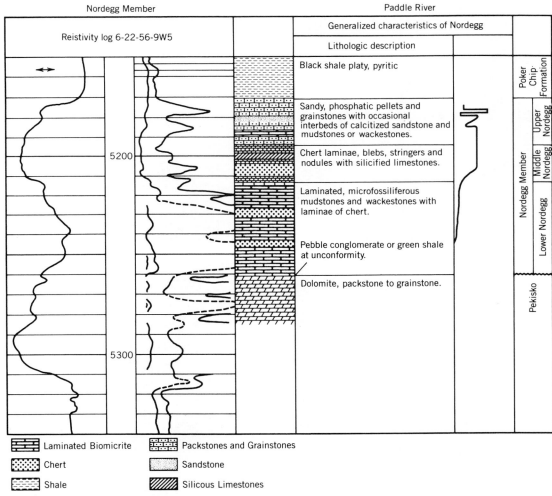

Figure 31.13. Typical resistivity log response and generalized lithology of the Nordegg Member. (After Bovell 1979, p. 18.)

Geologic Background

TABLE 31.2 Modal Analysis (100 points) of the Nordegg Detrital Member[a]

Component %	Depth					
	4836.5 ft (1474.5 m)	4838.7 ft (1475.2 m)	4847.5 ft (1477.9 m)	4866.4 ft (1483.7 m)	4868.5 ft (1484.3 m)	4872.7 ft (1485.6 m)
Quartz	60	3	7	10	36	69
Quartz cement	15	—	—	—	—	—
Chert	9	67	53	87	31	30
Isotropic grains	12	1	—	—	25	—
Calcite	3	29	40	—	—	—
Rock fragments	1	—	—	3	7	—
Clay	—	—	—	—	1	—
Dolomite	—	—	—	—	—	1
Total Points	100	100	100	100	100	100

[a] From cores of the Honolulu et al. Paddle River 6-34-56-8-W5M, Paddle River Field, Alberta Canada.

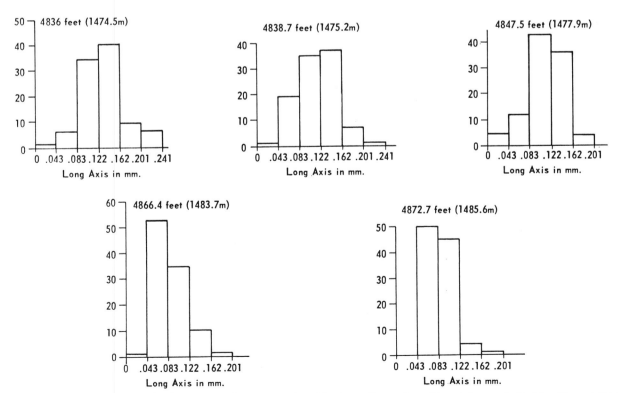

Figure 31.14. Grainsize for long axes of quartz grains, Nordegg–Formation, Honolulu et al. Paddle River 6-34-56-8-W5M, Paddle River field, Alberta, Canada.

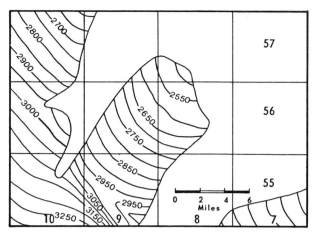

Figure 31.15. Structure contour map of the upper surface of the Nordegg Member for the Paddle River field and vicinity constructed on a sea level datum. (After Bovell 1979.)

solutions that cemented the sandstones and filled many of the vugs.

The pre-Cretaceous period, when the unconformity occurred, subjected the Nordegg to subaerial erosion along its subcrop edge. The drainage system that developed provided a surface with considerable relief including channels and a number of promontories along the erosional edge. It is these promontories that are the locus of gas fields such as the Paddle River field. (See Figures 31.15 and 31.16).

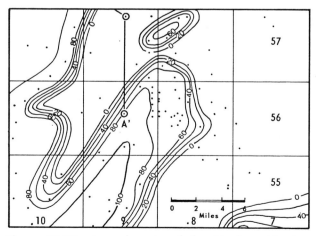

Figure 31.16. Isopach map of the Nordegg Member showing the promontory where the Paddle River field is located. (After Bovell 1979, p. 19A.)

The characteristic electric log profile is provided in Figure 31.17.

It is probable that much of the porosity and permeability that is present in the Paddle River field is caused by the leaching that took place during the time when the Nordegg was exposed to erosion. The subaerial exposure also appears to have masked some of the finer details of the original sediment.

The complexity of the events that led to the accumulation of gas at Paddle River makes it impossible to cite an analogous Recent sediment example that seems plausible. This very complexity also suggests that numerous exploration opportunities still remain.

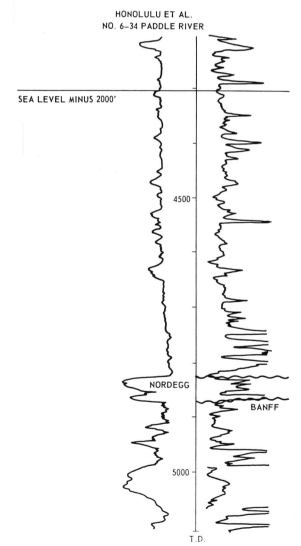

Figure 31.17. Electric log profile of the Honolulu et al. 6-34-56-8W5.

References

Bovell, G. R. L., 1979, *Sedimentation and Diagenesis of the Nordegg Member in Central Alberta*, Masters Thesis, Queen's University, Kingston, Ontario, Canada.

Deere, R. E. and P. Bayliss, 1969, Mineralogy of the Lower Jurassic in West Central Alberta, *Bulletin of Canadian Petroleum Geology*, Vol. 17, No. 2, pp. 133–154.

Prather, R. W. and G. B. McCourt, 1968, Geology of Gas Accumulations in Paleozoic Rocks of Alberta Plains, in *Natural Gases of North America*, Vol. 2, Memoir 9, American Association of Petroleum Geologists, Tulsa, OK, p. 1238.

32

Pembina Oil Field, Basal Belly River Formation, Alberta, Canada

Large fields such as the Pembina field in Alberta, Canada, are commonly multipay fields. In this chapter, the Cretaceous basal Belly River Formation in the Pembina oil field will be examined. This productive zone is of interest for two reasons. First, it has been a very erratic and confusing exploration target and, second, it has provided confusing data from the cores analyzed by the usual core laboratory methods (Iwuagwu and Lerbekmo 1981).

General Geologic Setting

The Pembina field is located approximately 60 miles southwest of Edmonton and 280 miles north of the U.S. border (Figure 32.1). The major producing zone is the Cardium Sand, which by itself has made this a giant field. The field is about 30 miles east of the edge of the Foothills belt in an area where homoclinal beds dip southwest at 45 ft/mile. The surface sediments are sands and shales of early Tertiary age, which are underlain by Cretaceous, Jurassic, Mississippian, Devonian, and Cambrian sediments. The formations that occur in southwestern Alberta are listed in Figure 32.2. The Upper Cretaceous basal Belly River sands occur approximately 2000 ft (610 m) above the major Cardium production at a depth of 3500 ft (1067 m). (See Figure 32.2.)

Petrologic Detail

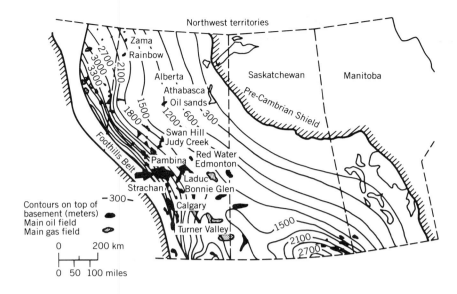

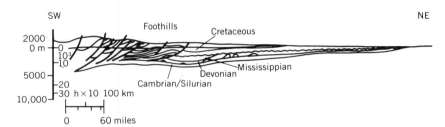

Figure 32.1. Map and cross section of western Canada showing location of the Pembina field in relation to the major structural elements. (After Perrodon 1983, p. 276.)

Petrologic Detail

The Belly River Sandstone is described by Williams et al. (1954, p. 305) as a lithic arenite with "fragments of chert and silicious volcanics . . . andesite, quartz . . . and a little feldspar. Rock fragments greatly predominate." This description provides a very brief idea of the sediment type but does not cover the clay mineralogy where most of the difficulties arise.

The reservoir rock with relatively good porosity and permeability will be emphasized because the important features are more readily viewed in these coarser specimens.

Table 32.1 provides a summary of the point count data for various mineralogic constituents as they appear in thin sections from six different depths in a core from the Cities Service Winfield, Pembina No. 10-12-4-7-4W5M. Rock framgents from sedimentary metamorphic and igneous and sources are all represented, suggesting that several source areas may contribute to a complex mixture.

The components that vary the most are chert and chert cement. Obviously, the source area high in chert is dominant up to the sample at 3501 ft (1067 m). The sample from 3501 ft (1067 m) has the most chert, the most quartz, the lowest clay content, and the best porosity and permeability. It also exhibits the largest mean grainsize as expressed by the measurement of the long axes of quartz grains (Figure 32.3).

All of these parameters appear to fit the appropriate model of a good oil reservoir sand. However, examination of the point count data from the sample below 3504 ft (1068 m) shows that the combined clay mineral content (kaolinite plus montmorillonite–illite and chlorite) is twice that of the sample from 3501 ft (20 vs. 10%), even though the mean value for long axes of the quartz grains is only slightly reduced (Figure 32.3). The result is that the permeability is reduced from 260.00 to 8.60 millidarcies while the porosity is lowered from 19.6 to 17.0%.

		FORMATION	LITHOLOGY	THICKNESS	AV. VELOCITY (MAX. DEVIATION) IN FEET/SEC.	RANGE OF VELOCITIES	REMARKS
TERTIARY CLASTICS	PLEISTOCENE-RECENT		GRAVEL, SAND, SILT, GLACIAL DRIFT	0-5000'	11,000 (1000)	10,000-12,000	
		unconformity					
	MIOCENE	ST. EUGENE (CONFINED TO CRANBROOK AREA)	GRAVEL, SAND, SILT				
		unconformity					
	EOCENE AND OLIGOCENE	KISHENENA (FLATHEAD VALLEY)	NON-MARINE MUDSTONE, MARL, SILTSTONE, SANDSTONE, CONGLOMERATE	6000'+			
		unconformity					
	PALEOCENE	PASKAPOO - PORCUPINE HILLS	NON-MARINE SANDSTONE, SHALE, COAL, BASAL CONGLOMERATE	4000'+			
		unconformity					
MESOZOIC CLASTICS (FOREDEEP) / EXOGEOSYNCLINAL	PALEOCENE AND/OR UPPER CRETACEOUS	WILLOW CREEK	NON-MARINE SANDSTONE, SHALE, MUDSTONE	350-2700'		12,000-14,000	
	UPPER CRETACEOUS	EDMONTON (BLOOD RESERVE, ST. MARY RIVER)	NON-MARINE SANDSTONE, SHALE, COAL, BASAL CONGLOMERATE	1000-3100'	11,400 (400)		
		BEARPAW	MARINE BLACK SHALE	0-60'			
		BELLY RIVER	NON-MARINE SANDSTONE, MUDSTONE, SHALE	1200-4000'	12,600 (1700)		
		WAPIABI	MARINE SHALE, SILTSTONE	1100-1800'	12,800 (1000)		
		CARDIUM	MARINE SANDSTONE, SILTSTONE, SHALE	30-450'	13,400 (2400)		
		BLACKSTONE	MARINE SHALE, SILTSTONE, BASAL GRIT	400-1000'	13,100 (1900)		
	LOWER CRETACEOUS	CROWSNEST VOLCANICS (CONFINED TO CROWSNEST AREA)	VOLCANIC AGGLOMERATE, TUFF	0-1800'		13,000-15,000	BASAL MESOZOIC REFLECTION EVENT "NEAR MISS."
		INTRUSIVES (CONFINED TO FLATHEAD AREA)	TRACHYTE, SYENITE 95-112 M.Y. OLD				
		BLAIRMORE	NON-MARINE SANDSTONE, SILTSTONE, SHALE, BASAL CONGLOMERATE	1000-6500'	14,700 (1900)		
		unconformity					
	LOWER CRETACEOUS & UPPER JURASSIC	KOOTENAY	NON-MARINE SANDSTONE, SILTSTONE, SHALE, COAL	0-4000'	14,800 (2600)		
	JURASSIC	FERNIE	MARINE SHALE, SILTSTONE, LIMESTONE, SANDSTONE	100-1000'+	13,700 (3000)		
		unconformity					
	TRIASSIC	WHITEHORSE	DOLOMITE, SANDSTONE	0-800'			
		SULPHUR MOUNTAIN	LAMINATED ARGILLACEOUS, SILTSTONE, SANDSTONE	0-1000'			
		unconformity					
PALEOZOIC CARBONATES MIOGEOSYNCLINAL	PERMIAN	ISHBEL	QUARTZITIC SANDSTONE, SILTSTONE, CHERT	0-2000'		20,000-21,000	
		unconformity					
	PENNSYLVANIAN	SPRAY LAKES GROUP — KANANASKIS	MARINE CHERTY DOLOMITE	0-170'			
		TUNNEL MTN. (CONTAINS MINOR DISCONFORMITIES)	MARINE DOLOMITIC SILTSTONE, SANDSTONE	0-1800'			
	MISSISSIPPIAN	RUNDLE GROUP — ETHERINGTON	MARINE LIMESTONE, SILTY DOLOMITE, ANHYDRITE	0-850'	20,400 (2500)		
		MOUNT HEAD	MARINE THIN-BEDDED LIMESTONE, SILTY DOLOMITE	0-1000'			
		LIVINGSTONE (TURNER VALLEY - SHUNDA - PEKISKO)	MARINE CRINOIDAL, CHERTY LIMESTONE	800-1400'			
		BANFF	MARINE, DARK ARGILLACEOUS, CHERTY LIMESTONE	500-1050'			
		EXSHAW	MARINE, BLACK SHALE	10-40'			
		unconformity					
	UPPER DEVONIAN	PALLISER	MARINE, MASSIVE LIMESTONE, DOLOMITE	900-1200'	20,000-21,000		
		ALEXO	MARINE, SILTY LIMESTONE, DOLOMITE, SILTSTONE	20-600'			
		FAIRHOLME	MARINE LIMESTONE, SHALE, DOLOMITE, DOLOMITIZED REEFS	950-1500'			
	LOWER AND/OR MIDDLE DEVONIAN	BASAL DEVONIAN CLASTICS	SILTY & SANDY DOLOMITE, RED BEDS	0-120'			
		unconformity					
	ORDOVICIAN	MONS & SARBACH (FRONT RANGES OF BOW VALLEY & NORTH)	LIMESTONE, DOLOMITE, PUTTY-COLOURED SHALE	0-1500'			
	UPPER CAMBRIAN	LYNX (ELKO) (FRONT RANGES)	MARINE DOLOMITE, SILTY DOLOMITE, SHALE, LIMESTONE	0-1800'+			
	MIDDLE CAMBRIAN	ARCTOMYS (FRONT RANGES)	SILTSTONE, SILTY DOLOMITE, SHALE (SHALLOW WATER)	0-200'+			CAMBRIAN REFLECTION EVENT "NEAR BASEMENT"
		PIKA (FRONT RANGES)	MARINE LIMESTONE, DOLOMITE	0-320'			
		ELDON	MARINE LIMESTONE, DOLOMITE	700-1000'			
		STEPHEN	MARINE LIMESTONE, SHALE	100-400'			
		CATHEDRAL (BURTON)	MARINE DOLOMITE, LIMESTONE	200-1000'			
		MOUNT WHITE (EAGER)	MARINE SHALE, SANDSTONE, QUARTZITE, LIMESTONE	500-1500'			
	LOWER CAMBRIAN	ST PIRAN - GOG (CRANBROOK)	QUARTZITE, SHALE	20-8000'			
		unconformity					
BELTIAN	MIDDLE-LOWER LATE PROTEROZOIC	UPPER PURCELL (CRANBROOK WATERTON AREA) NOT PRESENT IN FOOTHILLS	ARGILLITE, DOLOMITE, QUARTZITE, PURCELL LAVA	8000-12,000	17,000-18,000	17,000-18,000	
		LOWER PURCELL (CRANBROOK WATERTON AREA) NOT PRESENT IN FOOTHILLS BASE OF PURCELL NOT OBSERVED	QUARTZITE, ARGILLITE, LIMESTONE, DOLOMITE	8000-20,000'			
		unconformity					
BASEMENT	EARLY PROTEROZOIC	BASEMENT	IGN. & METAMORPHIC "SHIELD" TYPE ROCKS, CONSOLIDATED DURING HUDSONIAN (1600-1900 M.Y.)				

C.G.D.

Figure 32.2. General stratigraphic sequence in southwestern Alberta, Canada, showing the relative position of the Belly River Formation (after Bally et al. 1966, p. 346).

Petrologic Detail

TABLE 32.1 Modal Analysis (100 Points) of the Belly River Sandstone from Cores of the Cities Service Winfield, Pembina #10-12-47-4W5M, Pembina Field, Alberta, Canada

Component %	Depth					
	3472 ft (1058 m)	3490 ft (1064 m)	3501 ft (1067 m)	3504 ft (1068 m)	3506 ft (1069 m)	3513 ft (1071 m)
Quartz	30	30	34	23	24	26
Kaolinite	11	8	7	8	9	7
Montmorillonite/illite and chlorite	14	15	3	12	12	15
Orthoclase feldspar	5	3	3	7	6	6
Plagioclase feldspar	3	1	2	4	8	6
Microcline		1		1		
Biotite and muscovite		1		1		4
Metaquartzite rock fragments	6	6	5	8	5	4
Schist rock fragments	4	4		3		
Granite rock fragments			1	1	1	3
Volcanic rock fragments	3	13	10	6	8	8
Orthoquartzite fragments		2			1	
Shale rock fragments		2	3	3	1	2
Siltstone rock fragments		2		1		
Chert and chert cement	8	9	29	20	22	17
Quartz cement	5	1	3	2		1
Carbonate cement	4	2		2		
Glauconite pellets	1					1
Opaque organics	6				1	
Total points	100	100	100	100	100	100
Porosity, %	13.0	13.0	19.6	17.0	18.6	15.9
Permeability, millidarcies	0.38	1.30	260.00	8.60	49.00	1.32
Photomicrograph #	7	8	9	10	11	12

Iwuagwu and Lerbekmo (1981, p. 485) state the following:

Core analysis data would suggest that the reservoir is characterized by high porosity and high water saturation, and therefore would have a low oil/water production ratio, but this has not been the case. From the petrographic study, it is clear that this apparent discrepancy results at least partly from the presence of authigenic clay.

The authigenic clay is characterized by small crystal size, large inner surface area and submicroscopic porosity. The large surface area of the clay renders it capable of adsorbing a considerable quantity of water.

The overall importance of this authigenic clay to the production characteristics is illustrated by the capillary pressure curves for the two samples compared previously (Figures 32.4 and 32.5). These curves indicate that the estimated irreducible water saturation as a percentage of the pore volumn is 38% for the sample from 3501 ft (1067 m) (Figure 32.4) and 60% for the sample from 3504 ft (1068 m) (Figure 32.5).

The main points that concern petroleum reservoir engineers are summarized by Iwuagwu and Lerbekmo (1981, p. 491) as follows:

Engineering practices should recognize the inherent problems of clay-fluid systems in the basal Belly River

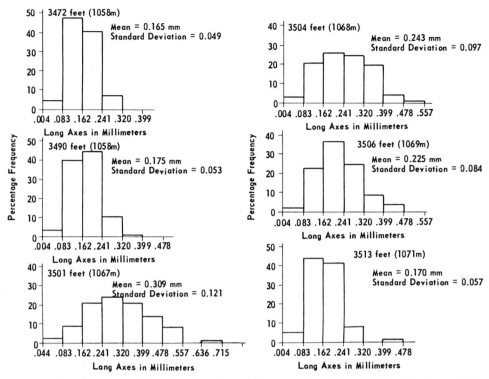

Figure 32.3. Grainsize for long axes of quartz grains, Belly River Sandstone, Cities Service Winfield, Pembina No. 1012-47-4W5M, Pembina field, Alberta, Canada.

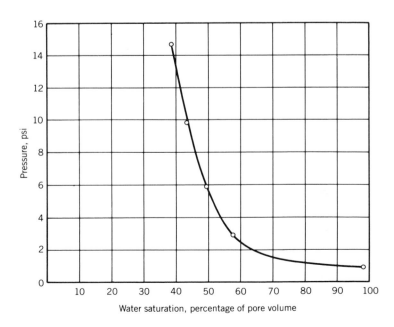

Figure 32.4. C. S. Winfield 10-12-47-4W5M, Alberta, Canada, capillary pressure curve. Depth, 3501 ft; air permeability, 260 millidarcies; porosity, 19.6%.

Geologic Interpretation

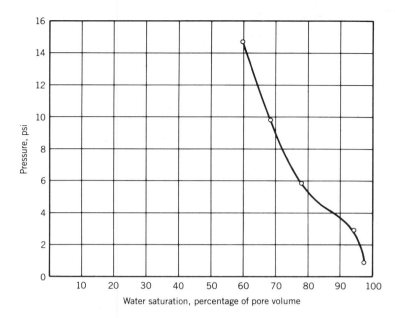

Figure 32.5. C. S. Winfield 10-12-47-4W5M, Alberta, Canada, capillary pressure curve. Depth, 3504 ft; air permeability, 8.6 millidarcies; porosity, 17.0%.

sandstone, and incorporate petrographic knowledge of this reservoir in drilling, completion and production programs. The petrography indicates the possibility of some undesirable formation damage caused by: (a) swelling of the authigenic clay linings due to fresh water sensitivity of the mixed-layer montmorillonite/illite clays, (b) migration of kaolinite "booklets" within the pore system, and (c) secondary mineral precipitation, such as gelatinous ferric hydroxide $Fe(OH)_3$ from chlorite due to poor acid treatment (Almon and Davis 1981). Such problems could be avoided or mitigated by adequately designed fluid systems based on an understanding of this reservoir's authigenic mineral suite.

More recent work by Longstaffe (1986) indicates that the chlorite crystallized from brackish water at low temperature. The Kaolinite and calcite crystallized later at high temperature indicating greater depth of burial.

Photomicrographs and Petrographic Description

The photomicrographs (Figures 32.6–32.11) are selected to illustrate certain important features such as grain size, clay mineral content, cementation, and so on. Taken together, they provide a representative sample of the basal Belly River Sandstone as it occurs in the Pembina field.

Geologic Interpretation

The basal Belly River oil production, like that from the Cardium Sandstone, involves sets of stratigraphic traps; however, exploration has been puzzling. An illustration of the confusion is provided by the following statement (Anonymous 1964, p. 7):

The sand is so erratic that drilling of 40-acre spacing yields insufficient data for a meaningful exploration map. The most that can be done is to establish likely trends of sand occurrence over a limited area.

More recent work by Shouldice (1979, p. 234) suggests that the environment of deposition for the basal Belly River at Pembina was deltaic and that the better producing sandstones are delta channel sands. The finer-grained sands with lower porosity are interpreted as "lagoonal" areas between channels. Figures 32.12 and 32.13 show typical electric logs of the basal Belly River sands in the Pembina field and the outline of the 10 ft net isopach for the B sands as reproduced by Shouldice (1979, p. 233). The electric log profile for the B sand resembles what Taylor (1977, p. 159) has called the "transgressive sand." The outline of the 10-ft net isopach does indeed resemble a complex deltaic sequence.

Figure 32.6. Basal Belly River Sandstone, 3472 ft (2058 m), Cities Service Winfield Pembina No. 10-12-47-4W5M, Pembina field, Alberta, Canada (0.44 in the photo equals 0.17 mm). The glauconite (green grain in the middle) is generally thought of as an indicator of a marine environment. The overall complexity of the mineral composition and the angularity of the individual grains is clearly illustrated. Several varieties of chert are present, which is characteristic of the basal Belly River Sandstone.

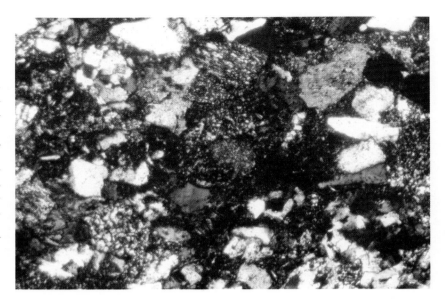

Figure 32.7. Basal Belly River Sandstone, 3490 ft (1064 m), Cities Service Winfield Pembina No. 10-12-47-4W5M, Pembina field, Alberta, Canada (0.44 in. in the photo equals 0.08 mm). Clay mineral coatings surrounding the grains are authigenic and consist of chlorite with subordinate amounts of montmorillonite–illite. Kaolinite crystallized in the pores at a later time. Reservoir engineers should be alerted to possible problems in launching secondary and tertiary recovery projects when such coatings are present. Similar coatings have been discussed by Mellon (1959) and Scholle (1979, p. 132).

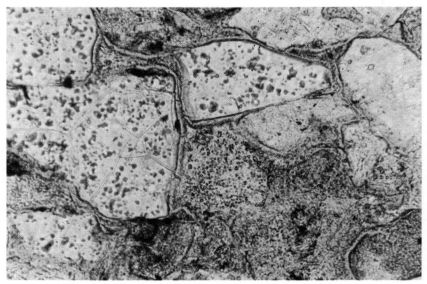

Figure 32.8. Basal Belly River Sandstone, 3501 ft (1067 m), Cities Service Winfield Pembina No. 10-12-47-4W5M, Pembina field, Alberta, Canada (0.44 in. in the photo equals 0.17 mm). Many volcanic rock fragments are present and one of these is illustrated as very dark grains with small feldspar laths. Fresh plagioclase feldspar is also present. The most obvious characteristic of this specimen, aside from the coarseness of the grainsize (mean for long axes of quartz grains is 0.309 mm), is the large amount of chert. In addition, quartz cement is important, as illustrated by the quartz overgrowths on top of the quartz grain (white) in the center.

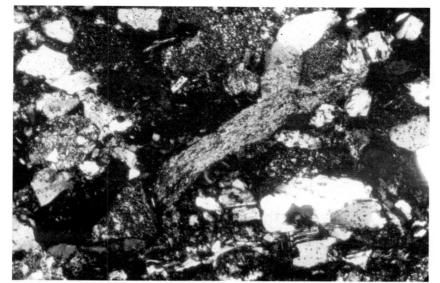

Figure 32.9. Basal Belly River Sandstone, 3504 ft (1068 m), Cities Service Winfield Pembina No. 10-12-47-4W5M, Pembina field, Alberta, Canada (0.44 in. in the photo equals 0.17 mm). As in Figures 32.7–32.9, rock fragments are important constituents. The grain in the center is a schist fragment. Chert composed of microcrystalline quartz is a dominant mineral component along with volcanic rock fragments. Although difficult to trace, authigenic kaolinite fills some of the pore space.

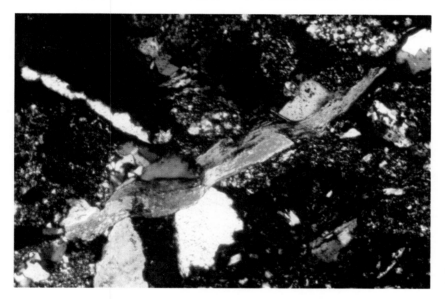

Figure 32.10. Basal Belly River Sandstone, 3506 ft (1069 m), Cities Service Winfield Pembina No. 10-12-47-4W5M, Pembina field, Alberta, Canada (0.44 in. in the photo equals 0.17 mm). The obvious feature is the muscovite and biotite (brown flakes), which are bent around the grains. Various varieties of chert form the dominant mineral species.

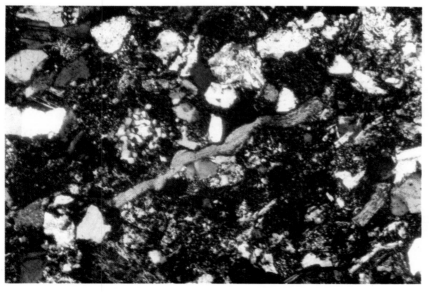

Figure 32.11. Basal Belly River Sandstone, 3513 ft (1071 m), Cities Service Winfield Pembina No. 10-12-47-4W5M, Pembina field, Alberta, Canada (0.44 in. in the photo equals 0.27 mm). A rather general view with twisted mica flakes, angular quartz grains (white and gray), and numerous chert grains exhibiting both microquartz and megaquartz varieties. Volcanic rock fragments showing small feldspar laths are also present. Although not visible, both glauconite and authigenic kaolinite are also present in the sample.

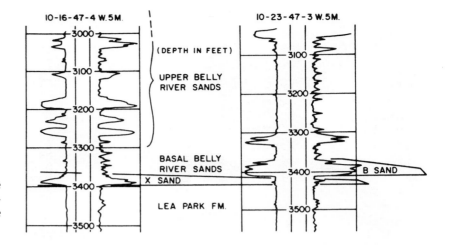

Figure 32.12. Updip pinchout of the Basal Belly River Sandstone in the Pembina–Keystone area. (After Shouldice 1979, p. 233.)

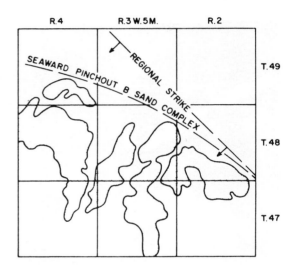

Figure 32.13. Outline formed by the 10-ft net sand isopach of the Pembina–Keystone basal Belly River "B" sands. (After Shouldice 1979, p. 233.)

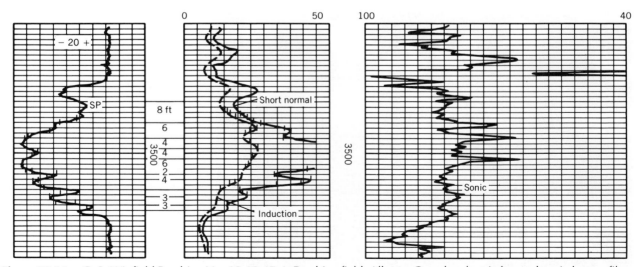

Figure 32.14. C. S. Winfield Pembina No. 10-12-47-4, Pembina field, Alberta, Canada, electric log and sonic log profiles.

Geologic Interpretation

Figure 32.14 shows the portion of the electric log for the basal Belly River interval in the Cities Service Winfield Pembina 10-12-47-4. The profile of the spontaneous potential indicates that the sand coarsens upward. However, it more closely resembles the pattern produced by sands deposited at a distributary mouth. The presence of a glauconite throughout the core for this well, on the other hand, tends not to support the distributary interpretation.

As Selly (1977, p. 208) observes, "A delta involves a wide variety of sedimentary processes, ranging from traction currents in fluvial channels, through beach processes at the shore-face, to grain-flow and turbidites at the foot of some delta slopes." Therefore, on balance, the deltaic interpretation appears tenable.

In searching for a modern example, it becomes clear that the literature provides illustrations that focus on the deltas of the major rivers of the world, such as the Mississippi and the Nile (Figures 32.15 and 32.16). Examination of the fluvially dominated (Mississippi) delta and the wave-dominated (Nile) delta indicates that the same distribution of the latter more closely resembles the sand distribution of the basal Belly River Formation as it is illustrated by Shouldice (1979) (Figure 32.13.) If this is the case, then both deltaic channels sands and beach sands

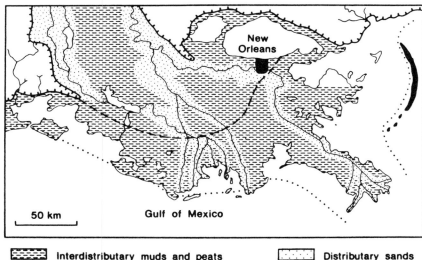

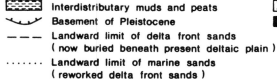

Figure 32.15. Distribution of major sand facies in the modern Mississippi delta (a fluvially dominated delta). (From Selley 1977, p. 201.)

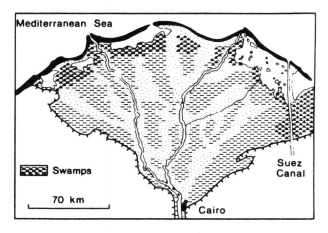

Figure 32.16. Distribution of the major sand facies in the modern Nile delta (a wave-dominated delta). (From Selley 1977, p. 203.)

associated with wave-dominated delta margins can be represented in the basal Belly River producing sand system. High-wave activity and vigorous longshore currents would provide an environment where sediments from different source areas could be mixed efficiently. In this case, the presence of a wide variety of rock fragments appears to be a likely result.

Exploration

The oil-prone area of the basal Belly River appears to be relatively limited (Shouldice 1979, p. 230); nevertheless, the possibility for additional deltaic complexes within the basal Belly River outside the limits of the Pembina–Keystone area appear to be worth additional examination.

References

Almon, W. R. and Davis, D. K., 1981, Formation Damage and the Crystal Chemistry of Clays, in *Short Course in Clays and the Resource Geologist,* F. J. Longstaff, Ed., Mineralogical Association of Canada, pp. 81–103.

Anonymous, 1964, Belly River Sands Oil-Rich But Erratic, Confusing and Coy, *Oil Week,* Vol. 15, No. 27, pp. 7–20.

Bally, A. W., P. L. Gordy, and G. A. Stewart, 1966, Structure, Seismic Data and Orogenic Evolution of Southern Canadian Rocky Mountains, *Bulletins of Canadian Petroleum Geology,* Vol. 14, No. 3, pp. 337–381.

Iwuagwu, C. J. and J. F. Lerbekmo, 1981, The Role of Authigenic Clays in Some Reservoir Characteristics of the Basal Belly River Sandstone, Pembina Field, Alberta, *Canadian Society of Petorleum Geologists Bulletin,* Vol. 29, No. 4, pp. 479–491.

Longstaffe, F. J., 1986, Oxygen Isotope Studies of Diagenesis in the Basal Belly River Sandstone, Pembina-I-Pool, Alberta, *Journal of Sedimentary Petrology,* Vol. 56, No. 1, pp. 78–88.

Mellon, G. B., 1959, The Petrology of the Blairmore Group, Alberta, Canada, Unpublished Ph.D Thesis, The Pennsylvanian State University, 279 pp.

Perrodon, A., 1983, *Dynamics of Oil and Gas Accumulations,* Memoir 5, Elf Aquitaine, Pau, France, 368 pp.

Selley, R. C., 1977, Deltaic Facies and Petroleum Geology, in *Developments in Petroleum* Geology, Vol. 1, G. D. Hobson, Ed., Applied Science Publishers, London pp. 197–224.

Shouldice, J. R., 1979, Nature and Potential of Belly River Gas Sand Traps and Reservoirs in Western Canada, *Canadian Society of Petroleum Geologists Bulletin,* Vol. 27, No. 2, pp. 229–241.

Taylor, J. C. M., 1977, Sandstones as Reservoir Rocks, in *Developments in Petroleum Geology,* G. D. Hobson, Ed., Applied Science Publishers, London, pp. 147–196.

Williams, H., F. J. Turner, and C. M. Gilbert, 1954, *Petrography: An Introduction to the Study of Rocks in Thin Section,* W. H. Freeman, San Francisco, 405 pp.

33

Alaska

The 365 million acres composing the land area of Alaska is so vast that it is possible to provide only a brief glimpse of the geologic provinces favorable to oil and gas accumulation (Figure 33.1). For the purpose of this particular work, the focus is on a single field on the North Slope, as illustrated in Figure 33.1. Specifically, the field of particular interest is the Prudhoe Bay field. As Williams (1984, p. 48) observed, "There is still a great deal of interest in what many explorationists believe is the primer U.S. frontier province, despite the high costs and harsh environment." The search for another elephant-class oil field continues, and it is appropriate to examine a small sample of the most important reservoir rocks from this northern frontier.

Geologic Background

The area involved in this discussion lies between the Brooks range on the south and the Beaufort Sea on the north. One recent estimate (Bird 1985, p. 656) indicates that during the 40 years of exploration on the North Slope, 21 accumulations have been found with an in-place total volume of 60 billion barrels of oil and 35 trillion ft^3 of gas. By any standard, this is an enormous amount of oil and gas. Figure 33.2 provides an index map of the North Slope and illus-

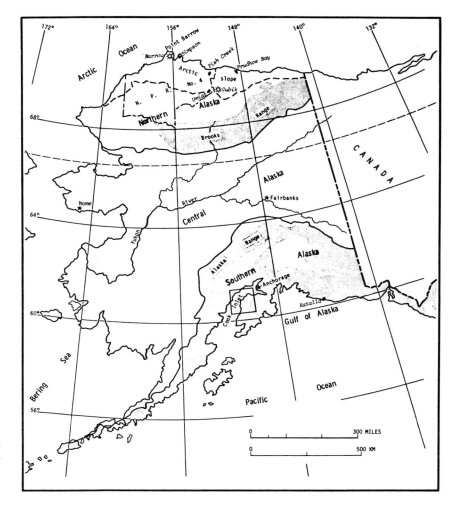

Figure 33.1. Index map of Alaska showing the major areas of petroleum exploration and development. (After Landes 1970, p. 491.)

trates the locations of the major structural features. The diagrammatic cross section in Figure 33.2 proceeds from northeast to southwest and shows the position of the Prudhoe Bay field in relation to the large-scale structure.

Landes (1970, p. 498) divides the northern province into two subprovinces, the Rocky Mountain extension on the south and the coastal plain to the north. The southern portion contains the Brooks range, which is composed of Paleozoic sandstones, shales, and limestones and metamorphic rocks of pre-Mississippian age.

The foothills to the north are made up of marine and nonmarine sandstones and shales of Mesozoic age including substantial amounts of bituminous coal. The North Slope has a thin covering a Quaternary deposits penetrated in a few places where windows of Tertiary clastics occur. This is mainly the area between the Colville River and the Canadian border.

The northern province contains two major tectonic features and these are the geosynclinal belt characterized by the mountain system and the Mesozoic Colville geosyncline to the north. Sediments deposited in the geosyncline have been folded in the foothills region to the south. Toward the eastern end of the Colville geosyncline, Tertiary sediments of both marine and nonmarine varieties are infolded into the Mesozoic formations.

Because of the oil-bearing sandstones and oil shales that crop out in the foothills belt, in 1923 the Naval Petroleum Reserve No. 4 was set aside by executive order. This includes 37,000 square miles south of Point Barrow.

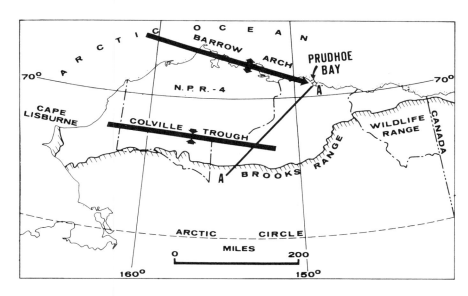

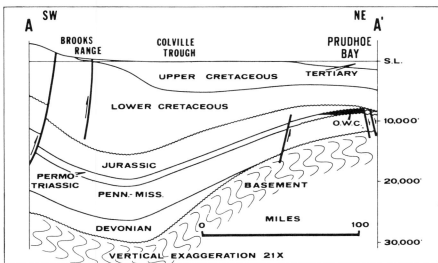

Figure 33.2. *Top:* Map showing the major structural features of the North Slope. *Bottom:* Generalized cross section of the North Slope. (After Jones and Speers 1976, p. 26.)

Between 1945 and 1955, 82 core drill and test wells were drilled by the Navy and this resulted in the discovery of the Umiat, Cape Simpson, and Fish Creek oil fields, all of which are not commercially attractive at this writing.

By any standard, the Prudhoe Bay field is a supergiant, and because of its size, it will continue to be of lasting interest to both geologists and petroleum engineers. The following discussion will attempt to include some information concerning both the geology and the current recovery projects.

Prudhoe Bay Field, Sadlerochit Formation, North Slope, Alaska

The Prudhoe Bay field is generally conceded to be the largest oil field in the United States. Jamison et al. (1980, p. 289) indicate that the Permian–Triassic reservoirs alone contain 9.6 billion barrels and 26 trillion ft^3 of gas. Beyond these productive zones, there are other oil and gas accumulations in the Mississippian and Pennsylvania Lisburne carbonates and the

Lower Cretaceous Kuparuk River sandstones (Figure 33.3); hence, this area will be the focus of exploration and production attention for some time to come. The thrust of this study is directed at the major producing horizon, the Sadlerochit Formation of Permian and Triassic age.

The major structural elements of the North Slope are the Barrow arch parallel to the coastline eastward from Point Barrow and the Colville trough to the south. From a thickness of 2500 ft (760 m) at Point Barrow, the sedimentary column thickens to more than 30,000 ft (9100 m) in the deep parts of the Colville trough.

For the purpose of this work, we will follow the work of Jamison et al. (1980, p. 297) and consider the pre-Mississippian rocks as the economic limit. Carbonates and sandstones of the Mississippian and Triassic lie unconformably on the pre-Mississippian surface. The rock units above this unconformity from oldest to youngest include the Endicott, Lisburne, and Sadlerochit groups, the Shubuik Formation, the Sag River Formation, the Kingak Shale, and the Kuparuk River Formation. All of these are believed to have a northerly source for their sediment, mainly from highlands once located near the present Barrow arch. It is also possible that they originated from a source area that is now detached from north Alaska (Jamison et al. 1980, p. 297).

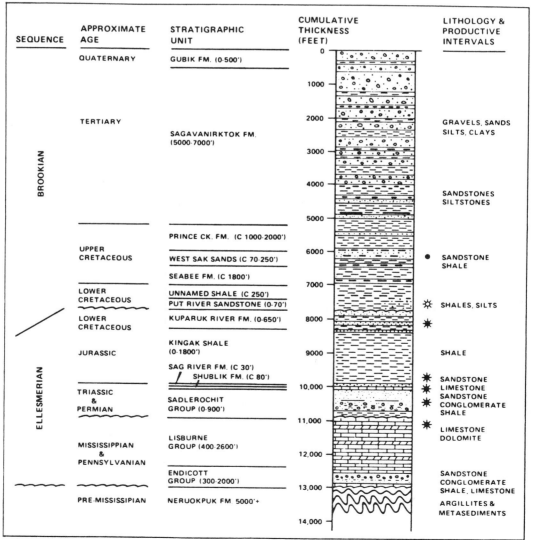

Figure 33.3. Generalized stratigraphic column of the Prudhoe Bay field. (After Jamison et al. 1980, p. 299.)

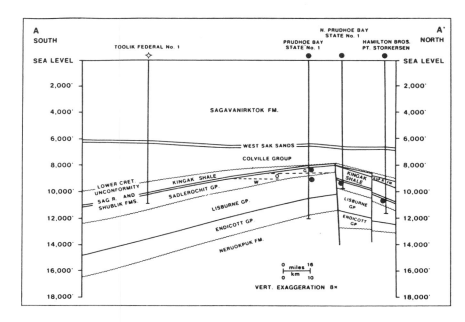

Figure 33.4. North-to-south cross section of the Prudhoe Bay complex. (After Jamison et al. 1980, p. 296.)

The Ellesmerian sequence (see Figure 33.3) is unconformably overlain by the Brookian sequence and is called the Lower Cretaceous unconformity by Jones and Speers (1976). This unconformity is most pronounced in the Prudhoe Bay area and is the major trapping feature of the field. The two cross sections illustrated in Figures 33.4 and 33.5 provide the general structural and stratigraphic relationships and also show how the Lower Cretaceous unconformity, where combined with good structural position, forms the trap. The generalized structure based on top of the Sadlerochit Formation is shown in Figure 33.6.

Test production rates for the discovery well (Figure 33.7) were 25.6 million ft^3/day of gas and 2025 barrels/day of oil from the Sadlerochit. However, later field well rates were much higher and ranged from 15,800 to 19,900 barrels/day (Jamison et al. 1980, p. 301). These figures suggest that some attention

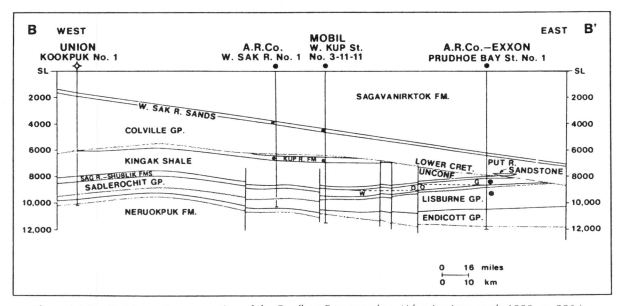

Figure 33.5. East-to-west cross section of the Prudhoe Bay complex. (After Jamison et al. 1980, p. 296.)

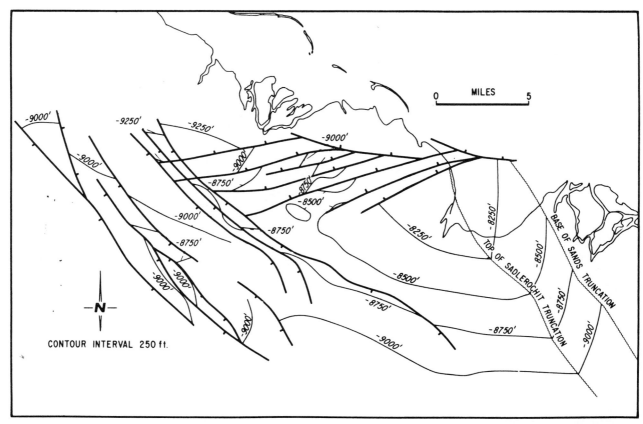

Figure 33.6. Generalized structure map based on the top of the Sadlerochit Formation, Prudhoe Bay field. (After Jones and Speers 1976, p. 45.)

should be paid to the pore space configurations. According to Jamison et al. (1980, p. 302), "In the Prudhoe Bay Field the Ivishak sandstone (Sadlerochit) consists primarily of two fine to medium-grained pebbly sandstone sequences separated by an interval dominated by massive conglomerates."

Although separated by shale intervals, sandstones of the lower sequence are clean and massive with some conglomerates and grade downward into finer-grained sandstones. The isopach map for the sadlerochit sand is provided in Figure 33.8. This lower sandstone sequence commonly has porosities in the 20–30% range and permeabilities from 75 to over 3000 millidarcies (Jamison et al. 1989, p. 302).

The photomicrographs in Figures 33.9–33.15 were taken from the Sadlerochit and illustrate to some degree the typical mineralogy. Figure 33.16 illustrates a typical well log response to pyritic intervals within the Sadlerochit Group.

Mineral Composition and Grainsize

Probably the most obvious feature of the Sadlerochit Sandstone is the presence of large chert pebbles dispersed among the finer-grained quartz and other chert. In the thin-section samples from wells 83-109-H715 (Figure 33.12) and 83-109-H517 (Figure 33.17), the means for the long axes of the 10 larger chert pebbles in each section were 4.7 and 4.5 mm, respectively. On the other hand, the mean values for the long axes of quartz grains from the same thin sections were 0.316 and 0.356 mm, respectively.

A detailed discussion of the various chert types present is beyond the scope of this work; however, it should be noted that some of the chert grains and pebbles are highly fractured and are reminiscent of the Monterey Chert from southern California.

Mineral Composition and Grainsize 317

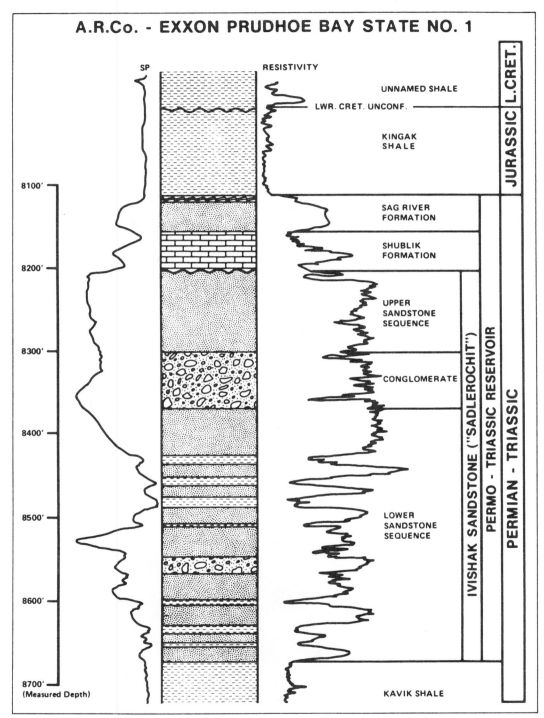

Figure 33.7. Electric log of the Permo-Triassic reservoir encountered in the discovery well at Prudhoe Bay. (After Jamison et al. 1980, p. 300.)

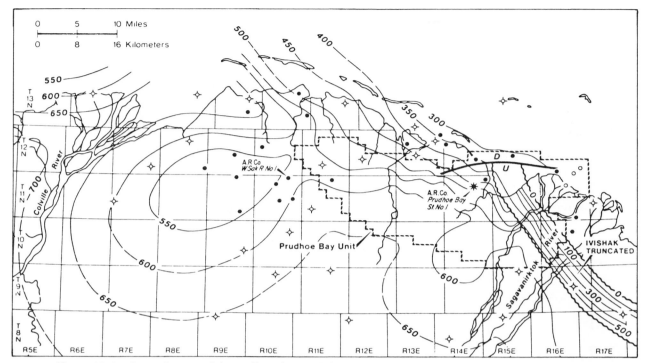

Figure 33.8. Isopach map of the Sadlerochit (Ivishak) Sandstone. (After Jamison et al. 1980, p. 302.)

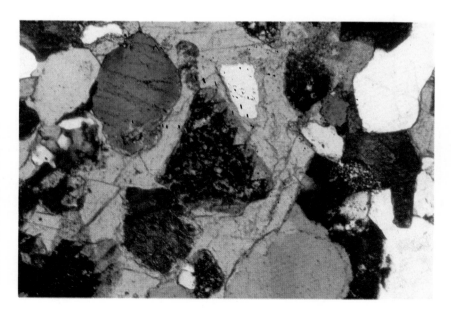

Figure 33.9. Sadlerochit Group, 11,295 ft (3443.6 m), Arco 83-109-J-290, Prudhoe Bay field, North Slope, Alaska. (0.44 in. in the photo equals 0.17 mm). Porosity, 17.2%; permeability, 63 millidarcies. Examples of poikilotopic calcite cement replacing both chert and quartz grains. Particularly striking is toothlike replacement of the chert grain in the center. Quartz overgrowths are common and mask some of the well-rounded quartz grain surfaces.

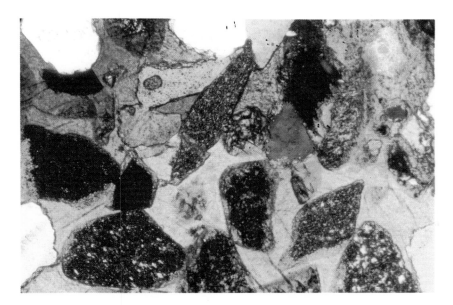

Figure 33.10. Sadlerochit Group, 11,295 ft (3443.6 m), Arco 83-109-J290, Prudhoe Bay field, North Slope, Alaska (0.44 in. in the photo equals 0.17 mm). Porosity, 17.2%; permeability, 63 millidarcies. Another example of poikilotopic calcite cement that replaces both chert and quartz grains. In this instance, both the chert and quartz appear to be angular.

Figure 33.11. Sadlerochit Group, 10,104 ft (3080.5 m), Arco 83-109-H290, Prudhoe Bay field, North Slope, Alaska (0.44 in. in the photo equals 0.17 mm). Porosity, 18.8%; permeability, 79 millidarcies. The specimen has been impregnated with blue plastic in order to show the pore configurations. Brownish grains are chert, whereas the white grains are quartz. Much of the cementation is in the form of quartz overgrowths.

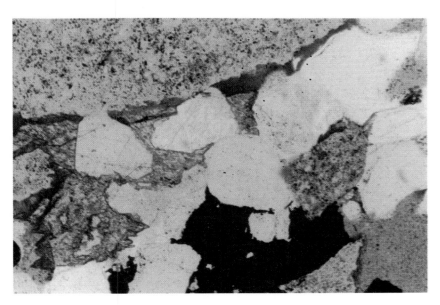

Figure 33.12. Sadlerochit Group, 10,521 ft (3207.6 m), Arco 83-109-H715, Prudhoe Bay field, North Slope, Alaska (0.44 in. in the photo equals 0.17 mm). Porosity, 24.9%; permeability, 199 millidarcies. Blue plastic injected into the specimen shows the open pore space. Large grain (upper left) is part of a chert pebble. Measured in thin section, these pebbles ranged from 5 to 7 mm in diameter. Both quartz and calcite cement are present. Dark material at the bottom is pyrite. As Jones and Speers (1976, p. 37) indicate, pyrite forms a zone that can be observed on various well logs. Typical well log response to pyrite zones in the Sadlerochit Group is illustrated in Figure 33.16.

Figure 33.13. Sadlerochit Group, 10,323 ft (3147.3 m), Arco 83-109-H517, Prudhoe Bay field, North Slope, Alaska (0.44 in. in the photo equals 0.17 mm). Porosity, 16.6%; permeability, 151 millidarcies. Fractured chert pebble has been healed with macroquartz. Overall fracturing of the chert resembles that of the Monterey Formation as it occurs in the Santa Maria basin in California.

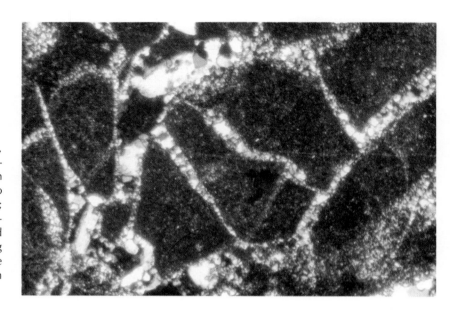

Figure 33.14. Sadlerochit Group, 11,386 ft (3471.3 m), Arco 83-109-J380, Prudhoe Bay field, North Slope, Alaska (0.44 in. in the photo equals 0.17 mm). Porosity, 21.9%; permeability, 121 millidarcies. Fractured zone in the sand has been filled in with calcite. Although some replacement of both chert and quartz is observable, it has not occurred as "aggressively" as illustrated in Figures 33.9 and 33.10.

Figure 33.15. Sadlerochit Group, 10,323 ft (3147.3 m), Arco 83-109-H517, Prudhoe Bay field, North Slope, Alaska (0.44 in. in the photo equals 0.04 mm). Porosity, 16.6%; permeability, 151 millidarcies. High magnification view of pore spaces. Top grain is chert, whereas the two side grains are quartz with overgrowths. The pore at the bottom appears to be lined with clay mineral matter.

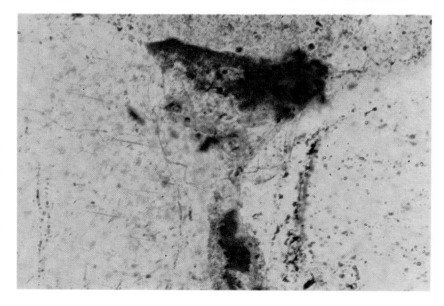

Mineral Composition and Grainsize

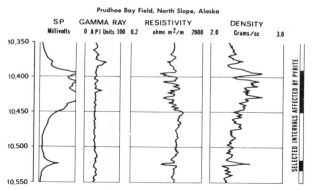

Figure 33.16. Typical well log response to pyritic intervals within the Sadlerorchit Group in BP well 13-11-13. (After Jones and Speers 1976, p. 37.)

Additional observations that deserve mention include the rather high degree of quartz cementation in the sample from well 83-109-J507 (see Table 33.1) and the accompanying large permeability of 1231 millidarcies. The reason for the high-permeability figure is not obvious.

Another feature that deserves discussion is the presence of highly rounded quartz grains that have numerous quartz overgrowths. The way in which some of these grains have been eroded suggests that they are derived from a second-cycle othoquartzite.

Geologic Interpretation

The interpretation of the environment of deposition for the Sadlerochit Group is essentially the same as that proposed by Jamison et al. (1980, p. 302). The fundamental geologic unit involves a delta and the numerous subenvironments associated with deltaic deposition. Sandstones from the lower sequence are separated by shales that were deposited in bays between the major distributary channels. This lower sandstone–shale sequence is overlain by massive, nonmarine conglomerates marking the southernmost advance of the Sadlerochit delta.

Overlying the conglomeratic phase are fine to medium-grained sandstones of the so-called upper sand-

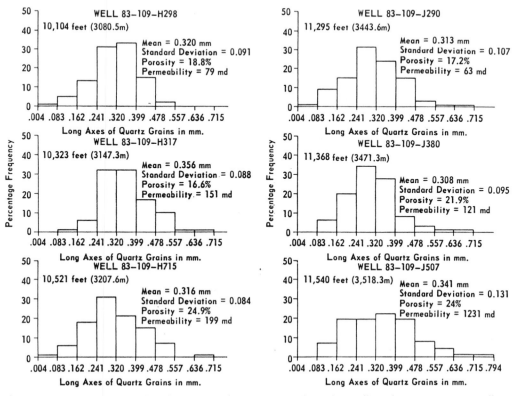

Figure 33.17. Grainsize of the long-axes of quartz grains from the Sadlerochit Formation, Prudhoe Bay field, North Slope, Alaska.

TABLE 33.1 Modal Analysis (100 Points per Thin Section) of the Producing Sandstone from the Sadlerochit Formation, Prudhoe Bay Field, Alaska[a]

Component	J 290, 11,295 ft (3443.6 m)	J 380, 11,386 ft (3471.3 m)	J 507, 11,540 ft (3518.3 m)	H 298, 10,104 ft (3080.5 m)	H 517, 10,323 ft (3147.3 m)	H 715, 10,521 ft (3207.6 m)
Quartz	35	48	45	48	46	46
Chert (all varieties)	27	20	32	21	31	24
Calcite	13	9	3	4	—	—
Dolomite	3	—	—	—	—	—
Clay minerals	3	6	1	2	6	2
Pyrite	1	—	—	—	—	5
Phyllite and schist fragments	1	2	4	1	—	1
Pore space	17	12	14	20	12	20
Metamorphic quartz	—	2	—	4	5	2
Feldspar	—	1	—	—	—	—
Organic opaque	—	—	1	—	—	—
Total points	100	100	100	100	100	100
Porosity, %	17.2	21.9	24.0	18.8	16.6	24.9
Permeability, millidarcies	63	121	1231	79	151	199

[a] Thin sections were kindly provided by Arco Alaska, Inc.

stone sequence. The basal sandstone section of this sequence is attributed to deposition in braided streams, whereas the upper part is believed to reflect a nearshore marginal-marine environment. (Jamison et al. 1980, p. 302.) These interpretations appear to be based on numerous samples taken over the 10-year period of development.

Source Rocks

That the Permo-Triassic formations (including the Sadlerochit Group) were exposed to erosion during the Early Cretaceous means that the oil in the Prudhoe Bay structure could not have been in place earlier than the Early Cretaceous. The next step is to analyze the shales associated with the reservoir rocks both above and below. Morgridge and Smith (1972) carried out such an analysis for both the Jurassic shales and the Cretaceous marine shales. They determined both the total carbon and the C^{15+} hydrocarbon contents for a number of different shales. The results indicated that both the Jurassic and Cretaceous shales were rich in these two measurements; however, the Cretaceous marine shales appeared to be the better source rocks of the two with a value of 3000 ppm C^{15+} hydrocarbons and 5.4 wt. % organic matter. Furthermore, it was noted that only the Cretaceous marine shales are in contact with all the field reservoirs. On this basis, Morgridge and Smith (1972) believed that the Cretaceous shales are the most probable source rocks.

Additional organic geochemical studies of the crude oils from three reservoir sands indicate that the crudes have a common origin (Jones and Speers 1976, p. 44). Stable isotope work based on the oils from these sands suggests that they originated in a restricted marine environment that contained a significant amount of terrestrially derived organic matter. All of the above tends to support the idea that the marine Cretaceous shales were the major source rocks for the Prudhoe Bay accumulation.

Table 33.2 provides a summary of the reserves and productive capacity for Prudhoe Bay Field as estimated by the Federal Energy Administration in 1974.

Source Rocks

TABLE 33.2 Summary Report of Natural Gas Reserves and Productive Capacity, December 31, 1974, Prudhoe Bay Field[a]

	Crude Oil (MMbbl)	Lease Condensate (Mbbl)	Associative[b] (BCF)	NonAssociative (BCF)	Liquids (MMbbl)
		Wet Basis			
Hydrocarbons originally in place	19,245	830	38,318	NA	—
Proved ultimate recovery	8,760	379	30,331	NA	—
Cumulative production[c]	1	0	6	NA	—
Proved reserves	8,759	379	30,325	NA	—
		Dry Basis			
Proved reserves	—	—	29,082	NA	949
Reserves in shutin reservoirs	8,759	379	29,082	NA	949
Indicated secondary and tertiary reserves	0	—	0	—	—
Production					
1973 (total)	NA	0	NA	NA	0
1974 (total)	NA	0	NA	NA	0
Long-term projection of production (annual total)					
1975	NA	0	NA	NA	—
1976	NA	0	NA	NA	—
1977	146.0	0	18.2	NA	—
1978	515.4	0	36.5	NA	—
1979	547.5	0	36.5	NA	—
1980	547.5	0	36.5	NA	—
1981	547.5	0	36.5	NA	—
1982	547.5	0	220.5	NA	—
1983	547.5	0	415.1	NA	—
1984	547.5	0	425.7	NA	—
	(Mbbl)	(Mbbl)	(MMCF)	(MMCF)	(Mbbl)
Daily averages					
December 1974 production	NA	0	NA	NA	0
Short-term productive capacity (60-day basis)	NA	0	NA	NA	—

MMbbl—million barrels, Mbbl—thousand barrels.
[a] From Office of Policy and Analysis and Office of Energy Resource Development, 1975, Final Report, Vol. II, *Oil and Gas Resources, Reserves, and Productive Capactities,* Federal Energy Administration, Washington, D.C., 151 pp.
[b] Production volumes for 1973–1976, as well as short-term productive capacities, do not include minor amounts of fuel usage.
[c] Cumulative production data from the State of Alaska.

Recovery Project

According to Williams (1984, p. 47), plans call for expansion of the miscible enriched gas flood now in progress to boost Prudhoe recovery beyond that recoverable from water flooding by another 115 million barrels of oil. This figure is 12% beyond the $2 billion Prudhoe Bay water flood that started up in mid-1984.

The plan calls for continuing injection of water alternating with hydrocarbon gas enriched with natural gas liquids into the Sadlerochit reservoir sand. The project will employ an inverted nine-spot on 80 acre spacing for at least 10 years, or until the miscible gas slug pore volume reaches 10%.

Exploration

Although there have been exploration failures such as the apparent dry hole at Mukluk Island in the Beaufort Sea, other projects are reporting success. Specifically, the presence of a giant gas–condensate reservoir in the Point Thomson area has been confirmed, Anonymous (1984, p. 30). There may be 5 trillion ft^3 of recoverable gas and 350×10^6 bbl of condensate present in the Point Thomson field. Continued drilling in the whole area will undoubtedly find additional large accumulations during the years ahead. It is indeed one of the last frontiers with an "elephant" potential.

References

Anonymous, 1984, Exxon: N. Slope Gas/Condensate Field is a Giant, *Oil and Gas Journal,* Vol. 82, No. 11, p. 30.

Bird, K. J., 1985, North Slope Oil and Gas: The Barrow Arch Paradox, Abstract, *American Association of Petroleum Geologists Bulletin,* Vol. 6, No. 4, pp. 656.

Jamison, H. C., L. D. Brockett, and R. A. McIntosh, 1980, *Prudhoe Bay—A 10-Year Perspective in Giant Oil and Gas Fields of the Decade 1968–1978,* M. T. Halbouty, Ed., American Association of Petroleum Geologists, Memoir 30, Tulsa, OK.

Jones, H. P. and R. G. Speers, 1976, Permo-Triassic Reservoirs in Prudhoe Bay Field, North Slope, Alaska, in *North American Oil and Gas Fields,* J. Braustein, Ed., American Association of Petroleum Geologists, Memoir 24, p. 23–50.

Landes, K. K., 1970, *Petroleum Geology of the United States,* Wiley-Interscience, New York, 571 pp.

Morgridge, D. L. and W. B. Smith, Jr., 1972, Geology and Discovery of Prudhoe Bay Field, Eastern Artic Slope, Alaska, in *Stratigraphic Oil and Gas Fields—Classification, Exploration Methods, and Case Histories,* R. E. King, Ed., American Association of Petroleum Geologists, Memoir 16, pp. 489–501.

Williams, R., 1984, Apparent Mukluk Wildcat Failure Doesn't Dim North Slope Outlook, *Oil and Gas Journal,* Vol. 82, No. 3, pp. 45–48.

34

Dineh-Bi-Keyah Oil Field, Apache County, Arizona

Oil reservoirs in igneous rocks are comparatively rare, and even more unusual are oil occurences in unweathered fresh igneous rock; nevertheless, this is the case in the Dineh-Bi-Keyah field in Apache County, Arizona. This author was able to obtain only one specimen from this reservoir; however, it is of such interest that it deserves special consideration and discussion even at this extremely sparse sampling level.

General Geologic Setting

The Dineh-Bi-Keyah field is located in the Chuska Mountains on the Toadlena anticline in eastern Apache County, Arizona, about midway between the Black Mesa and San Juan basins, (Figures 34.1 and 34.2). The Paleozoic rocks are overlain by Triassic and Jurassic nonmarine red beds and sandstones. In the Black Mesa basin, these sediments are in turn overlain by Cretaceous marine beds and locally by Tertiary and Quaternary sediments. The overall relationship of the oil-producing sill to various stratigraphic zones is shown in the general south-to-north structural cross section in Figure 34.3. For a more detailed discussion of the stratigraphy, the reader is referred to McKenny and Masters (1968).

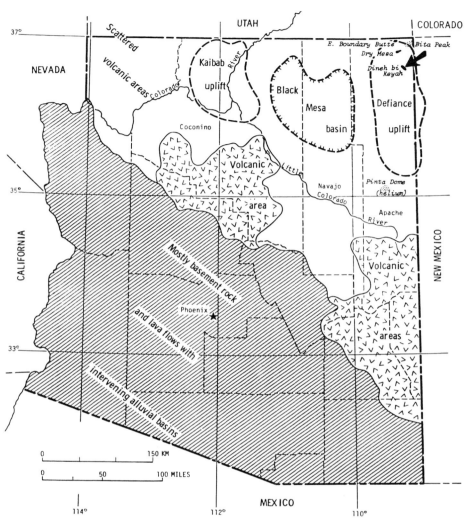

Figure 34.1. Generalized geologic map of Arizona showing the location of the Dineh-Bi-Keyah field and its relation to the major geologic features. (After Landes 1970, p. 448.)

In the Black Mesa basin, there are many diatremes, explosive vents, and other volcanic phenomena with dikes and sills penetrating rocks in various ways. In addition to showing the location of the field, Figure 34.1 also shows a generalized geologic map of Arizona where the major areas of Quarternary and Tertiary age volcanics and intrusives are illustrated.

The oil reservoir for the Dineh-Bi-Keyah field is situated in what at first was described in the literature (Pye 1967, p. 81) as a syenite sill that ranges from less than 100 to more than 125 ft in thickness. The electric log of the discovery well is illustrated in Figure 34.4.

The porosity ranges from 10 to 15% and the permeability from 0 to 25 millidarcies. The sill appears to have been injected into Pennsylvanian limestones and shales of the Hermosa Formation in late Oligocene time. According to McKenny and Masters (1968), several laboratories have dated samples from the sill by the potassium–argon method and have determined an age ranging from 31 to 35 million years.

The producing well that confirmed the discovery was completed, pumping 2860 barrels of oil per day of 45° API gravity oil from perforations of the zone of 3050–3114 ft (932.9–949.4 m). The photomicrographs in Figures 34.5–34.10 were obtained from a piece of the core taken from the producing zone in

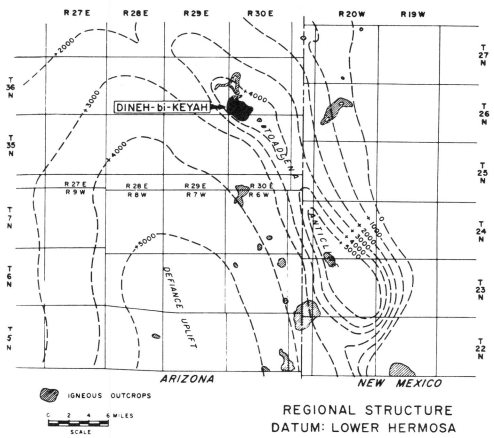

Figure 34.2. Regional structure and igneous outcrops (cross-hatched) in the vicinity of the Dineh-Bi-Keyan field. (After O'Sullivan and Beikman 1963.)

the Kerr–McKee Navajo No. 2. The cored interval is marked in Figure 34.11.

Mineralogy

The mineralogy of the sample differs from that reported in the early literature (Kornfeld and Travis 1967; McCaslin 1967) because much of the ground mass appears to be sanidine and not nepheline. Good optical figures were obtained on a number of mineral grains that could be either nepheline (uniaxial negative) or sanidine (biaxial negative), and in every case the separation of the isogyres indicated that the mineral was biaxial (i.e., sanidine). Other tests indicated that most of the ground mass material was biaxial negative (sanidine) and not uniaxial negative (nepheline). Kenny and Masters (1968) also identified the major constituent as sanidine.

The mineral composition as determined by counting 100 randomly selected points is summarized by Table 34.1.

TABLE 34.1 Mineral Composition (%), producing sill, Dinah-Bi-Keyah Field, Apache County, Arizona[a]

Constituent	Percent
Sanidine	40
Diopside and augite	20
Phlogopite	29
Opaque minerals	2
Pore space	9
Total	100

[a] Based on 100 points.

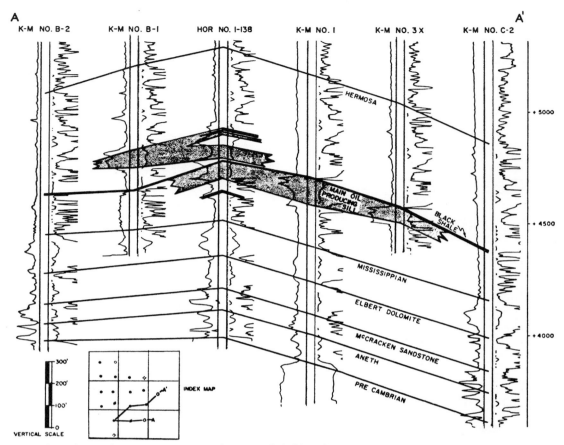

Figure 34.3. General south–north cross section, Dineh-Bi-Keyah field. (After McKenny and Masters 1968, p. 2055.)

General Comments

The reasons permeability and porosity developed in an unweathered sill are not at all obvious. Some workers have proposed that the sill came into contact with a pre-existing Pennsylvania limestone reservoir. This appears unlikely because only tight limestone occurs around the margin of the field. McKenny and Masters (1968) proposed the following possibilities.

1. Oil originally trapped in low-porosity limestone migrated into the sill along fractures.
2. Oil moved up a fault plane and into the sill from the Devonian McCracken Sandstone.
3. Oil was forced out of the black shale by the heat associated with the emplacement of the sill and then migrated back into the sill when cooling occurred.

Based on the mineralization that took place in both the fractures and the pores, it seems plausible to suggest that the fractures occurred first. These may have been associated with additional uplift that probably continued along the Toadlena anticline. The fractures allowed what appear to have been warm, mineralized fluids to leach out or alter the minerals originally present, in some cases increasing the porosity and permeability. Apparently, the coarse-grained rocks near the center of the flow were more susceptible to both leaching and crystal growing. The oil itself may have entered and porous zone along with the flow of fluids during the latter stages when the sill was cold.

This particular field illustrates the opportunities that can exist within areas that contain numerous intrusions into marine rocks with source bed potential. Figure 34.12 provides a thermal-maturation map for the Paleozoic rocks of Arizona based on conodont

General Comments

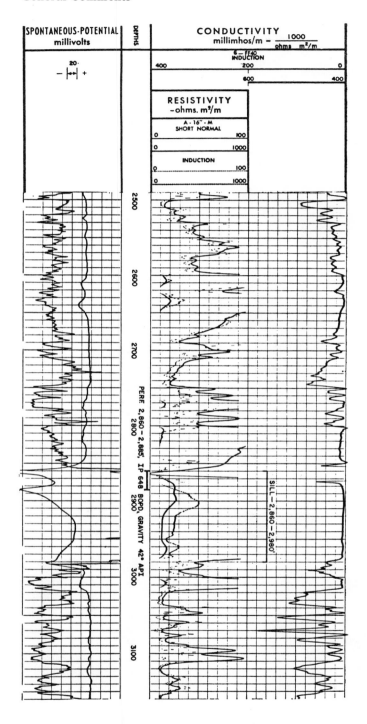

Figure 34.4. Part of the field copy of the electric log of Kerr-McGee Corp, Navajo No. 1 before the well was plugged and subsequently re-entered for completion as an oil discovery. The sill is between 2,860 ft. (871.9 m) and 2980 ft. (908.5 m). (After W. D. Pye, "How Unique Arizona Syenite Oil Reservoir Formed-World Oil," 1967, p. 83.)

Figure 34.5. Oligocene sill, 3100 ft (945 m), Dineh-Bi-Keyah oil field, Apache County, Arizona (0.44 in. in the photo equals 0.07 mm). Close examination reveals several fan-shaped groups of apophyllite needles that have grown into the open pore space. According to McKenny and Masters (1968), both large vertical and horizontal fractures occur in some of the cores. These fractures frequently are filled with apophyllite crystals, suggesting that the fluids circulated through the fractures and into the pores of the reservoir rock. The augite crystal jutting into the pore also shows alteration to chlorite (light green) caused by circulating fluids at the end of the crystal.

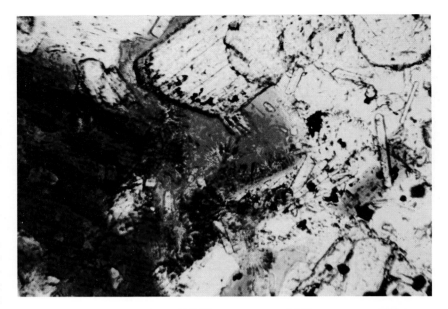

Figure 34.6. Oligocene sill, 3100 ft (945 m), Dineh-Bi-Keyah oil field, Apache County, Arizona (0.44 in. in the photo equals 0.26 mm). The plane light picture shows a large pore with the phlogopite (brown) and other crystals of augite in random orientation. The pore space is relatively clear of growing crystals. A few crystals of magnetite can also be seen at top right center. This phlogopite or biotite-rich zone suggests that this local zone is a "minette."

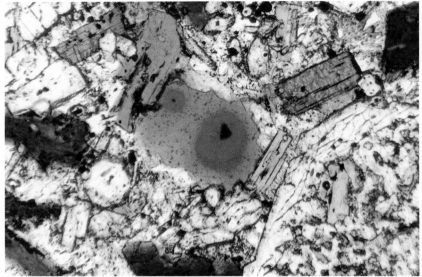

Figure 34.7. Oligocene sill, 3100 ft (945 m), Dineh-Bi-Keyah oil field, Apache County, Arizona (0.44 in. in the photo equals 0.26 mm). Under crossed nicols at relatively low magnification, the size ranges of the augite (yellowish brown) and diopside (pink) become clear. As McKenny and Masters (1968) observed, the larger pores and vugs are associated generally with medium or coarsely crystalline rock. The freshness of the minerals suggests that no weathering has occurred.

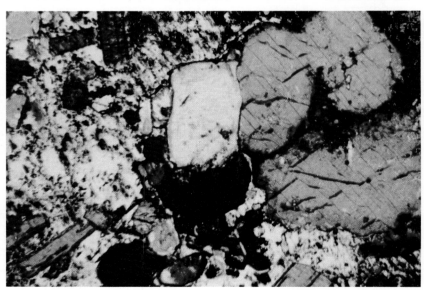

Figure 34.8. Oligocene sill, 3100 ft (945 m), Dineh-Bi-Keyah oil field, Apache County, Arizona (0.44 in. in the photo equals 0.26 mm). The photomicrograph shows the distribution of phlogopite and black opaque minerals—commonly magnetite or titaniferous magnetite. The pores (pink) again show fine apophyllite crystals growing into them. At the lower left, the phlogopite crystal shows alteration and crystals growing in the altered zone.

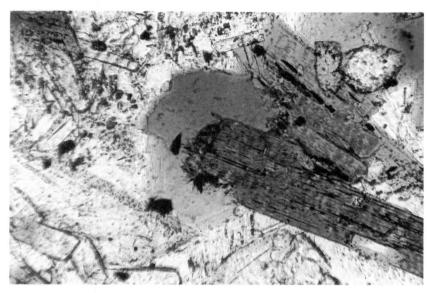

Figure 34.9. Oligocene sill, 3100 ft (945 m), Dineh-Bi-Keyah oil field, Apache County, Arizona (0.44 in. in the photo equals 0.07 mm). Another large pore with altered phlogopite and small crystals of apophyllite growing into the pore space.

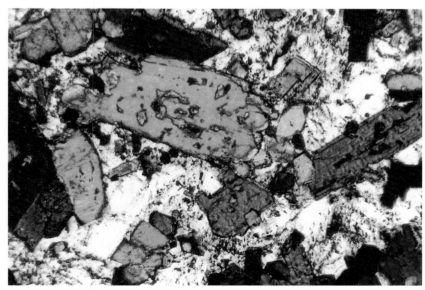

Figure 34.10. Oligocene sill, 3100 ft (945 m), Dineh-Bi-Keyah oil field, Apache County, Arizona (0.44 in. in the photo equals 0.25 mm). A general view of the reservoir rock under crossed nicols with dark brown phlogopite and augite (blue). Again, the freshness of the minerals indicates that no weathering has taken place.

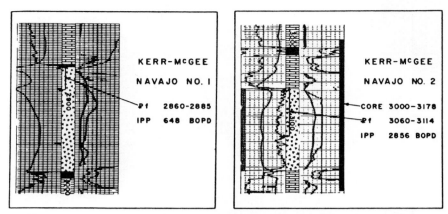

Figure 34.11. Parts of the electric logs of the Kerr-McGee No. 1 and No. 2 Navajo. The No. 2 log shows the cored interval from which the sample was obtained. (After McKenny and Masters 1968, p. 2049.)

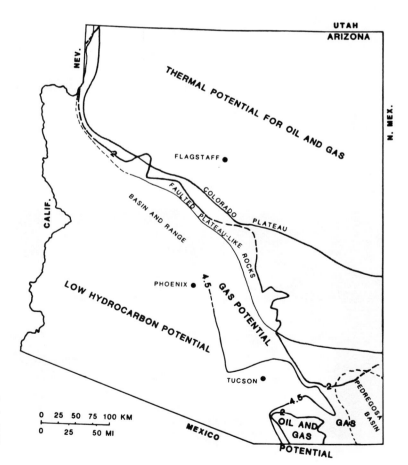

Figure 34.12. Map of Arizona showing potential for oil and gas occurrence in Paleozoic rocks based on thermal-maturation studies of conodonts. (After Wardlaw and Harris 1984, p. 1105.)

color studies. It is clear from this map that the whole northeast half of the state has "thermal potential" for oil and gas, suggesting that the local igneous intrusions in that area may deserve additional study. It is clear from the Dineh-Bi-Keyah example that unweathered igneous rocks can provide the kind of porosity and permeability that favors oil and gas accumulations. Exploration activity that continues in this area (McCaslin 1982) will probably provide additional insights into how and why such fields occur. Worldwide explorationists might well keep this example in mind.

References

Kornfeld, J. A. and M. M. Travis, 1967, Arizona's Spectacular Oil Strike Tops Rocky Mountain Field Interest *World Oil,* Vol. 164, No. 6, pp. 180–190.

Landes, K. K., 1970, *Petroleum Geology of the United States,* Wiley-Interscience, New York, 571 pp.

McCaslin, J. W., 1967, New Arizona Oil Field An Eye-Opener *Oil and Gas Journal,* Vol. 80, No. 37, pp. 103–104.

McCaslin, J. W., 1982, Remote Arizona Wildcat to Test State's Potential, *Oil and Gas Journal,* Vol. 80, No. 37, pp. 103–104.

McKenney, J. W. and J. W. Masters, 1968, Dineh-Bi-Keyah Field, Apache County, Arizona, *American Association of Petroleum Geologists Bulletin,* Vol. 52, No. 10, pp. 2045–2057.

O'Sullivan, R. B. and H. M. Beikman, 1963, Geology, Structure, and Uranium Deposits of the Shiprock Quadrangle, New Mexico and Arizona, *U.S. Geological Survey Misc. Geol. Inv.,* Map I-345.

Pye, W. D., 1967, How Unique Arizona Syenite Oil Reservoir Formed, *World Oil,* Vol. 165, No. 1, pp. 82–83.

Wardlaw, B. R. and A. G. Harris, 1984, Conodont-Based Thermal Maturation of Paleozoic Rocks in Arizona, *American Association of Petroleum Geologists Bulletin,* Vol. 68, No. 9, pp. 1101–1106.

35

West Coast

Any study of the oil production from the far western part of the United States of necessity focuses on California. Even though the state has ranked first in the country in annual yield (i.e., 1926), a relatively small amount of its geology is represented by sedimentary rocks favorable to oil and gas (see Figure 35.1).

As Landes points out (1970, p. 461), almost every kind of hydrocarbon trap "except for those peculiar to carbonate rock reservoirs" has been discovered in California. Anticlinal accumulations are most common, accounting for about two-thirds of the volume of oil. Faulting is common, and combination traps involving plunging anticlines and faults are observed in some fields. Sedimentary traps involving lensing and wedgeouts resulting from truncation and sealing are also part of the array of oil-accumulating mechanisms.

For the purpose of this study, additional attention will be paid to the fracture-related production that is beautifully illustrated in a number of California reservoir rocks.

General Geologic Background

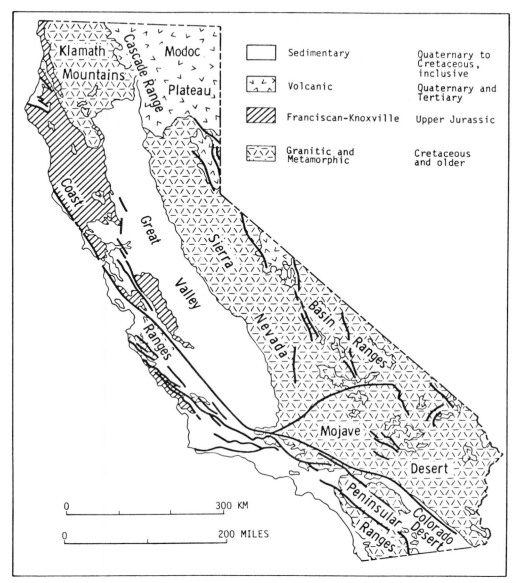

Figure 35.1. Major rock units and fault systems in California. Sedimentary basins are confined to the areas shown in white. (From California Division of Mines and Geology Bulletin 176, Mineral Commodities of California, edited by Lauren A. Wright 1957, "Petroleum," by Charles F. Jennings, Fig. 2, p. 411.)

General Geologic Background

The oil and gas fields are either in the Great Valley or in sedimentary basins closely related to the Pacific shoreline. The latter includes both onshore and offshore production. The remainder of the state is underlain by igneous and metamorphic rocks.

Of the eight productive sedimentary basins, six border the Pacific or are a relatively short distance inland. From south to north, these basins are the Los Angeles, Ventura, Santa Maria, Cuyama, Salinas, Santa Cruz, and El River (Figure 35.2).

The basins related to the Great Valley are the San Joaquin basin to the south and the Sacramento basin to the north. The ages of the major productive rocks and the relative volumes of hydrocarbons produced for these basins are shown in Figure 35.3.

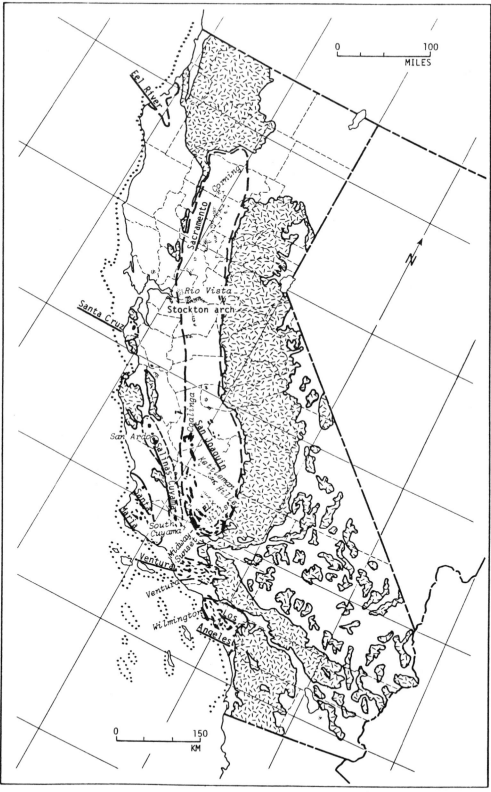

Figure 35.2. Major sedimentary basins in California (underlined). Oil fields are solid black; gas fields stippled. (After Landes 1970, p. 465.)

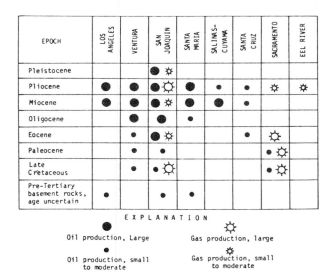

Figure 35.3. Chart illustrating ages of the oil and gas reservoirs and the relative volumes of production for the sedimentary basins of California. (After Landes 1970, p. 463.)

The reservoirs chosen for this study range in age from the Pliocene to the Upper Cretaceous; and although selected to be somewhat representative, it should be recognized that there is a wider range of possibilities including some production from a fractured schist.

The first reservoir to be discussed involves a diatomite of upper-Miocene–Pliocene age. This type of reservoir is rather unusual and deserves careful study with an eye to exploration in other parts of the globe. The other factor that makes the diatomite particularly interesting is that a project to mine a deposit at the McKittrick field from the surface is in progress.

Lost Hills Field Monterey "Diatomite," Kern County, California

The Lost Hills field is located approximately 50 miles (80 km) northwest of Bakersfield in Kern County, California (see Figure 35.4). Primary production of the naturally fractured Monterey Shale in the southeastern Lost Hills field began in 1913. If production from all segments of the field are combined, the cumulative production is around 10×10^6 barrels of oil and about $40,000,000 \times 10^3$ ft^3 of gas. The point of particular interest is that we are dealing with what is called a "fractured shale pool."

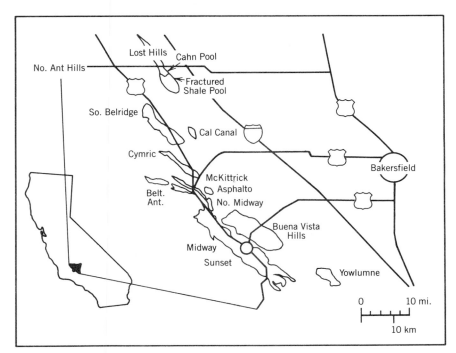

Figure 35.4. Location of the Southeast Lost Hills Fractured Shale pool. (After McGuire et al. 1983).

Geologic Background

In late middle to late Miocene time, the San Joaquin Valley was a marine basin similar to the basins that exist off the present-day coast of southern California. The relative stratigraphic position of the Monterey is illustrated in Figure 35.5. Tectonic activity associated with the San Andres fault resulted in troughs and anticlinal banks being formed on this ancient sea floor. The glacial climate that began during this time produced coastal upwelling of the sea close to shore, which stimulated rapid reproduction in the diatom population, and their opaline frustules rained down on the seafloor, forming diatomite. According the McGuire et al. (1983, p. 3), the topography of the Miocene basin controlled two types of deposition: (1) synclinal troughs that provided channels for gravity flows and ended in the troughs and on the lower bank slopes and (2) anticlinal bank top diatomites that tended to be relatively pure. The latter graded downslope into less siliceous, coarser-grained sediments. The facies changes from diatomaceous mudstone upslope into less permeable diatomites provided the trapping mechanism for oil and gas once the source rocks containing diatomites and other organic matter were buried deeply enough to produce hydrocarbons. Examples of diatomites from the Lost Hills field are shown in Figures 35.6 and 35.7.

According to McGuire et al. (1983, p. 3), the burial diagenesis of the diatomites and diatomite mudstones helped produce a permeability contrast across the facies change to form a trap. The uppermost diatomaceous mudstones apparently kept their primary porosity and permeability due to clay inhibition of opal diagenesis.

GENERALIZED STRATIGRAPHIC COLUMN

EPOCH	SALINAS VALLEY	SANTA MARIA BASIN		SO. SANTA BARBARA CO.	
Upper Pleistocene	Alluvium	Orcutt		Marine Terrace	
Lower Pleistocene	Paso Robles and Pancho Rico	Paso Robles		Santa Barbara	■
Upper Pliocene		Careaga			
Middle Pliocene		Foxen	■	Absent	
Lower Pliocene		Sisquoc	■	Sisquoc	
Upper Miocene	Santa Margarita	Sta. Margarita (?)		Monterey	
Middle Miocene	Monterey	■ Monterey	■		
		Point Sal	■		
Lower Miocene	Sandholdt	Sespe		Tranquillon	
				Rincon	■
	Vaqueros			Vaqueros	▨
Oligocene	Absent		■	Alegria-Sespe	▨
				Gaviota	
Upper Eocene	Absent	Absent		Sacate(Coldwater)	■
				Cozy Dell	
				Matilija	
Middle Eocene	Absent	Absent		Anita and Sierra Blanca	
Lower Eocene	Absent	Absent			
Paleocene	Absent	Absent			
Upper Cretaceous	Absent	Absent		Jalama	
Middle Cretaceous	Absent	Absent			
Lower Cretaceous	Absent	Absent		Espada	
Jurassic and older	Granodiorite Basement complex	Knoxville		Hondo	
		Franciscan		Franciscan	

■ Denotes oil production   ▨ Denotes dry gas production

Figure 35.5. Stratigraphic column and reservoir rocks for central coastal California showing relative position of the Monterey Formation. (After Landes 1970, p. 476.)

Mineralogy

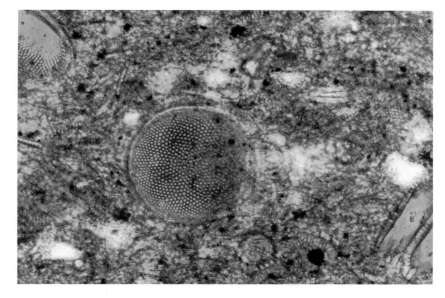

Figure 35.6. Diatomite, 1886 ft (575 m), Chevron "Vulcan" 4B, Lost Hills field (Upper Miocene–Pliocene), Kern County, California (0.44 in. in the photo equals 0.04 mm). The sample has been impregnated with blue plastic to show the open pores. Most of the porosity is associated with the diatom tests. The circular diatom near the center is *Coscinodiscus*, Enranburg, one of the genera that make up the bulk of assemblage.

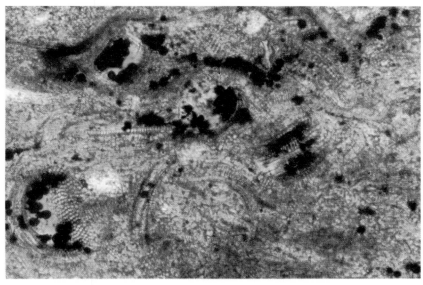

Figure 35.7. Diatomite, 1886 ft (575 m), Chevron "Vulcan" 4B, Lost Hills field (Upper Miocene–Pliocene), Kern County, California (0.44 in. in the photo equals 0.04 mm). View of the pyrite framboids (black circular objects) associated with diatom test fragments indicating a reducing environment at the time of deposition.

The hydrocarbons generated in these beds were moved upward as the amorphous opal converted to cristobalite–tridymite crystals and diagenetic quartz.

According to Hedberg (1980), as the light hydrocarbons diffused upward along the bedding planes, overpressuring occurred, which resulted in microfracturing.

The generalized cross section (Figure 35.8) along the southeasterly plunging axis of the Lost Hills structure shows the producing zone of the Chevron Cahn lease, which is close to the Vulcan property discussed in this work.

Mineralogy

The average mineral composition of the reservoir rock produced by the "diagenetically enhanced, hy-

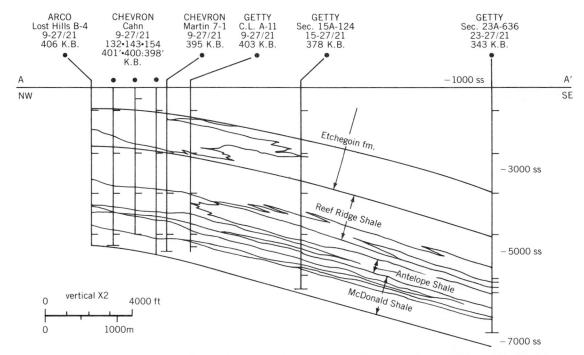

Figure 35.8. Generalized cross section along the southeasterly plunging axis of the Lost Hills anticline. (After McGuire et al. 1983.)

drocarbon trapping facies change" (McGuire et al. 1983) is as follows.

DIATOMACEOUS MUDSTONES	
Clay minerals	36%
Detrital quartz	26%
Diatom debris	38%

Structure

The Southeast Lost Hills fractured shale pool, as McQuire et al. name it, is located on the southwest limb and southeasterly plunging axis of the Lost Hills anticline. The fold apparently experienced structural growth during the last 13 million years while deposition of the sediments was taking place. Evidence for this is the thickening of strata off the southwest flank of the field that continued through Miocene time. Thinning and lensing out of strata over the axis of the Miocene fold also accounts for several stratigraphic traps within the diatomite.

The style of folding changed in the Pliocene to Recent interval resulting in a tilting of the fractured shale pool down to 15° to the northeast.

Fracturing

According to McGuire et al. (1983), (1) fractures are not limited to producing zones, (2) many of the fractures are mineralized and appear to provide little effective permeability, (3) fractures are not limited to certain lithologic types or stratigraphic zones, (4) most of the natural fractures are vertical or within 10° of vertical, and (5) most of the fractures are aligned in a northeast–southwest direction. This may not hold true along the axis or in unproductive zones.

Some wells have produced oil at elevated rates prior to hydraulic fracturing, which indicates that there is effective permeability present due to natural fracturing. Hedberg (1980, p. 185) concluded that overpressuring in the stratigraphic traps created the microfractures oriented perpendicular to the axis of least compressive stress. Fortunately, these microfractures were not sufficiently widespread to cause the loss of the oil from the traps.

Source Rocks

Almost all of the sediments from the middle and upper Miocene are potential source rocks. Of 35 samples analyzed, all have organic carbon values exceeding 1% (McGuire 1983). Furthermore, the major part of the organic matter is amorphous kerogen of algal origin, which is likely to produce oil (Clark and Clark 1982).

In addition, the thermal alteration index values of 2.0– are high enough to indicate that the sediment has been buried long enough and deep enough to produce petroleum.

McGuire et al. (1983) postulate a rather complicated series of events leading to the final accumulation of hydrocarbons in the diatomaceous mudstones and siliceous shales. In summary, these events are as follows:

1. Diatomite converts to procelanite at a depth corresponding to a temperature of around 35°C.
2. At an increased depth and a temperature of 50–55°C, early generation of oil and gas occurs accelerated by the catalytic action of smectic clays.
3. The compaction and porosity loss associated with the conversion to porcelanite results in higher capillary pressures.
4. Contact in capillary pressures within the mudstone forces the nonwetting phase updip into that part of the lense with primary porosity.
5. At a depth of burial corresponding to 70 °C, the cristobalite–tridymite within the mudstone converts to quartz, resulting in more compaction and greater brittleness as well as loss of porosity. This in turn causes increased capillary pressures that force more oil to move into the updip part of the lens.

Reservoir Characteristics

Much of the middle to upper Miocene sedimentation is "laminated" so that thin zones with high water saturations occur adjacent to zones with high oil saturations. This situation is compounded by the microfractures, which are intermittent in occurrence.

References

Clark, J. L. and N. M. Clark, 1982, Vitrinite Reflectance and Kerogen Analysis of Miocene Monterey Formation Rocks of the San Joaquin Valley, in *Monterey Formation and Associated Coarse Clastic Rocks; Central San Joaquin Basin, California,* Society of Economic Paleontologists and Mineralogists, Pacific Section, Volume and Guidebook, pp. 43–54.

Hedberg, H. D., 1980, Methane Generation and Petroleum Migration, in *Problems of Petroleum Migration,* American Association of Petroleum Geologists, Studies in Geology, No. 10, pp. 179–206.

Landes, K. K., 1970, *Petroleum Geology of the United States;* Wiley-Interscience, New York, 571 pp.

McGuire, M. D., J. R. Bowersox, and L. J. Earnest, 1983, *Diagenetically Enhanced Entrapment of Hydrocarbons—Southeastern Lost Hills Fractures Shale Pool, Kern County, California;* Society of Economic Paleontologists and Mineralogists, Pacific Section.

36

McKittrick Field, Diatomite Oil Mine, Kern County, California, and Fractured Monterey Chert

From the petroleum engineer's point of view, one of the most interesting projects involving the Miocene diatomite is the oil mine proposed by Getty Oil Company. The oil-bearing diatomite located in the McKittrick field in western Kern County (Figure 36.1) ranges from surface exposures to a depth of 1214 ft (370 m) with an average thickness of 800 ft (244 m) to 900 ft (274 m). Getty plans to mine about 400 million barrels of oil from 680 ha.

According to Earnest (1981, p. 17), during middle Pleistocene time the diatomite was offset along the McKittrick thrust. The latter involves a large-scale zone of reverse faulting with the fault zone being roughly vertical on the southwest edge of the field. In the main area of the McKittrick field to the northeast, the fault zone flattens and becomes irregular. It seems likely that the diatomite was emplaced by a combination of plastic flow along the McKittrick thrust and the gravity sliding of large blocks over the central and northeastern parts of the field. The cross section in Figure 36.2 provides a schematic view of the geology associated with the diatomite at McKittrick.

As in the case in the Lost Hills field, the oil in the diatomite is probably syndepositional with the diatomite being both its own source and reservoir rock. The pore space is composed of hydrocarbon-filled diatom tests and

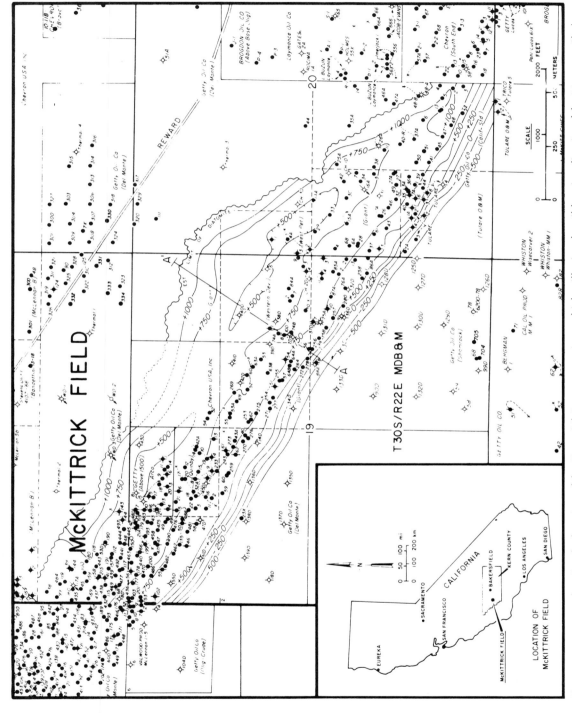

Figure 36.1. Map showing the location of the Getty Oil Company project, which will mine the oil-bearing diomite of the Monterey Shale as it occurs in the McKittrick field, Kern County, California. (After Earnest 1981, p. 17.)

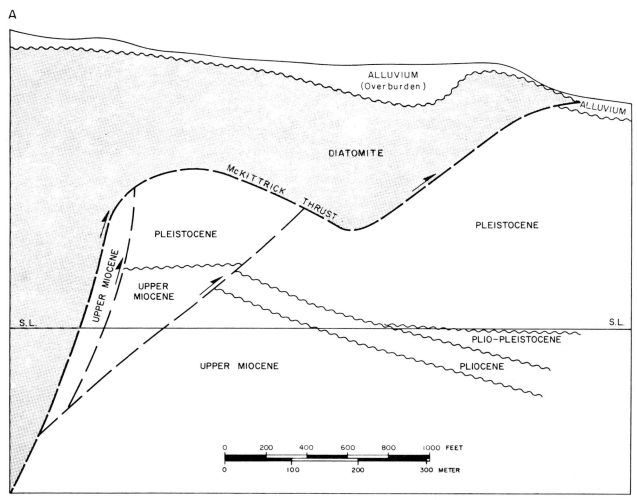

Figure 36.2. A schematic cross section showing the geology associated with the diatomite portion of the Miocene (Mohnian Stage) facies of the Monterey Shale, Kern County, California. (After Earnest 1981, p. 18.)

intergranular pores. Permeability values of usually less than 1 millidarcy allowed little migration although some light hydrocarbons have been lost in surface exposures. In the lower intervals, the oil saturation frequently is as high as 60% of the pore volume.

Earnest (1981, p. 18) observes that although the outcrop appears to be homogeneous, cores and electric logs show sections of chert, sand, and clay beds included in the diatomite. Even with these lithologic inhomogeneities, the ore body averages 15% oil by weight, or 0.88 barrels (140 liters) per ton.

The Dravo Process

The Dravo extraction process used in one of the pilot plants involves slurrying crushed diatomite ore with a solvent composed of heptane, cyclohexane, and ethanol. The slurried ore is pumped through six extractor tanks in countercurrent flow until the solvent removes the oil from the ore (Williams 1982). Once the oil–solvent–water mix is clarified, it is transported to a multieffect evaporation system where the spent residue is removed by stream stripping. The evaporation step produces four streams: a solvent–water mixture, an oil and water-free clean solvent, a solvent–oil mixture, and an oil–solvent mixture. These are separated or stripped to provide product oil or solvent and water for reuse.

The second pilot plant employs Lurgi technology where crushed diatomite ore is steam heated in a rotary dryer to remove most of the water and some votalile hydrocarbons. The cooling and condensation of the product gases from the rotary dryer and those produced from the retort mixer yield by-products gas, naptha, middle oil, heavy oil, oily aqueous liquor, and sour phenolic gas liquor. The oil liquor is also used to moisten the spent solids and for injection tempering of flue gases. The heavy oil is diluted with naptha and processed for dust removal and then distilled to recover the naptha.

The Lurgi oil is a cracked stock and therefore requires rehydrogenation in any commercial-scale unit so that the final product is capable of being transported in existing pipelines.

Oil mining followed by conventional oil recovery processes is being investigated at a number of sites both in California and elsewhere (Parkinson 1985). Whether or not the oil mine approach will become a model for other shallow hydrocarbon accumulations remains to be seen.

Fractured Monterey Chert (California)

The lessons learned about the Monterey Formation from the onshore studies in places such as the Lost Hills field and the McKittrick field have been applied to the California offshore. The discovery of the Point Arguello field and the logic that guided the exploration is worth a brief review. The beginning clues to the production possibilities of the Monterey offshore were obtained from wells drilled on the Santa Barbara Channel acreage where, as Cox (1984, p. 1) indicates, most of the wells "tested surprisingly well from the Monterey formation, a hard, laminated rock of cherts, siliceous shales and dolomites which many geologists believed would hold little producible oil." The problem for the explorationists was twofold: heavy oil and low producibility. If the prospect were to be commercial, the oil had to be free flowing; therefore, the question became, under what conditions would free-flowing oil be formed. If the temperature gradient were sufficiently high, it was reasoned that hot spots should provide the conditions for the formation for free-flowing oil. The temperature gradient maps that were developed indicated that reservoir temperatures in excess of 200°F existed at Point Arguello, which meant that if permeability existed, flowable oil was likely to be present. Cox (1984, p. 18) summarizes the additional rationale that, as will be recognized, was developed for the Lost Hills field as follows:

Diatoms raining on the sea floor formed the bulk of the Monterey. The cell walls of the micro-organisms were mainly made of amorphous silica. Heated and pressed during burial, amorphous silica molecules can recrystallize in two stages. At around 100°F, the silica may enter a cristobalite–tridymite-like phase. At over 174°F, it probably goes into a quartz phase if other subsurface mineral conditions are right.

It was realized that the Monterey could be in the quartz phase at Point Arguello, which would be the most brittle and fracture–prone condition.

The next question was, where were the best fractures formed? In this case, it was reasoned that the "brittle, quartz-rich rocks" would most likely be shattered at or near the crest of the most tightly folded anticlines. Thus the discovery well was located on the crest of the Point Arguello anticline.

The results of the tests from this well indicated that five intervals in a 1000-ft oil column flowed at a combined rate of 6480 barrels of oil per day and 1.68×10^6 ft^3 gas per day.

The photomicrographs in Figures 36.3–36.5 show the naturally fractured Monterey chert. Blue-colored plastic has been impregnated into the samples in order to provide a clear view of the fracture system.

Comments

At present, there is a resurgence in studies carried out on the Monterey Formation largely as the result of continuing success stories such as the discoveries at Point Pedernales and Point Sal. It is furthermore clear that considerable attention will continue to be devoted to the exploration potential of the Santa Maria basin, where it will take years to exhaust the exploration possibilities.

346

Figure 36.3. Monterey Formation, Chert Upper Miocene, California (0.44 in. in the photo equals 0.17 mm). E. W. Christensen, staff petrographer for Chevron's Exploration Department in the western region, makes the following comment concerning this chert sample: "At first glance, one might conclude that most of the fracturing is the result of sample preparation—but this is not the case in much of the slide—similar fracturing occurs on a megascopic scale in many outcrops."

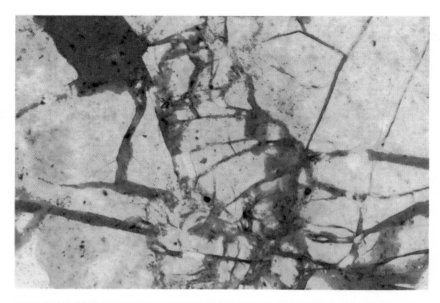

Figure 36.4. Monterey Formation, Chert, Upper Miocene, California (0.44 in. in the photo equals 0.17 mm). Varying types of chert, that is, dark versus light in thin section, appear to reflect different origins. As in the previous slides, blue plastic fills the fractures.

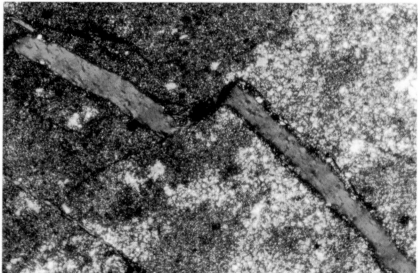

Figure 36.5. Monterey Formation, Chert, Upper Miocene, California (0.44 in. in the photo equals 0.17 mm). The extreme brittleness of the Monterey Chert is reflected in the intense fragmentation. It is this type of fracturing that appears to provide the reservoir pore space in such fields as Point Arguello.

From a broader perspective, the whole concept involving the reservoir potential of fractured cherts needs to be examined and applied to other areas both inside and outside the United States.

References

Christensen, E. W., 1984, personal communications.

Cox, B., 1984, Unlocking the Door at Arguello, *American Association of Petroleum Geologists Explorer,* Vol. 5, No. 7, pp. 18–20.

Earnest, L. J., 1981, Diatomite May Yield Petroleum, *Geotimes,* March, pp. 17–19.

Parkinson, G., 1985, New Ways to Recover Hard-to-reach Petroleum, *Chemical Engineering,* Vol. 92, No. 8, pp. 31–33.

Williams, R., 1982, Getty to Tap Heavy Oil in Diatomite. *Oil and Gas Journal,* Vol. 80, No. 3, pp. 66–67.

37

West Elk Hills Field, Stevens Sandstone, Kern County, California

The Stevens Sandstone of the San Joaquin Valley has produced over 400 million barrels from 15 fields (MacPherson 1978, p. 2243) and continues to be an attractive exploration target. The general locations of the Elk Hills and Kettleman Hills fields are shown in Figure 37.1. MacPherson uses the term *Stevens Sandstone* to include more than 4000 ft (1200 m) of interbedded deep-marine sandstone and shale that originated from two submarine fans toward the eastern end of the Bakersfield arch as shown in Figure 37.2. This particular sandstone provides a good example of a reservoir that originated as a turbidite deposit. As exploration moves into deeper waters offshore, the quest for turbidite traps will undoubtedly become more important. It is with this prospect in mind that the following discussion is focused on the Stevens Sandstone reservoir in the West Elk Hills field.

Geologic Background

The San Joaquin Valley is the southernmost end of the narrow elongate marine basin that lies between the Sierra Nevada Mountains toward the east and the California coast ranges on the west. The basement rock for the eastern half of the valley is composed of Sierra Nevada granite where the typical structure is

Geologic Background

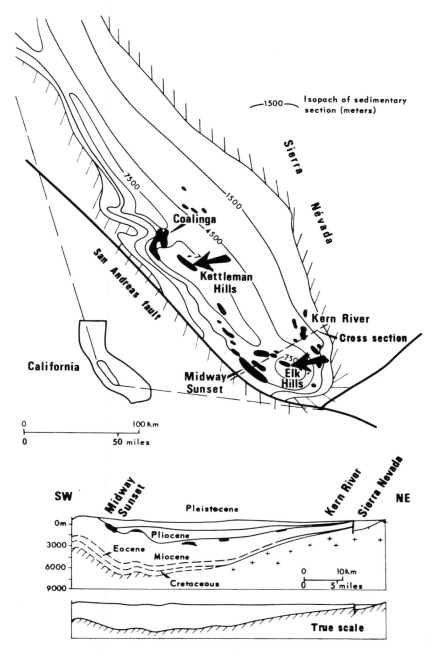

Figure 37.1 Cross section and map of the San Joaquin basin, California, showing the locations of the Kettleman Hills and Elk Hills fields. (After Perrodon 1983, p. 288.)

that of block faulting and open folds. The western half of the valley is underlain by Franciscan metamorphics where the structure is characterized by thrust faulting and tight folding.

Throughout the Cenozoic, the basin was a marine seaway that was continually being filled with clastics from the Sierras. This situation allowed the shorelines to move back and forth near the present-day mountain ranges. The deep-marine environments lay a short distance west of the present valley axis. The climax of the marine sedimentation phase occurred in the late Pliocene, after which the region was ultimately filled in by fluviolacustrine sediments. The generalized stratigraphic column for both sides of the San Joaquin basin is shown in Figure 37.3.

The major feature of the southern part of the valley

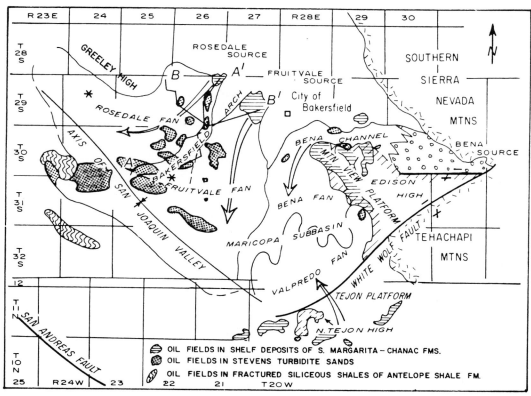

Figure 37.2 Map of the important turbidite fans in the southern San Joaquin Valley, California. The Rosedale–Fruitvale fan complex includes the Stevens Sandstone. (After MacPherson 1978, p. 2245.)

is the Bakersfield arch, or Kern arch, a broad subsurface structure that plunges west from the community of Bakersfield toward the center of the valley.

Four major turbidite cycles occurred in the late Miocene on the Bakersfield arch, and these compose what is called the Stevens Sandstone (see Figure 37.4).

The source area for the fan sediments are thought to be two submarine canyons that cut through the nearshore shelf and allowed the turbidites to funnel down and spread out as sea floor fans. As each sand influx occurred, fan built out away from the mouth of its canyon, and then the next influx would move out over the first and be deposited farther into the deep part of the basin. Shales that can be correlated over a wide area indicate that at times there was little or no sand supply. At the deepest end of these fans, siliceous shales occur (Figure 37.4).

The Stevens Sandstone is a group name used to include a total thickness of over 4900 ft (1500 m) of marine sandstone and shale. This whole sequence reflects several cycles of turbidite deposition. In general, these sands range from fine to coarse grained and show "fining" upward in some areas near the source. An example of the coarse arkosic sand is shown in Figure 37.5. More distant areas from the source show coarsening of the grainsize upward (MacPherson 1978, p. 2248).

The Elk Hills field appears to have been in what MacPherson (1978, p. 2264) calls the distal-margin facies. In this environment, diatom blooms flourished in areas where upwelling currents provided a continuous supply of nutrients. According to MacPherson (1978, p. 2266), the areas of upwelling also are preferred areas of turbidite deposition on the outer continental shelves, where gravity flows supply abundant clay minerals. At Elk Hills, local flexures have created open oil-filled fracture systems updip from the sandstone reservoirs.

The most striking feature of the Stevens Sandstone is the secondary porosity caused by the dissolution of the feldspars. The thin-section photomicrographs in Figures 37.6, 37.7, and 37.8 show this phenomenon quite clearly. In this respect, the

Epoch	San Joaquin Valley			
	West side–Central		East Side	
Pleistocene	Tulare	■	Kern River	■
Upper Pliocene	San Joaquin	▨	Kern River	■
Lower Pliocene	Etchegoin	▨	Etchegoin	■
Mio-Pliocene	Beef Ridge	■	Chanac	■
Upper Miocene	Antelope–McLure Stevens McDonald	■	Santa Margarita Fruitvale	■
Middle Miocene	McDonald Devilwater Gould Button Beds	■	Round Mountain Olcese	■
Lower Miocene	Media Carneros Santos–Vaqueros Phacoides Salt Creek	■	Olcese Freeman–Jewitt Pyramid Hill Vedder Walker	■
Oligocene	Tumey–Wagonwheel Oceanic	■	Tumey Walker	■
Eocene	Kreyenhagen Point of Rocks Canoas Domengine–Avenal	■	Walker–Famoso	■
Paleocene	Martinez	■	Absent	
Upper Cretaceous	Moreno	■	Absent	
	Panoche	▨		
Lower Cretaceous	Part of Panoche and other units		Absent	
Jurassic and older	Basement complex		Schist Granite	■

■ Denotes oil production   ▨ Denotes dry gas production

Figure 37.3 Stratigraphic column for the east and west sides of the San Joaquin basin. ■, Oil production; ▨, dry gas production. (After Landes 1970, p. 481.)

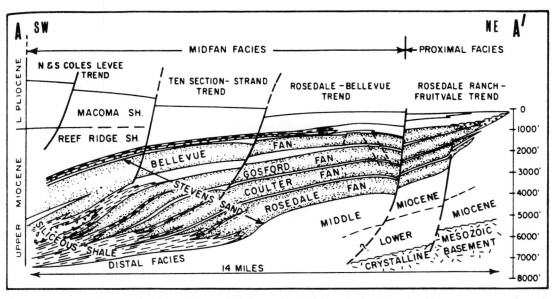

Figure 37.4 Cross section along the axis of the Bakersfield arch showing the Upper Miocene fans that comprise the Stevens Sand, Kern County, California. (After MacPherson 1978, p. 2247.)

Figure 37.5 Stevens Sandstone, 8765 ft (2672.3 m), Chevron No. 335-7R, 7R pool, West Elk Hills field, Kern County, California (0.44 in. in the photo equals 0.17 mm). Coarse arkosic sandstone with large unweathered feldspar grains (note the cross-hatched microcline). Some leaching and replacement with kaolinite seems to have occurred (grain at lower left corner). Also some fracturing across both the feldspar and quartz grains has occurred.

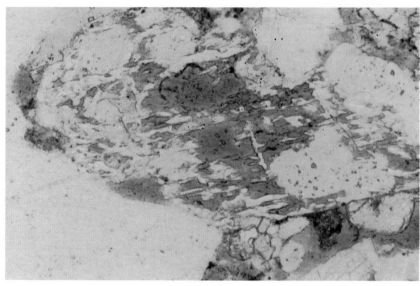

Figure 37.6 Stevens Sandstone, 8765 ft (2672.3 m), Chevron No. 335-7R, 7R pool, West Elk Hills field, Kern County, California (0.44 in. in the photo equals 0.17 mm). Specimen has been impregnated with blue plastic and shows a classic example of secondary porosity caused by the dissolution of feldspar. In this case, the secondary porosity is interconnected by other porosity types.

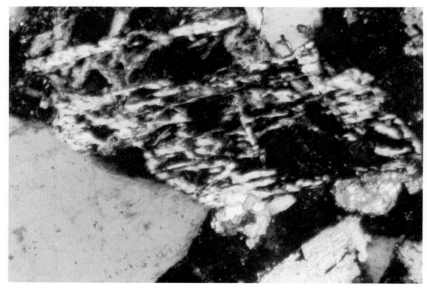

Figure 37.7 Stevens Sandstone, 8765 ft (2672.3 m), Chevron No. 335-7R, 7R pool, West Elk Hills field, Kern County, California (0.44 in. in the photo equals 0.17 mm). The same view as in Figure 37.6; however, it is taken with crossed polarizers so that the remaining unleached mineral shows up more clearly.

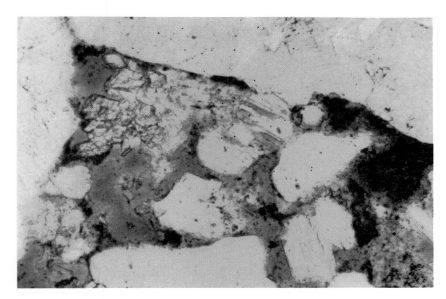

Figure 37.8 Stevens Sandstone, 8765 ft (2672.3 m), Chevron No. 335-7R, 7R pool, West Elk Hills field, Kern County, California (0.44 in. in the photo equals 0.17 mm). Another view showing various types of porosity. Secondary porosity of feldspar (center) has resulted from incipient stages of dissolution. The primary porosity appears to have been maintained with only a few dolomite grains filling in some of the pore space.

Stevens Sandstone resembles the Frio sands of the Gulf Coast, which also exhibit feldspar dissolution and secondary porosity as a result. It is also interesting to note that the Stevens Sand shows structural similarities with growth faults occurring at various intervals as shown in Figure 37.3. It is conceivable that such faults provide the pathway from the basement for the solutions that remove the feldspars. Shanmugam (1985, p. 379) indicates that the chemical breakdown of feldspars, particularly plagioclase, occurs because of acid attack by CO_2-charged waters. This may have been the case with the Stevens sandstone.

Recent Equivalents

The California coast provides a number of examples of submarine fans that continue to be studied as keys to the turbidite fans present in the San Joaquin Valley and elsewhere. The work by Normark (1978) in particular has focused on improving our understanding of submarine fans from the Recent of California.

Although the offshore fans of California are as close to the onshore oil reservoirs as possible in space and time, other types of turbidites are also being studied. Weirich (1984, p. 384) has worked with the occurrence, flow pattern, and internal dynamics of both surge and continuous turbidity currents as they occur in a glacial lake in southeastern British Columbia. It appears likely that more studies of this type will continue to be made and that this whole subject will be the subject of increasingly active research.

Walker (1978, p. 946) provides an excellent diagram of the many subenvironments encountered in the formation of submarine fans as shown in Figure 37.9. The identification and separation of the sediment types will remain a geologic challenge for some time to come.

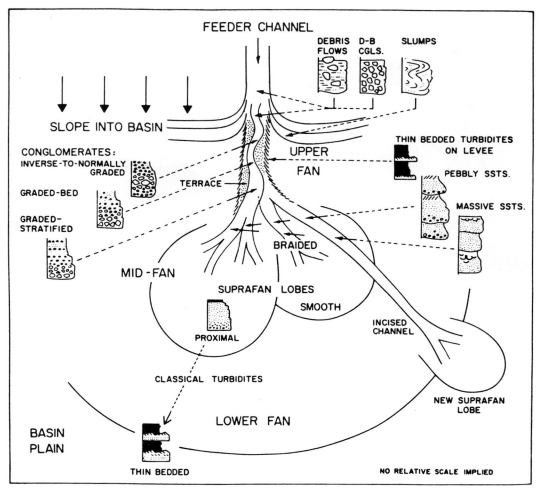

Figure 37.9 Model of submarine-fan deposition showing facies, fan morphology, and environment of deposition. (After Walker 1978, p. 946.)

References

Landes, K. K., 1970, *Petroleum Geology of the United States,* Wiley-Interscience, New York, 571. p.

MacPherson, B. A., 1978, Sedimentation and Trapping Mechanisms in Upper Miocene Turbidite Fans of Southeastern San Joaquin Valley, *American Association of Petroleum Geologists Bulletin,* Vol. 62, pp. 2243–2274.

Normark, W. R., 1978, Fan Valleys, Channels, and Depositional Lobes on Modern Submarine Fans: Characters for Recognition of Sandy Turbidite Environments, *American Association of Petroleum Geologists Bulletin,* Vol. 62, No. 6, pp. 912–931.

Perrodon, A., 1983, *Dynamics of Oil and Gas Accumulations,* Memoir 5, Elf Aquitaine, Pau, France, 368 pp.

Shanmugam, G., 1985, Significance of Secondary Porosity in Interpreting Sandstone Composition, *American Association of Petroleum Geologists Bulletin,* Vol. 69, No. 3, pp. 378–384.

Walker, R. G., 1978, Deep-Water Sandstone Facies and Ancient Submarine Fans: Models for Exploration for Stratigraphic Traps, *American Association of Petroleum Geologists Bulletin,* Vol. 62, No. 2, pp. 932–966.

Weirich, F. H., 1984, Turbidity Currents: Monitoring Their Occurrence and Movement with a Three-Dimensional Sensor Network, *Science,* Vol. 224, No. 4647, pp. 384–387.

38

Guijarral Hills Field, Gatchell–McAdams Sandstone, Fresno County, California

The Gatchell–McAdams Sandstone as it occurs in the Guijarral Hills field represents a permeability change type of oil accumulation. This field is interesting because it many instances electric log techniques do not distinguish between the "plugged," or impermeable, zone in the sand and the more permeable oil-productive zone. The reasons for this will become apparent as the mineralogic composition is discussed.

Geologic Background

The Gatchell Sandstone is middle Eocene in age and lies between below a carbonaceous siltstone and above another siltstone, both of which are middle Eocene in age. The cumulative production from the Gatchell is approximately 500×10^6 barrels of oil and 340×10^9 ft^3 of gas. Most of this production has come from three fields, and these are Coalinga East Extension, Pleasant Valley, and Guijarral Hills (Figure 38.1). The latter is the focus of this work; however, both the Pleasant Valley field and the Guijarral Hills field are believed to have similar traps. In discussing the Pleasant Valley field, Weddle (1951, p. 623) indicates that the upper 200 ft or so of Gatchell Sand is filled with interstitial white kaolinite "to the extent that this part of the sand forms an impermeable

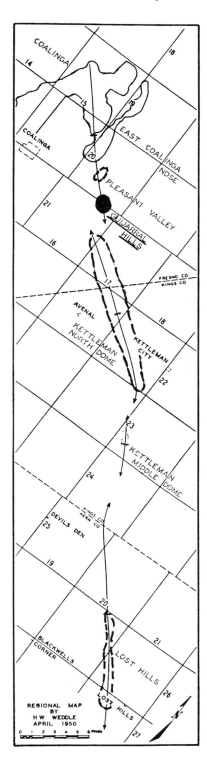

Figure 38.1 Map showing the location of the Guijarral Hills field in relation to the Coalinga–Kettleman Hills line of folding. (After Weddle 1951, p. 620.)

barrier to the upward migration of oil and gas" (Figure 38.2). The upper part of the sand is water wet and kaolinite filled, exhibiting permeabilities ranging from 0 to as high as 50 millidarcies. However, permeabilities in the oil-saturated reservoir rock below range up to 400 millidarcies. The photomicrographs in Figures 38.3–38.5 are from the clean reservoir sand in the lower part of the Gatchell Sand and show relatively little of the authigenic kaolinite.

Mineral Composition

The fact that the oil in the Guijarral field is trapped in the bottom of the sand is a direct reflection of the mineral composition. According to Schneeflock (1978, p. 850), electric logs show no continuous shale break between the lower oil-productive sand and the upper water-wet section. In addition, sonic logs indicate that the sand is "almost uniform in porosity throughout."

Electric logs do not reveal the difference between the upper and lower zones because the ion exchange potential of kaolinite produces almost the same spontaneous potential and resistivity responses as a clean wet sand. Sonic logs, on the other hand, give a response to the upper kaolinite-plugged sand zone that resembles "shale porosity."

Work by Merino (1975) indicates that the kaolinite crystallized in the pore space after the sand was deposited, According to Schneeflock (1978, p. 852), the kaolinite in the Gatchell is an alteration product of formation water, feldspars, and micas. In the case of the oil reservoir sand, the formation waters were displaced at an early stage by the oil accumulation; therefore, the kaolinization did not occur. Apparently, oil-saturated Gatchell sands with permeabilities in the 50–75-millidarcy range resulted from oil accumulation after some kaolinization had taken place.

Figures 38.3–38.5 are from reasonably good Gatchell reservoir sands where large amounts of kaolinite are not likely to have formed.

Recent Equivalents

The problem of locating Recent sands that have an updip equivalent of kaolinite plugging the pores is not

Recent Equivalents

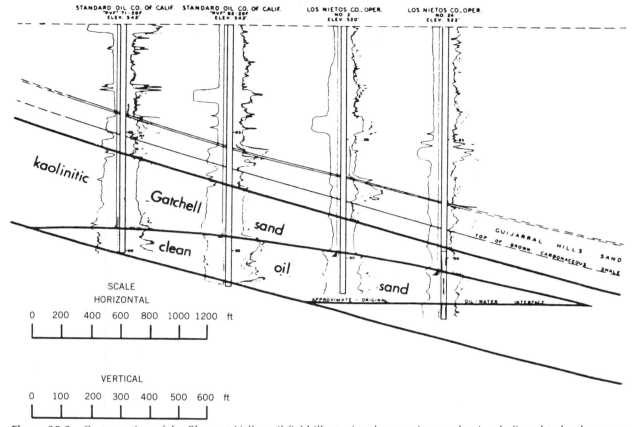

Figure 38.2 Cross section of the Pleasant Valley oil field illustrating the trapping mechanism believed to be the same as that for the Guijarral Hills field. (After Schneeflock 1978, p. 850.)

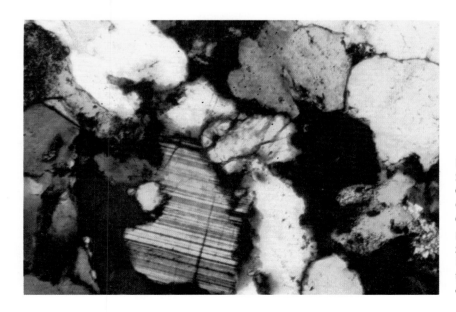

Figure 38.3 Gatchell–McAdams Sandstone, 10,340 ft (3,152.4 m), Chevron Bourdiew No. 1A–1H, Guijarral Hills field, Fresno County, California (0.44 in. in the photo equals 0.17 mm). Fresh plagioclase feldspar with fractures. Blue plastic has been injected into the pore space in order to get some concept of the morphology.

Figure 38.4 Gatchell–McAdams Sandstone, 10,340 ft (3,152.4 m), Chevron Bourdiew No. 1A–1H, Guijarral Hills field, Fresno County, California (0.44 in. in the photo equals 0.17 mm). Porous sand that has been impregnated with blue plastic to show the nature and shape of the pore space. Note the angular nature of the quartz grains.

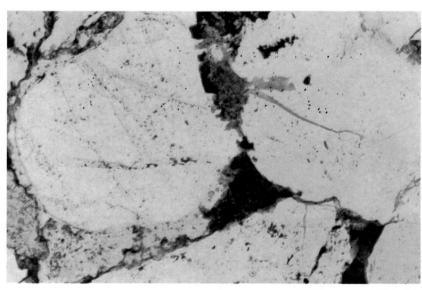

Figure 38.5 Gatchell–McAdams Sandstone, 10,340 ft (3,152.4 m), Chevron Bourdiew No. 1A–1H, Guijarral Hills field, Fresno County, California (0.44 in. in the photo equals 0.17 mm). Large quartz grains with crenulated contacts as well as quartz overgrowths show some dissolution effects. Blue plastic has been impregnated into the pore space; however, the lighter blue color (center, upper half) indicates some authigenic kaolinite is present.

likely to be located; however, the chemistry of why and how feldspars are changed to kaolinite is well documented and, in time, more research will be forthcoming on this topic as it applies to petroleum reservoirs. The subject of authigenic clays and their occurrence was discussed by Wilson and Pittman (1977, p. 27), who observed the following:

Our petrographic observations indicate most fluvial channel sandstones as well as most turbidite, beach, dune and tidal sands contain little, if any, detrital clay (excluding bioturbation).

On the other hand, sandstones deposited in any depositional environment may contain significant amounts of authigenic clay, regardless of compositional maturity.

In the case of the Guijarral Hills field reservoir sand, emplacement of the oil at an early stage stopped the formation of kaolinite.

References

Merino, E., 1975, Diagenesis in Tertiary Sandstones from Kettleman North Dome, California, I. Diagenetic Mineralogy, *Journal of Sedimentary Petrology,* Vol. 45, No. 1, pp. 320–336.

Schneeflock, R., 1978, Permeability Traps in Gatchell (Eocene) Sand of California, *American Association of Petroleum Geologists Bulletin,* Vol. 62, No. 5, pp. 848–853.

Weddle, H. W., 1951, Pleasant Valley Oil Field, Fresno County, California, *American Association of Petroleum Geologists Bulletin,* Vol. 35, No. 3, pp. 619–623.

Wilson, M. D. and E. D. Pittman, 1977, Authigenic Clays in Sandstones: Recognition and Influence on Reservoir Properties and Paleoenvironmental Analysis, *Journal of Sedimentary Petrology,* Vol. 47, No. 1, pp. 3–31.

39

Kettleman Hills Field, Gatchell—McAdams Sandstone, Kings and Fresno Counties, California

The Kettleman Hills field provides an interesting example of a reservoir sand in which numerous diagenetic minerals have been identified. According to Merino (1975, p. 320), at least 15 such minerals are present in the sandstones of the Temblor Formation and the McAdams sandstone. The samples discussed in this chapter are from the Gatchell–McAdams Sandstone where the diagenetic variability is not as pronounced as in the Temblor Formation; however, some unique features are of considerable interest.

Geologic Background

The Kettleman Hills field includes an area about 30 miles long and 4 miles wide and lies along the west side of the San Joaquin Valley, roughly 75 miles northwest of Bakersfield and 10 miles east of Reef Ridge (see Figure 37.1). The Kettleman Hills are elongate, slightly offset anticlines that trend northwest–southeast as shown in Figure 39.1. In 1938, oil was discovered in the McAdams Sandstone of Eocene age and is entrapped by thick shales known as the McLure and Kreyenhagon formations.

Geologic Background

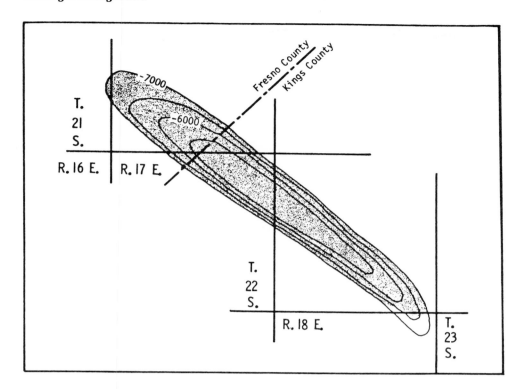

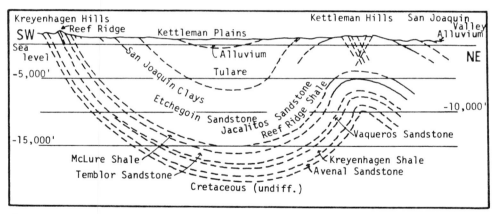

Figure 39.1 *Upper:* Structure map of North Dome and Kettleman Hills field, Kings and Fresno counties, California. *Lower:* Southwest–northeast cross section for Kreyenhagen Hills through the Kettleman Hills anticline. (After Landes 1970, p. 484.)

The sediments are alternating sandstones, siltstones, and shales that were deposited in varying environments and depths of water. MacPherson (1978, p. 2263) interprets the sequence at Kettleman N. dome as a series of turbidite deposits wedging out at the basin margin (Figure 39.2).

Merino's estimate of the mineral composition of the McAdams Sandstone is provided in Table 39.1. It will be noted that the bottom half of the lower McAdams Formation tends to have more quartz and fewer rock fragments than the upper McAdams, and this includes both the *lithic fragments* and the *polycrystalline quartzose grains*.

Merino (1975, p. 325) noted that calcite occurs in the McAdams and in some places includes dolomite rhombs. As the photomicrographs in Figures 39.3–39.6 illustrate, at least in one instance, fluids returned after the dolomite rhombs had formed and dissolution

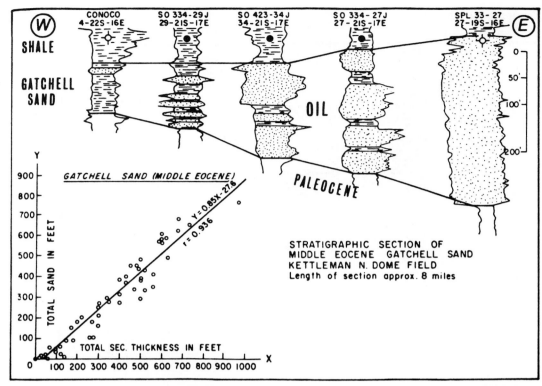

Figure 39.2 Typical electrical log cross section of Kettleman North Dome field showing basin–margin wedging of the Gatchell turbidite. (After MacPherson 1978, p. 2263.)

TABLE 39.1 Generalized Mineral Composition of the McAdams Sandstones at Kettleman North Dome (%) (Merino 1975, p. 323)[a]

	Matrix	Quartz	Feldspar	Lithic Fragments	Polycrystalline Quartzose Grains	Plagioclase	Other
Upper McAdams	≈5	<85	3–4	4	Trace to 5%	0–3	1–3% quartz overgrowths, glauconite, biotite, zircon, kaolinite, pyrite, sporadic calcite ± dolomite
Lower McAdams	0	88	≈10	2	2–4	2–5	Quartz overgrowths, kaolinite, iron oxide, pyrite, biotite
Sandy micritic biosparites with dolomite glauconite, biotite, and foraminifera							
	0	89	≈5	1	0	2–3	Quartz overgrowths, glauconite, pyrite, mica, calcite, dolomite, biotite

[a] Reprinted with permission of *Journal of Sedimentary Petrology*, Vol. 45, No. 1, pp. 320–336.

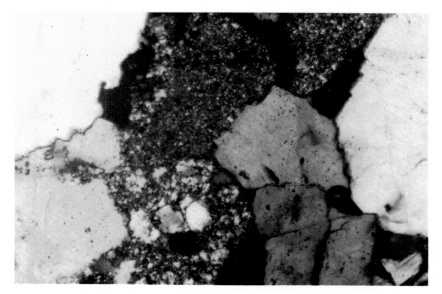

Figure 39.3 Gatchell–McAdams Sandstone, 9811 ft (2991.2 m), Chevron No. E423-34J, Kettleman Hills field, Kings and Fresno counties, California (0.44 in. in the photo equals 0.17 mm). Various types of chert grains and megaquartz grains are cemented together at the contact areas. In some areas, crenulated boundaries occur, suggesting considerable pressure. However, pore spaces (black) have remained remarkably clear from fine-grained debris.

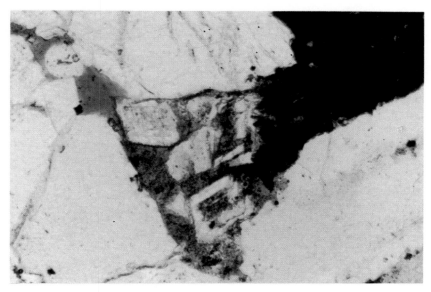

Figure 39.4 Gatchell–McAdams Sandstone, 9811 ft (2991.2 m), Chevron No. E423-34J, Kettleman Hills field, Kings and Fresno counties, California (0.44 in. in the photo equals 0.17 mm). Blue plastic impregnated specimen shows dolomite crystals that have grown in the pore space. In at least some cases, circulating solutions have dissolved the center zone of the zoned dolomite crystals, leaving a diamond-shaped hollow shell.

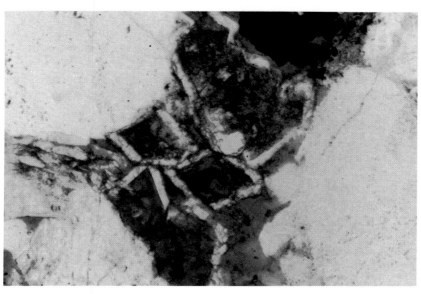

Figure 39.5 Gatchell–McAdams Sandstone, 9811 ft (2991.2 m), Chevron No. E423-34J, Kettleman Hills field, Kings and Fresno counties, California (0.44 in. in the photo equals 0.17 mm). As in Figure 39.4, dolomite crystals that initially grew into the pore space have had the center zones removed by dissolution, leaving only the outer less soluble zone behind. It appears that the remaining hollow shells have been crushed to some extent by increasing rock pressure.

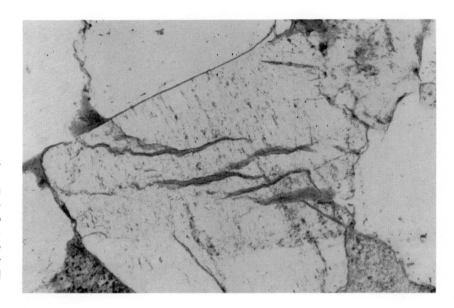

Figure 39.6 Gatchell–McAdams Sandstone, 9811 ft (2991.2 m), Chevron No. E423-34J, Kettleman Hills field, Kings and Fresno counties, California (0.44 in. in the photo equals 0.17 mm). Blue plastic injected into the pore space reveals the open fractures within a large feldspar grain. Quartz overgrowths have filled in some of the remaining pore space.

of the centers of the zoned rhombs occurred, leaving open network. This latter fluid emplacement phase appears to have enhanced the permeability and porosity of this zone significantly.

Merino (1975, p. 324) also notes that numerous quartz and plagioclase grains in the Temblor Formation are fractured, and he attributes these to the forces of compaction. This same kind of fracture occurs in the McAdams as illustrated in Figure 39.6; however, as Figure 39.6 shows, the blue plastic has filled the fractures and, therefore, there has been no recementation in this particular specimen.

Discussion

The McAdams Sandstone as illustrated in Figure 39.3 shows considerable quartz cementation, which in many cases takes the form of sutured boundaries. The McAdams also has experienced dolomite-forming solutions that left zoned crystals. At a later time, the solutions apparently became more acidic and were able to remove the center zones from the dolomite crystals, thereby creating a loose dolomite framework in some of the pore spaces. This seems to have enhanced the porosity considerably; however, the overburden pressures were enough to collapse some of the dolomite framework.

The across-the-grain fractures have remained open with no evidence of cementing materials being deposited. In brief, many of the diagenetic minerals deposited in the Temblor Formation are not observed in the McAdams. It is possible that the solutions that came along later and removed the center zones from the dolomite crystals were also capable of removing some of the diagenetic minerals present in the Temblor.

Recent Sediment Equivalents

If MacPherson's idea is correct, we are again dealing with a basin-margin wedging turbidite. As mentioned in previous discussions of the Stevens sandstones in the Elk Hills field the offshore areas of southern California have a number of examples where submarine fans and turbidites occur in the submarine valleys.

From the explorationist's point of view, the deposition of numerous diagenetic minerals in the pore space is a problem. The fact that the McAdams shows evidence of having a number of these materials removed suggests that where this dissolution has taken place, reservoir conditions are more favorable for both primary and secondary recovery purposes.

Crude Oil Maturation and Source Rocks

Initially, it appeared that the crude-oil analyses from the California fields carried out by the Bureau of

Crude Oil Maturation and Source Rocks

Mines would not provide enough information to be worthy of a special discussion; however, examination of the samples and their overall origins stimulated additional work. In Figures 39.7–39.9, it becomes clear that we are dealing with a sequence of three crude oils that starts with a very young, shallow oil in the Pleistocene–Pliocene Tulare Formation from the McKittrick field and progresses to the deeper and older oil illustrated by the Miocene Stevens sands from Elk Hills.

Specifically, the correlation index values for the shallow Pleistocene–Pliocene oil are quite high, in the range 42–52, indicating a fairly high degree of aromaticity that begins to provide an appreciable amount of distilled product at a distillation temperature 200°C (Figure 39-7).

The next oil in the sequence is from the Etchegoin Sand (1810–2368 ft) of Pliocene age in the Lost Hills field where a diatomite is the source material. It will be noted that the distillation fractions begin at 100°C, and the correlation index is within the 28–41 range, indicating a more paraffinic character than that from the Tulare (Figure 39-8).

Lastly, the crude from Stevens sand in the Elk Hills field (Miocene) is much more paraffinic and shows distillation fractions beginning at 50°C with correlation index values extending from 10 to 36 (Figure 39-9). All of the above suggest that time and intensity (temperature) have the major influence on the crude oil so that other factors such as source materials may be masked. In this instance, all three oils appear to have diatomite as a likely source material. If one examines the API gravities for the shallow sands (less than 2000 ft) and the deep sands from deeper than 10,000 ft in California (see Figure 39.10), it becomes strikingly clear that there are major differences in these two classes of crude oils (Biederman 1965, p. 79). Furthermore, if we compare the individual components of these relatively young oils with those from older strata in other parts of the country, differences

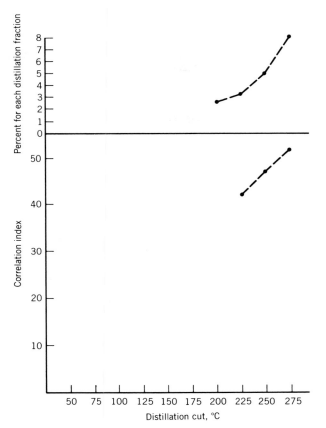

Figure 39.7 Correlation index and distillation fractions for crude oil from McKittrick field, Tulare Formation, Pleistocene–Pliocene, 1000–1050 ft, Kern County, California. (Data from McKinney et al. 1966, p. 57.)

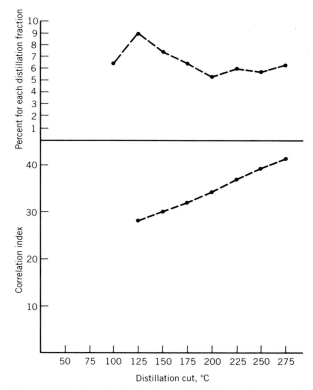

Figure 39.8 Correlation index and distillation fractions for crude oil from Lost Hills field, Etchegoin Sand, 1810–2368 ft, Pliocene, Kern County, California. (Data from McKinney and Garton 1957, p. 52).

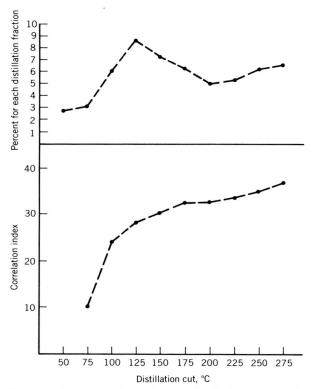

Figure 39.9 Correlation index and distillation fractions for crude oil from Elk Hills field, Stevens Sand, 5199–6379 ft, Upper Miocene, Kern County, California. (Data from McKinney and Garton 1957, p. 45.)

are observed in the amounts of components such as n-pentane, n-butane, n-hexane, and n-heptane (Biederman 1965, p. 80). In general, as Tissot and Welte (1978, p. 173) have observed, at greater depths cracking becomes important and the distribution of hydrocarbon types "becomes more and more dominated by lighter molecules in all types of sediments.

In recent years, the subject of petroleum maturation and its subsequent migration has received considerable attention. The general sequence of events is discussed by Hunt (1984) and is summarized briefly as follows: Biological production of methane begins in near-surface sediments and extends down to a depth of 300 m. This can occur in both freshwater lake and marine deposits. Some C_{2+} light hydrocarbons are also present in Recent near-surface sediments in trace amounts (parts per billion) and appear to be the result of both biological and low-temperature (<50°C) chemical reactions. These include such low-molecular-weight species as ethane, ethylene, propane, propylene, butane, pentane, butenes, pentenes, hexenes, and toluene.

With increasing depth, important changes begin to take place when the subsurface temperature reaches the 100–110°C range. Thermal cracking of the organic matter occurs, which produces hydrocarbons of the type found in most petroleums. In this zone, hydrocarbons occur in fine-grained shale sediments in the parts per thousand range. The bulk of the hydrocarbon generation probably takes places within a subsurface temperature range of 100–150°C. More straight-chain hydrocarbon species are formed at the high-temperature end of this range, which shows as increased paraffinicity in the correlation index curves in Figures 39.7–39.9. Below the main oil-generating zone and with increasing temperature up to 200°C, methane again becomes the major hydrocarbon species generated.

The exploration implications of this sequence are important because although we have drilling technology capable of drilling to a depth of 15 km (Hunt 1984, p. 1269), the deepest commercial gas well to date is the Chevron No. 1 Ledbetter in Wheeler County, Texas, which produces from strata at 8 km. In the United States during the last 5 years, about 400 wells have been drilled below 6.1 km, and none have found commercial quantities deeper than 8 km. The suggestion is that odds of encountering significant gas discoveries below this level decrease sharply.

Hunt (1984, p. 1269) indicates several additional implications concerning petroleum migration. Methane and ethane apparently migrate vertically over tens of meters through fine-grained shales by means of a transport mechanism that is thought to involve a combination of diffusion and solution. In contrast to this mechanism, C_{3+} hydrocarbons do not show evidence of significant vertical migration through fine-grained shales. As evidenced in a number of places in this book, hydrocarbons obviously do migrate from the source rocks to the reservoir rocks in many sedimentary basins. As Hunt (1984, p. 1269) states, "This suggests that the migration pathways are not through the fine-grained rock but are along more permeable avenues such as faults, fractures, bedding planes, unconformities, sheet sands, and other large pore openings that permit migration as an oil or gas phase." This observation coincides with this author's views concerning the mechanism by which of the oil and gas reached the reservoirs discussed in this volume.

At this point, some concluding thoughts concerning how this geochemical data relate to the subsiding basin. In keeping with the work of Ejedawe and

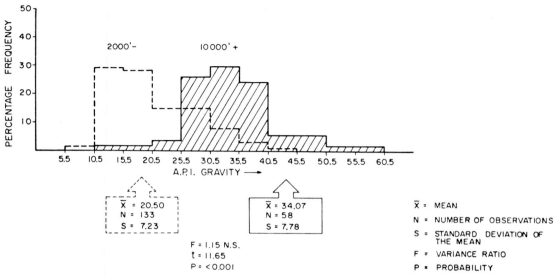

Figure 39.10 Distribution of API gravities of California crude oils of Cretaceous age and younger.

Coker (1984), it is appropriate to consider what happens to the *oil-generative window* through the basinal subsidence and postsubsidence phases. As these authors point out, the dynamics of sediment deposition and the resultant downwarping cause the oil-generative window to bend downward in times of rapid subsidence. In time, the limits of oil generation recover upward. The postsubsidence phase allows for the geothermal gradients to even out and the oil-generative window to spread to its maximum extent as time now becomes an important variable. Thus, old basins that have suffered only slight to moderate tectonic wrenching have the possibility of long-range migration along fracture patterns, which can result in very shallow production, such as that associated with the Rodney field (Chapter 6). This can occur even though the surrounding sediments provide little evidence of ever being elevated to even moderate temperatures.

With the dynamic model in mind, it becomes much easier to envision the reason there is such variation in the API gravities from field to field along the south Texas Gulf Coast. The depositional centers associated with deltas that have persisted through time produce their own downwarps superimposed on the regional setting. These features have their own fault block patterns that provide channels for oil and gas migration from deeper sources at irregular time intervals. These faults also extend laterally into the zones between the deposition centers and enable hydrocarbons to reach the clean sand bodies associated with the between-delta areas as well as the updip, nonmarine, river-derived channel sands.

In summary, the dynamic model proposed by Ejedawe and Coker (1984) appears to come closer to what actually takes place.

References

Biederman, E. W., Jr., 1965, Crude Oil Composition—a Clue to Migration, *World Oil*, Vol. 161, No. 7, pp. 78–82.

Ejedawe, J. E. and S. J. L. Coker, 1984, Dynamic Interpretation of Organic Matter Maturation and Evolution of Oil-Generative Window *American Association of Petroleum Geologists Bulletin*, Vol. 68, No. 8, pp. 1024–1028.

Hunt, J. M., 1984, Generation and Migration of Light Hydrocarbons, *Science*, Vol. 226, No. 4680, pp. 1265–1270.

Landes, K. K., 1970, *Petroleum Geology of the United States*, Wiley-Interscience, New York, 571 pp.

MacPherson, B. A., 1978, Sedimentation and Trapping Mechanism in Upper Miocene Stevens and Older Turbidite Fans of Southeastern San Jaoquin Valley, California, *American Association of Petroleum Geologists Bulletin*, Vol. 62, No. 11, pp. 2243–2274.

McKinney, C. M. and E. L. Garton, 1957, *Analyses of*

Crude Oils from 470 Important Oilfields in the United States, Bureau of Mines Report of Investigations 5376, 276 pp.

McKinney, C. M., E. P. Ferrero, and W. J. Wenger, 1966, *Analyses of Crude Oils from 546 Important Oilfields in the United States,* Bureau of Mines Report of Investigations 6819, 345 pp.

Merino, E., 1975, Diagenesis in Tertiary Sandstones from Kettleman North Dome, California. I. Diagenetic Mineralogy, *Journal of Sedimentary Petrology,* Vol. 45, No. 1, pp. 320–326.

Galloway, J. O., 1943, Kettleman Hills Oil Fields, in *California Division of Mines and Geology Bulletin,* Vol. 118, Pt. 3 Descriptions of Individual Oil and Gas Fields, p. 491 (Fig. 208).

Tissot, B. P. and D. H. Welte, 1978, *Petroleum Formation and Occurrence,* Springer-Verlag, Berlin, Heidelberg, and New York, 538 pp.

40

Thoughts and Conclusions

After completing such an odyssey involving hundreds of cores and over 5000 thin sections, it seems only fitting that an attempt be made to pull some of the information together and, if possible, to draw some conclusions concerning the conditions for oil accumulation. Initially, a few thoughts on the mineralogy and related structures and textures are in order.

It is increasingly clear that fractures can be very important to the formation of many oil fields. Specifically, fractures provide conduits leading from source rock shales to the more porous zones that form the reservoirs. This overall impression is illustrated by such great accumulation areas as the Central basin platform of west Texas where the basinal shales from both the Delaware basin and the northern Permian basin have supplied the porous zones surrounding the platform edge. Fractures would appear to be key to the emplacement of the oil. Similarly, fractures in the Spraberry are very important to the producibility of the individual wells in that field.

The fact that the Thornwood field in West Virginia produces any gas at all appears to be attributable to fractures that provide the reservoir space.

Moving to the more current areas of exploration, it is clear that the fractured chert of the Monterey Formation is key to the presence of producible oil and gas in the recent California offshore discoveries. The exploration strategy that led to this discovery and others in the same area is based on the search for fracture formation within the brittle chert that is high in organic matter and can act as its own source rock. For a more detailed discussion of fractures and how they affect various phases of oil and gas exploration and production, the reader is referred to the work of Aquilera (1980).

Looking at the mineralogic side, again it is clear that chert is an important constituent of a number of famous reservoir rocks. The whole Rocky Mountain belt ranging from Wyoming on up through Alberta and northwest to Prudhoe Bay is noteworthy for the high percentage of chert grains, both in conglomeratic phases as well as in sandstones. Even the Gulf Coast Frio in south Texas contains a significant amount of chert. Whether the chert has its origin in the siliceous beds associated with upwellings of the sea or whether there is a significant contribution from volcanic sources, the fact remains that chert is an important constituent of many reservoir rocks, particularly those in the western part of North America.

Another environment that looks as though it will become more important in the future is that of the turbidite fan. As our search moves out beyond the continental shelf, the turbidite fan will become an increasingly important target. It is also obvious that more work on Recent sediments, particularly turbidites, will probably pay off in terms of understanding the mechanisms and timing for sand deposition.

From the petroleum engineer's point of view, the problems encountered with tertiary recovery projects will require close examination of the actual core materials and the mineralogy involved.

The initial thoughts concerning the structure of this final chapter focused on producing a mathematical statistical summary; however, in the course of examining the actual samples and the data, it became clear that most of the comparisons that could be made would rely on very small samples of the whole population. Furthermore, it became obvious that lumping reservoir types together into classes and then seeking out statistically significant relationships would involve numerous assumptions, most of which are likely to be invalid.

The problem boils down to how can the reader of this volume best be served. The answer to this appears to involve bringing together some of the published estimates of future exploration and production potential for the various regions covered in this text. Specifically, the work of Dolton et al. (1981) from the U.S. Geological Survey is an appropriate place to begin a thoughtful review of the areas where the potential for new discoveries seems best. The overall volumes of sediment that are included in each region (Table 40.1) provide some feeling for the exploration probabilities, although places such as the Pacific Coast, particularly in the offshore, are obviously good places for future development. Part of the problem involves the depths of water in which production platforms or undersea collection systems can be maintained. The technologies are obviously moving the hunt for hydrocarbons into deeper waters, but questions concerning, for example, the maturity of the source materials and the limits imposed by the temperature should be reviewed before it is assumed that great quantities of reserves are available (see Figure 40.1).

Dolton et al. (1981, p. 30) have provided estimates of where the undiscovered resources of crude oil are likely to be, and it is interesting to note that the Colorado plateau and Basin and Range region is the most significant contributor. (See Figure 40.2). The illustration becomes more clear in Figure 40.3, where the ranges of the estimates as well as the mean values are provided. In dealing with the ranges, it should be observed that the low value corresponds to a 95% probability or more than that amount, and the high value corresponds to a 5% probability of more than the high figure.

According to Dolton et al. (1981, pp. 27, 28), the reason for the high potential of the Colorado plateau and Basin and Range Region (Figure 40.4) has to do with the gas presumed to be present in the Wyoming–Utah–Idaho overthrust belt. It is unfortunate that only one field from this region is covered in this book; on the other hand, the field that is discussed is an unusual one that should stimulate ideas about what might be the case in other areas where igneous sills are sandwiched into the sedimentary rock column.

From the exploration point of view, the worldwide comparison of the various oil provinces is valuable. Perrodon (1983, p. 320) has provided an interesting chart of exploration yield versus potential and intensity of exploration. It will be noted from Figure 40.5 that of the American provinces only west Texas, Oklahoma, and North Dakota are above the 100,000-m^3 line of oil discovered per wildcat. In terms of percentage of success, only west Texas, Wyoming, north Texas, and the Denver basin are above the 10% line.

The second of Perrodon's diagrams (1983, p. 322) (Figure 40.6) charts exploration success ratio versus yield and average discovery for various provinces on a worldwide basis. In these comparisons, only the Arctic slope would appear to have highly significant potential with the Cook inlet and offshore Louisiana some distance behind. The message that comes from this work is that the major U.S. basins have been more heavily explored than the remainder of the world.

TABLE 40.1 Estimates of Areas and Volumes of Sedimentary Rock by Region[a] (Dolton et al., 1981, p. 13.)

Region			Area		Volume	
			mi²	km²	mi³	km³
Onshore						
1	Alaska[b]		163	421	481	2,006
2	Pacific Coast		142	367	228	949
3	Colorado plateau—Basin and Range		414	1,072	958	3,994
4	Rocky Mountains—northern Great Plains		415	1,074	591	2,463
5	West Texas and eastern New Mexico		172	446	294	1,225
6	Gulf Coast		237	614	783	3,263
7	Mid-continent		401	1,039	332	1,382
8	Michigan basin		122	316	109	455
9	Eastern Interior		193	501	240	1,001
10	Appalachians		193	500	493	2,054
11	Atlantic Coast		154	399	141	586
	Total onshore		2,606	6,749	4,649	19,377
Offshore						
1A	Alaska	Shelf	527	1,364	657	2,740
		Slope	170	441	277	1,154
2A	Pacific Coast	Shelf	19	49	35	145
		Slope	64	165	78	327
6A	Gulf of Mexico	Shelf	124	32635	2,647	
		Slope	95	246	485	2,022
11A	Atlantic Coast	Shelf	106	274	245	1,020
		Slope	83	216	367	1,531
		Total shelf	776	2,010	1,572	6,552
		Total slope	412	1,067	1,208	5,034
		Total offshore	1,188	3,077	2,780	11,586
Onshore and offshore						
Total United States			3,794	9,826	7,429	30,963

[a] Area is in thousands of square miles (mi²) and square kilometers (km²). Volume is in thousands of cubic miles (mi³) and cubic kilometers (km³). All tabulated values are rounded and may not be precisely additive due to rounding.

[b] Areas and volumes of sedimentary rocks in parts of interior Alaska were excluded because distribution of such rocks is too poorly known to estimate at this time.

In other words, the exploration costs within the United States in general are not favorable as elsewhere. Although Perrodon's estimates may be disputed, it seems clear that efforts directed toward tertiary recovery techniques and methods for obtaining oil from tar sands and oil shales will not be wasted, if our country is to keep its in-house petroleum supplies at levels which are relatively safe from abrupt cutoffs. Meyerhoff (1985) pointed out that although the North Slope fields are huge by any standard, one must keep in mind that at the U.S. 1984 consumption rate of 15.7 million barrels a day, a field with 300 million

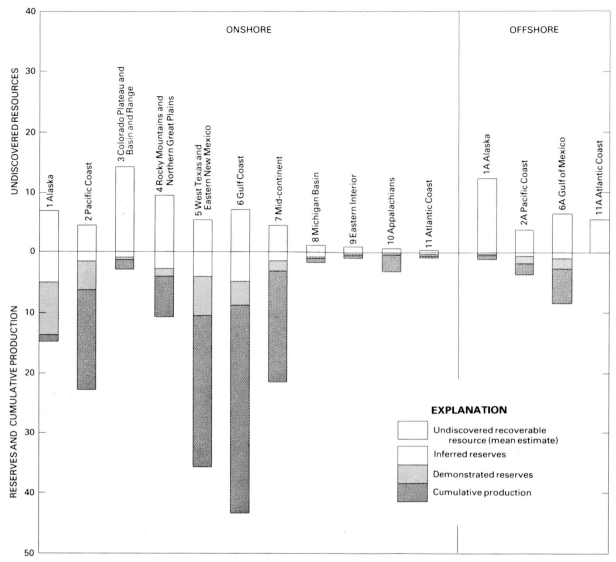

Figure 40.1 Graph of cumulative production and demonstrated reserves plus estimates of inferred reserves and as yet undiscovered amounts of crude oil in billions of barrels by region. (After Dolton et al. 1981, p. 30.)

Thoughts and Conclusions

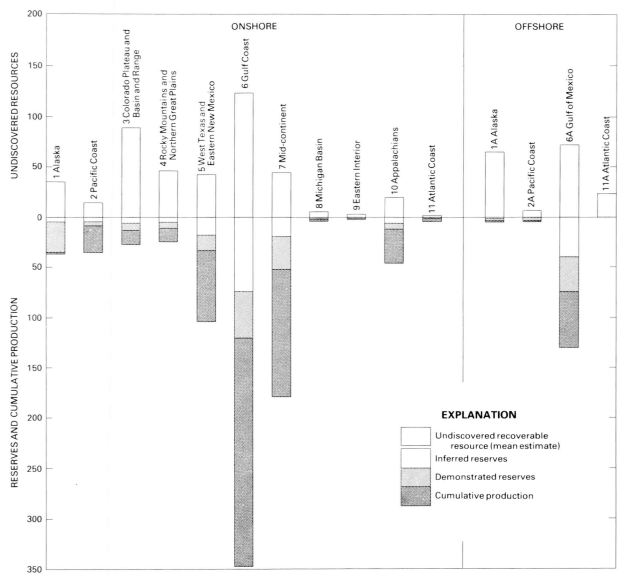

Figure 40.2 Graph of cumulative production and reserves plus estimates of inferred reserves and as yet undiscovered amounts of natural gas in trillions of cubic feet by region. (After Dolton et al. 1981, p. 31.)

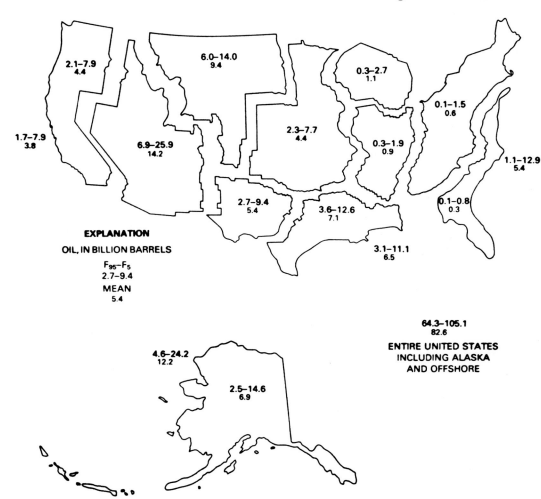

Figure 40.3 Estimates of undiscovered recoverable resources of crude oil showing ranges and means for each region. (After Dolton et al. 1981, p. 26.)

Thoughts and Conclusions

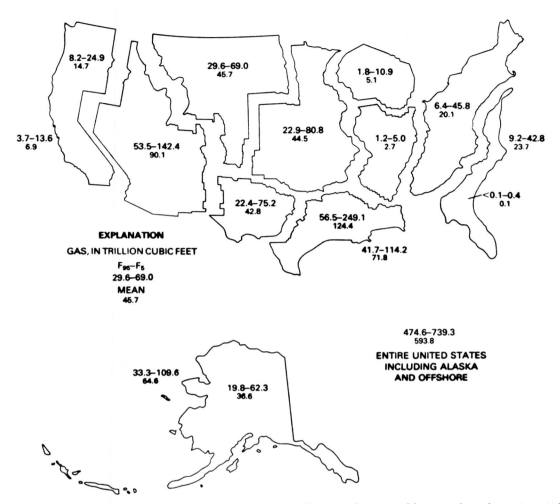

Figure 40.4 Estimates of the ranges and means of undiscovered recoverable natural gas by region. (After Dolton et al. 1981, p. 27.)

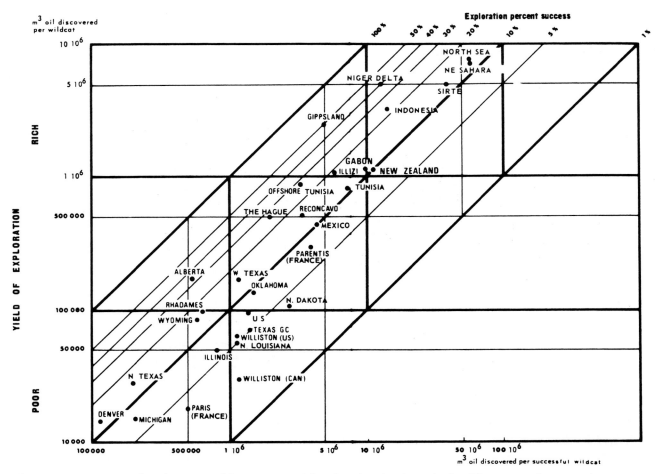

Figure 40.5 Diagram of exploration yield versus potential and exploration intensity for different petroleum provinces on a worldwide basis. (After Perrodon 1983, p. 320.)

References

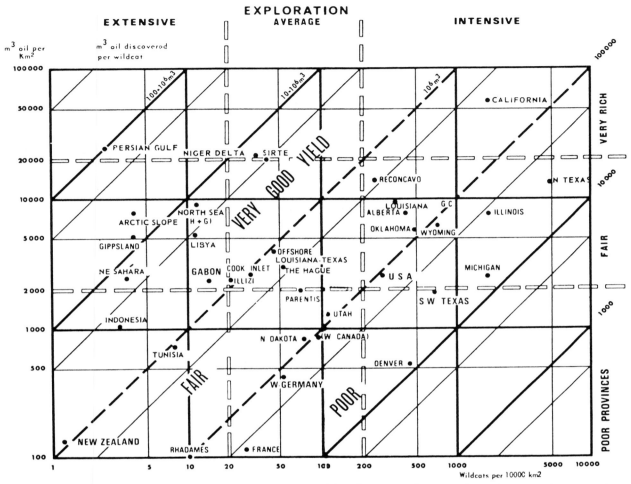

Figure 40.6 Diagram of exploration success ratio versus yield and average discovery for various petroleum provinces worldwide. (After Perrodon 1983, p. 322.)

barrels would last only 19 days and one with 8 billion barrels would last 510 days. We must continue to support a long-term view, even when it is much easier to assume that additional drilling will always be able to fill our needs.

References

Aguilera, R., 1980, *Naturally Fractured Reservoirs*, Petroleum Publishing Company, Tulsa, Oklahoma, 703 p.

Dolton, G. L., K. H. Carlson, R. R. Charpentier, A. B. Coury, R. A. Crovelli, S. E. Frezon, K. S. Khan, J. H. Lister, R. H. McMullin, R. S. Pike, R. B. Powers, E. W. Scott, and K. L. Varnes, 1981, Estimates of Undiscovered Recoverable Conventional Resources of Oil and Gas in the United States, *Geological Survey Circular 860*, 87 p.

Meyerhoff, A. A., 1985, Mergers, Shaky Prices Hit Oil and Gas in '84, *Geotimes*, Vol. 30, No. 3, pp. 14–16.

Perrodon, A., 1983, *Dynamics of Oil and Gas Accumulations*, Memoir 5, Elf Aquitaine, Pau, France, 368 p.

APPENDIX A

Stratigraphic Charts

The charts in this appendix are designed to provide the reader with some idea of where a particular reservoir fits into the overall geologic column. The numbers of the chapters covering specific reservoirs are placed opposite their approximate stratigraphic positions. For instance, Chapter 2 discusses the Bradford Third Sand reservoir in the Bradford field. The number 2 is found down the column labeled "Pennsylvania–West Virginia," opposite "Bradford 3rd" in the Devonian System in the chart covering the Paleozoic.

Similarly, charts are also provided covering reservoirs from the Mesozoic and the Cenozoic. The reservoirs from California are omitted from the Cenozoic chart because of the complexitiy of the nomenclature; however, they are more clearly defined in the chapters covering the West Coast. The reader may find much to disagree with in the forced correlations implied in the charts, and it should be recognized that the relationships shown are only approximate.

SYSTEM	SERIES	ARIZONA NEW MEXICO	WYOMING	WEST TEXAS	OKLAHOMA	KANSAS AND TEXAS PANHANDLE	NEBRASKA	ILLINOIS-KENTUCKY	MICHIGAN S.W. ONTARIO	PENNSYLVANIA W. VIRGINIA
PERMIAN	OCHOA	Ochoa		Ochoa						Dukard
PERMIAN	GUADALUPE	Guadalupe Tansill Yates 22 Seven Rivers Queen Grayburg San Andres	Embar Phosphoria 30	Tansill Yates Seven Rivers Queen Grayburg San Andres	Quartermaster White Horse El Reno Hennessey Garber Ramsey Fortuna	Taloga Day Creek White Horse Nippewalla Sumner (Red Cave)	Wellington Chase Council Grove Admire		Red Beds Woodville Ionia	
PERMIAN	LEONARD	Leonard Glorieta Yeso		Delaware Mt. Leonard Spraberry Clearfork Wichita Albany 21	Noble-Olson Wellington Chase Council Grove Admire	Chase (Brown Dol.) Council Grove Admire (Granite W.) 9,15				
PERMIAN	WOLFCAMP	Wolfcamp Cutler		Wolfcamp						
PENNSYLVANIAN	VIRGIL-CISCO	Rico	Tensleep 29 Minnelusa Amsden	Cisco	Cisco Wabaunsee Shawnee Douglas	Wabaunsee Shawnee Douglas	Wabaunsee Shawnee Douglas	McLeansboro	Saginaw Parma	
PENNSYLVANIAN	MISSOURI-CANYON			Canyon	Lansing Kansas City Cleveland	Lansing 14 Kansas City Pleasanton	Lansing Kansas City Pleasanton	Carbondale		Monongahela Conemaugh Allegheny Pottsville
PENNSYLVANIAN	DES-MOINES STRAWN	Hermosa 34 Paradox		Strawn	Strawn Marmaton Cherokee	Marmaton Cherokee	Marmaton 27 Cherokee			
PENNSYLVANIAN	ATOKA-BEND			Atoka-Bend	Atoka	Atoka				
PENNSYLVANIAN	MORROW	Molas			Morrow	Morrow				
MISSISSIPPIAN	CHESTER	Chester	Brazer	Chester Barnett	Chester	Chester		Pottsville Chester (Hardinsburg)		Mauch Chunk Mattox Greenbrier Big Lime
MISSISSIPPIAN	MERAMEC	Leadville Meramec	Madison	Osage, Chappel Woodford	Meramec	Meramec	Meramec	Valmeyer	Meramec	Keener Big Injun 4 Squaw Berea Papoose
MISSISSIPPIAN	OSAGE-KINDERHOOK	Osage			Osage Woodford Chattanooga Misener	Osage Kinderhook Chattanooga Misener	Osage Kinderhook	Kinderhook Hardin Chattanooga	Osage Kinderhook Antrim	
DEVONIAN		Ouray Devonian	Three Forks Darby Jefferson	Devonian				Beachwood Silver Creek Speeds Corniferous Dutch Creek	Traverse Dundee 6,8 Detroit River Sylvania	Conewango Conneaut Canadaway Bradford 3rd 2 Onondaga Oriskany 5
SILURIAN		Siluro-Devonian Fusselman	Laketown Big Horn Harding	Hunton Fusselman	Hunton	Hunton 12	Hunton	Silurian	Silurian-Niagaran	Newburg Medina Niagara Salina-Clinton
ORDOVICIAN		Montoya Simpson		Simpson	Viola Simpson 13 (Wilcox)	Viola 10 Simpson 11	Viola Simpson	Maquoketa Trenton St. Peter	Trenton Black River	Trenton
CAMBRO-ORDOVICIAN		Ellenburger		Ellenburger	Arbuckle	Arbuckle	Arbuckle	Knox	Prairie du Chien Knox	Beekmantown Knox
CAMBRIAN		Nickory	Deadwood	Wilberns Nickory	Reagan	Reagan 27	La Motte Deadwood	Potsdam	Mt. Simon	Gatesburg

PRECAMBRIAN

PALEOZOIC

System	Series	Stage	East Texas	Mississippi	Wyoming
Upper Cretaceous	Gulf Series	Montana	Nacatoch Annona Ozan	Ripley Selma	Lance, Fox Hills Lewis Masaverde Mancos Steele
		Colorado	Tokio Austin	Eutaw	Niobrara Carlile— Mowry Graneros Frontier 28 Muddy
			Eagle Ford		
			Tuscaloosa Woodbine	Tuscaloosa	— Dakota —
Lower Cretaceous	Comanchean Series	Dakota	Washita	Washita Fredericksburg Dantzler Cuevas	Lakota
		Washita			
		Fredericksburg	Fredericksburg		
		Trinity	Paluxy	Paluxy	
			Mooringsport, Ferry Lake Rodessa James 26 Pettet	Mooringsport, Ferry Lake Rodessa Sligo Hosston	
			Hosston Travis peak		
Jurassic			Cotton Valley Bodcaw Buckner Smackover Eagle Mills	Cotton Valley (Cadeville) 25 Smackover Eagle Mills	Morrison Sundance Twin Creek Nugget
Triassic					

Appendix A: Stratigraphic Charts

System	Series	Stage	South Texas	
Quarternary	Recent Pleistocene	Houston		
Tertiary	Pliocene	Citronelle	Goliad	
Tertiary	Miocene	Fleming	Oakville (Catahoula)	
Tertiary	Oligocene		Anahuac / Frio 24 / Hackberry (Catahoula)	
Tertiary			Vicksburg	
Tertiary	Eocene	Jackson	Jackson	
Tertiary	Eocene	Claiborne	Upper: Pettus, Yegua 23	
Tertiary	Eocene	Claiborne	Lower: Cook Mountain, Sparta, Mount Selman, Queen City	
Tertiary	Eocene	Claiborne	Carrizo	
Tertiary	Eocene	Wilcox Midway	Wilcox	
Tertiary	Eocene	Wilcox Midway	Wills Point, Kincaid	

APPENDIX B

Conversion Factors

To Convert From	To	Multiply By
Acres	ha	0.40469
Acres	m^2	4,046.9
Barrels (42 U.S. gal)	m^3	0.15899
Barrels/day	m^3/day	0.15899
Barrels/U.S. ton	m^3/tonne	0.17525
Barrels	gal	42
Btu	ft-lb	777.98
Btu/sec	W	1055
Btu/min	hp	0.0236
Btu/lb	cal/g	0.55556
cm	ft	0.032808
cm	in.	0.393700
cm	μm	10,000
cm$^-$	Å	1×10^8
cm^2	m^2	1×10^4
cm^2	in^2	0.1550
cm^2	ft^2	0.00108
cm^3	in.^{3}	0.061
cm^3	ft^3	3.531×10^{-5}
cm^3	ml	0.99997
cm^3	oz (U.S. fl.)	0.0338

Appendix B: Conversion Factors

To Convert From	To	Multiply By
cm^3	gal (U.S.)	2.6417×10^{-4}
cm	ft/sec	0.0328
cm/sec	ft/min	1.9685
cm/sec	m/min	0.06
cm/sec	mi/hr	0.02237
cm/sec	km/hr	0.03600
Erg	Btu	9.4805×10^{10}
ft	cm	30.48
ft	m	0.3048
H_2O (4°C)	$lb/in.^2$	0.4335
ft H_2O (4°C)	lb/ft^2	62.427
ft H_2O (4°C)	kg/m^2	304.79
ft^2	yd^2	0.111
ft^2	mi^2	3.587×10^{-8}
ft^2	acre	2.296×10^{-5}
ft^2	cm^2	929
ft^2	m^2	0.0929
ft^3	$in.^3$	1728
ft^3	yd^3	0.037
ft^3	cm^3	28,316
ft^3	m^3	0.02832
ft^3	l	28.316
ft^3	gal (U.S.)	7.4805
ft^3 H_2O	lb (4°C)	62.42
ft^3 H_2O	lb (15°C)	62.36
ft^3 min	l/sec	0.4720
ft^3/min	cm^3/sec	472
ft^3/min	gal/min	448.831
ft/min	ft/sec	0.01667
ft/min	mph	0.0114
ft/min	cm/sec	0.508
ft/min	m/sec	0.005
ft/min	m/min	0.3048
ft/min	km/hr	0.0183
ft/sec	cm/sec	30.48
ft/sec	m/min	18.288
ft/sec	km/hr	1.097
gal (U.S.)	gal (British)	0.8327
gal (U.S.)	oz	128
gal (U.S.)	cm^3	3785.4
gal (U.S.)	l	3.785
gal (U.S.)	m^3	0.00379

Appendix B: Conversion Factors

To Convert From	To	Multiply By
gal (U.S.)	in.^{3}	231
gal (U.S.)	ft^3	0.1337
gal (U.S.)	yd^3	0.00495
gal (U.S.)	lb H$_2$O (60°F)	8.3378
gal/min (U.S.)	ft^3/hr	8.0208
gal/min (U.S.)	l/sec	0.06308
g/l	ppm	1000
g/l	lb/ft^3	0.0624
Gravity	ft/sec^2	32.174
Gravity	cm/sec^2	980.665
ha	acre	2.471
ha	ft^2	107,640
ha	m^2	10,000
hp	Btu/min	42.418
hp	ft-lb/min	33,000
hp	ft-lb/sec	550
hp	kW	0.7457
in.^{2}	cm^2	6.4516
in.^{2}	m^2	6.4516×10^{-4}
in.^{2}	ft^2	0.0069
in.^{3}	cm^3	16.387
in.^{3}	ft^3	5.7870×10^{-4}
in.^{3}	m^3	1.6387×10^{-5}
in.^{3}	gal (U.S.)	0.004329
in.^{3}	l	0.01639
in.^{3}	oz (U.S. fl.)	0.5541
kg	oz (avoirdupois)	35.274
kg	lb	2.2046
kg	Ton (long)	9.842×10^{-4}
kg-m	ft-lb	7.2330
kg-m	lb/ft	0.67197
kg/m^2	lb/in.^{2}	0.00142
kg/m^2	lb/ft^2	0.20482
kg/m^3	lb/ft^3	0.06243
kl	ft^3	35.317
km	ft	3280.8
km	mi	0.62137
km^2	ft^2	1.076×10^7
km^2	mi^2	0.386
km^2	m^2	1×10^6
km^2	acre	247.1
kW	Btu/min	56.884

Appendix B: Conversion Factors

To Convert From	To	Multiply By
kW	hp	1.3410
kW-hr	Btu	3413
kW-hr	hp-hr	1.3410
l	in.^{3}	61.025
l	ft^3	0.035
l	gal (U.S.)	0.2642
l	oz (U.S. fl.)	33.814
l/sec	gal/min	15.8507
m	ft	3.2808
m	in.	39.370
m	mi	6.2137×10^{-4}
m/min	ft/sec	0.05468
m/min	km/hr	0.06
m/min	mph	0.03728
m^2	Acre	2.471×10^{-4}
m^2	ft^2	10.7639
m^2	in.	1550
m^3	ft^3	35.314
m^3	in.^{3}	61023
m^3	gal (U.S.)	264.73
m^3	l	999.973
Mile (statute)	in.	63360
Mile (statute)	km	1.60935
Mile (statute)	m	1609.35
Mile (statute)	Rod	320
mi/hr	cm/sec	44.704
mi/hr	ft/min	88
mi/hr	m/min	26.82
mg	oz (avoir)	3.5274×10^{-5}
mg	lb	2.2046×10^{-6}
mμ	Å	10
mμ	cm	1×10^{-7}
oz (avoirdupois)	oz (troy)	0.91146
oz (avoirdupois)	g	28.35
oz (fluid)	cm^3	29.5737
oz (fluid)	in.^{3}	1.8047
oz (fluid)	l	0.029573
ppm	Grain/gal	0.0584
lb	g	453.5924
lb	lb (troy)	1.2153
lb/ft	kg/m	1.48816
lb/ft^2	g/cm^2	0.48824

Appendix B: Conversion Factors

To Convert From	To	Multiply By
lb/ft^2	lb/in.^{2}	6.9445×10^{-3}
lb/in.^{2}	atm	0.068046
lb/in.^{2}	g/cm^2	70.307
lb/in.^{2}	mm Hg (0°C)	51.715
lb/in.^{3}	g/cm^3	27.68
lb/ft^3	g/cm^3	0.016018
lb/ft^3	kg/m^3	16.018
lb/ft^3	lb/in.^{3}	5.787×10^{-4}
Tonne	Barrels	7.31
Tonne	kl	1.16
Tonne	1000 U.S. gal	0.308
Tonne	lb	2204.62
Tonne	kg	1000
Ton (long)	kg	1016.047
Ton (long)	Ton (short)	1.12
W	ft-lb/min	44.254
W	hp	1.34×10^{-3}
W	Btu/hr	3.41304

Author Index

Adler, F. J., 146, 147, 148
Aguilera, R., 369, 377
Almon, W. R., 225, 233, 305, 310
Al-Sari, A. M., 57, 108
Anderson, J. H., 155, 161
Anonymous, 305, 310, 324
Avary, K. L., 31

Bally, A. W., 302, 310
Baker, D. R., 97, 108
Barlow, J. A., Jr., 266, 267, 268, 270, 271, 275
Barnes, C. R., 57
Bayliss, P., 292, 299
Bebout, D. G., 233
Behling, M. A., 31
Beikman, H. M., 327, 333
Bendler, E. P., 31
Biederman, E. W., Jr., 89, 90, 91, 92, 93, 94, 98, 99, 108, 124, 129, 182, 188, 365, 367
Bird, K. J., 311
Bovell, G. R. L., 292, 296, 298, 299
Bowersox, J. R., 341
Bowles, J. P. F., Jr., 114, 121
Brockett, L. D., 324

Callender, P. L., 222, 232
Calhoun, T. G., 241, 247, 248, 250
Caplan, W. M., 148
Cardwell, D. H., 26, 31, 42, 49
Carlson, K. H., 5, 377
Carlson, M. P., 148
Charpentier, R. R., 5, 377
Chen, P., 25, 30, 31
Chenoworth, P. A., 96, 106, 108
Choquette, P. W., 124, 129
Christensen, E. W., 346, 347
Clark, J. L., 341
Clark, N. M., 341
Claypool, G. E., 282, 283, 285, 286, 287, 288
Clisham, T. J., 57, 108
Cloud, P. E., Jr., 129
Cobban, W. A., 270, 271, 272, 273, 275
Cohee, G. V., 253, 262
Coker, S. J. L., 367
Coniglio, M., 191, 194
Cooper, C. G., 154, 156, 157, 158, 161
Cordiner, F. S., 276, 281
Coury, A. B., 5, 377
Cox, B., 345, 347
Crovelli, R. A., 5, 377

387

Danielson, H. H., 14, 15, 16, 17, 22
David, M., 155, 161
Davies, G., 233
Davis, D. K., 305, 310
Debrosse, T. A., 31
Deere, R. E., 292, 299
Dembicki, H., Jr., 182, 189
Denison, R. E., 137
Diecchio, R. J., 45, 49
Dolton, G. L., 4, 5, 370, 371, 372, 373, 374, 375, 377
Dott, R. H., Sr., 22
Dunbar, C. O., 19
Dutton, S. P., 133, 137, 172, 173, 179

Earnest, L. J., 341, 342, 343, 344, 347
Ejedawe, J. E., 366, 367
Exum, F. A., 235, 236, 239, 240

Fath, A. E., 77, 82, 86, 98, 108
Fenstermaker, C. D., 23, 24, 31
Ferrero, E. P., 18, 22, 148, 233, 250, 262, 275, 368
Ferris, B. J., 154, 156, 157, 158, 161
Fettke, C. R., 7, 8, 22
Folk, R. L., 222, 232
Frezon, S. E., 5, 377
Fryberger, S. G., 56, 57, 106, 108
Frye, J. C., 87, 123

Galloway, J. O., 368
Galloway, W. E., 203, 204, 205, 206, 222, 225, 227, 229, 231, 232, 233
Garton, E. L., 18, 22, 83, 86, 104, 105, 108, 120, 121, 148, 159, 161, 179, 228, 233, 365, 366, 368
Gatewood, L. E., 113, 114, 115, 117, 118, 121
Gilbert, C. M., 310
Gillett, S. L., 73
Goebel, E. D., 148
Goodell, H. G., 270, 275
Gordy, P. L., 310
Goth, J. H., Jr., 22
Gwinn, V. E., 43, 49

Halbouty, M. T., 7, 22, 137, 141, 148, 324
Hale, L. A., 275
Halley, R. B., 74, 141
Ham, W. E., 136, 137
Hantzschel, W., 79, 86
Harper, J. A., 31
Harris, A. G., 332, 333
Harrison, R. S., 191, 194
Haun, J. D., 266, 267, 268, 270, 271, 275
Hedberg, H. D., 339, 340, 341
Heinze, B., 182, 188
Henslee, H. T., 148
Hicks, I. C., 148
Hills, H. M., 170, 179
Hilpman, P. L., 77, 86, 124, 129
Hobday, D. K., 233
Holmquest, H. J., 153, 161
Houlik, C. W., Jr., 277, 281
Howard, J. D., 87

Hunt, J. M., 227, 233, 268, 275, 366, 368
Hurford, G. T., 241, 247, 248, 250

Illing, L. V., 124, 129
Isley, A. M., 19, 21, 22
Iwuagwu, C. J., 300, 303, 310

James, N. P., 124, 129
Jamieson, G. W., 268, 275
Jamison, H. C., 313, 314, 315, 317, 318, 321, 322, 324
Jennings, C. F., 335
Jewett, J. M., 87, 123
Jones, H. P., 313, 315, 316, 321, 322, 324
Jones, T. S., 155, 161, 173, 179

Kelley, D. R., 22
Kelly, W. W., Jr., 31
Kepferle, R. C., 45, 49
Khan, K. S., 5, 377
King, R. E., 22
Klemme, H. D., 22
Kline, P. C., 31
Kornfeld, J. A., 31, 327, 333
Krynine, P. D., 7, 12, 13, 19, 22, 24, 27, 31
Kuenen, P. H., 121

Land, L. S., 133, 137
Landes, K. K., 8, 22, 59, 60, 66, 67, 68, 69, 74, 77, 78, 86, 114, 116, 121, 132, 137, 150, 151, 161, 162, 169, 197, 198, 202, 252, 254, 255, 262, 264, 269, 275, 312, 324, 326, 333, 336, 337, 338, 341, 351, 354, 361, 368
Lane, E. C., 83, 86, 104, 105, 108
Larson, T. G., 148
Larson, W. S., 252, 253, 255, 256, 262
Legall, F. D., 56, 57
Lerbekmo, J. F., 300, 303, 310
Levorsen, A. I., 32, 40, 107, 108, 131, 136, 137, 166, 167, 169
Lindley, J., 22
Lindquist, S. J., 225, 233
Lindsay, D. R., 87
Linville, B., 14, 16, 22
Lister, J. H., 5, 377
Livingston, A. R., 276, 281
Longacre, S. A., 139, 141, 192, 193, 194
Longstaffe, F. J., 305, 310
Loucks, R. G., 74, 155, 161, 214, 222, 225, 227, 233
Love, A. H., 288
Lundegard, P., 49
Lytle, W. S., 9, 14, 22

Mackey, F. L., 114, 121
MacPherson, B. A., 348, 350, 351, 354, 361, 362, 364, 368
MacQueen, R. W., 57
Magara, K., 233
Masters, J. W., 325, 326, 327, 328, 330, 332, 333
Matthews, R. K., 196, 197, 202, 232, 233
Maughan, E. K., 288
Maynard, J. B., 49
McBride, E. F., 141
McCaslin, J. W., 327, 333
McCormac, M., 31

Author Index

McCourt, G. B., 290, 291, 299
McCracken, M. H., 148
McCublin, D. G., 275
McGlade, W. G., 22
McGuire, M. D., 337, 338, 340, 341
McIntosh, R. A., 324
McKenney, J. W., 325, 326, 327, 328, 330, 332, 333
McKinney, C. M., 18, 22, 120, 121, 148, 159, 161, 169, 179, 227, 228, 233, 247, 250, 262, 275, 365, 366, 368
McMullin, R. H., 5, 377
Mellon, G. B., 306, 310
Merewether, E. A., 270, 271, 272, 273, 274, 275
Merino, E., 356, 359, 360, 362, 364, 368
Merriam, D. F., 89, 96
Merritt, C. E., 137
Meyerhoff, A. A., 195, 202, 371, 377
Meyerhoff, H. A., 22
Milton, W. H., 22
Moore, C. H., 233
Moore, R. C., 78, 87, 123
Moore, S. B., 32, 40
Morgan, J. T., 276, 277, 278, 279, 280, 281
Morgridge, D. L., 322

Nanz, R. H., Jr., 206, 207, 222, 233
Nelson, R. A., 71, 74
Newmann, M. L., 120, 121
Normark, W. R., 353, 354
Nosow, E., 66

O'Connor, H. G., 87, 123
Ondrusek, P. S., 17, 22
O'Sullivan, R. B., 327, 333

Pan, C. H., 97, 108
Parker, M. C., 148
Parkinson, G., 345, 347
Patchen, D. G., 25, 29, 30, 31, 42, 46, 49
Paynter, W. T., 17, 22
Perrodon, A., 76, 87, 252, 262, 301, 310, 349, 354, 370, 371, 376, 377
Petta, T. J., 69, 74
Pettijohn, F. J., 121, 220, 233
Pike, R. S., 5, 377
Pippin, L., 131, 132, 137, 138, 139, 141, 143, 148
Pittman, E. D., 358, 359
Potter, P. E., 49, 59, 60, 66, 121
Powers, R. B., 5, 377
Prather, R. W., 290, 291, 299
Pratt, B. R., 73, 74
Pryor, W. A., 49
Pye, W. D., 326, 329, 333

Rainwater, E. H., 200, 202
Ramondetta, P. J., 170, 171, 172, 173, 178, 181, 189
Rascoe, B., Jr., 146, 148
Reeves, J. R., 77, 87, 89, 96
Reiss, B., 200, 202
Roliff, W. A., 51, 52, 53, 57

Samuels, N., 49
Sax, N., 152, 161

Scharef, F. J., 49
Schmitt, G. T., 162, 164, 166, 169
Schneeflock, R., 356, 357, 359
Scholle, P. A., 47, 49, 103, 108, 233, 237, 240, 274, 275, 294, 306, 310
Schramm, M. W., Jr., 148
Schulz, J., Jr., 202
Schwalb, H. R., 61, 65, 66
Schwarz, K. A., 31
Scott, E. W., 5, 377
Selley, R. C., 309, 310
Shanmugam, G., 353, 354
Sharp, E., 202
Shelton, E. M., 169
Shouldice, J. R., 305, 308, 309, 310
Sidwell, R., 162, 169
Siever, R., 121
Smith, C. M., Jr., 9, 22
Smith, H. M., 155, 161, 173, 179
Smith, N. M., 66
Smith, W. B., Jr., 322
Speers, R. G., 313, 315, 316, 321, 322, 324
Spencer, C. W., 270, 271, 272, 273, 275
Stenzel, W. K., 185, 189, 191, 194
Stewart, G. A., 310
Sutton, A. H., 66
Swann, D. H., 66

Taylor, J. C. M., 305, 310
Terriere, R. T., 242, 243, 244, 249, 250
Tetra Tech, Inc., 20, 22
Tillman, R. W., 225, 233
Tissot, B. P., 366, 386
Todd, R. G., 170, 179
Toucks, R. G., 141
Travis, M. M., 327, 333
Tucker, M. E., 98, 102, 103, 108
Turner, F. J., 310

Van Horn, L. E., 84, 85, 86, 87
Varnes, K. L., 5, 377
Velde, B., 133, 137

Waage, K. M., 19
Wagner, W. R., 22
Walker, F. H., 66
Walker, R. G., 353, 354
Wallace, L., 87, 123
Wardlaw, B. R., 332, 333
Wardlow, N. C., 233
Warn, G. F., 162, 169
Weddle, H. W., 355, 356, 359
Weimer, R. J., 79, 82, 87, 275
Weirich, F. H., 353, 354
Weller, J. M., 60, 66
Wells, J. S., 148
Welte, D. H., 366, 368
Wenger, W. J., 18, 22, 148, 233, 250, 262, 275, 368
West, T. S., Sr., 199, 200, 202
Whaling, J., 22
Wilkinson, W. M., 163, 169

Williams, H., 301, 310
Williams, R., 311, 324, 344, 347
Williamson, N. L., 40
Wilson, M. D., 358, 359
Wilson, W. W., 19, 21, 22
Wise, J. C., 202
Wood, G. V., 41, 49

Woodrow, D. L., 19, 21, 22
Wright, L. A., 335
Wulf, G. R., 275

Young, A., 180, 181, 182, 189

Zeller, D. E., 78, 87

Subject Index

Acid attack, CO_2-charged waters, 353
Admire Sand, photomicrographs, 80, 81
Alaska, 311–324
Alberta, Canada, 289–310
Algae, 244
Algal debris, 172, 173
Algal fenestral structure, 192
Algal pistolites, 192
Alkanes, normal, 286
Alternate gas–water miscible recovery, 247
Amarillo uplift, 131, 132, 137
Amorphous kerogen, algal origin, 341
Amorphous materials, 133, 135
Amorphous opal, 339
Amorphous sapropel, 172
Amorphous silica, 345
Anastomosing, thrombolites, 71
Ancestral Mississippi, 196
Anderson Co., Texas, 241
Andesite, 301
Angular fractures, 186, 188
Angularity, 19
Anhydrite, 140, 141, 180, 181

Anhydrite cement, 259, 261, 277, 278, 280
Anhydrite-cemented sandstone, 142, 144, 145
Anhydrite-effective seal, 172
Anomalous porosity vs. permeability, 124
Anoxic bottom conditions, 292
Anticline, 155, 157
Anticlinal Entrapment, 45
Apache Co., Arizona, 325
API gravities, regional plot, 185
API gravity distribution, California crudes, 367
Apophyllite needles, 330, 331
Appalachian basin, 6, 41
Appalachian Highlands, 6
Appalachian thrust belt, 3
Arbuckle Formation, 114
Arctic Slope, 370, 377
Areas and sedimentary rock volumes, 371
Arkose, coarse, 258
Ash falls, 206, 221, 222
Asphaltic residue, 120
Atlantic Coastal Plain, 6
Atlantic offshore, 6
Augite, 327, 330, 331

Authigenic chlorite, 133
Authigenic clay, 303, 305, 306
Authigenic minerals, 236

Back-reef facies, 180
Bahama Islands, 124
Bakersfield arch, 348, 350
Banff Formation, 292
Barrier beach environment, 236
Barrow arch, 314
Basal Belly River Formation, 300–310
Basal Belly River Sandstone, photomicrographs, 306, 307
Basal sand, 252–258
   Blackwood Creek field, photomicrographs, 258, 259
   Sleepy Hollow field, photomicrographs, 256
   Wildcat, photomicrographs, 259, 261
Basinal shales, 173
Basin-margin wedging turbidite, 362
Between delta sands, 367
Big Horn Basin, 285
   location, map, 277
Big Injun, 32
Bimodal grainsize, 256
Bimodal grainsize distribution, 48
Bingham test, 15, 17
Bioclastic fragments, 193
Biodegradation, 85, 173, 181
Biogenous opal, 292
Bioherm, 53
Biomicrite, 70
Biopolymer, 17
Biostrome, 53
Biotite flakes, 307
Black opaques, 135
Black powdery organic material, 244
Blackwood Creek field, 258, 259
Blanket sand, 24, 41
Blocky calcite, 124, 126
Blow-out, Wild Mary Sudik, No. 1, 114, 115, 117, 118
Blue-green algae, 73
Boehm lamellae, 47
Brachiopods, 70
Bradford Oil field, 7
Bradford Third Sand, 7
Braided streams, 322
Brine pre-flush, 15
Brittle quartz-rich rocks, 345
Brown Dolomite, 138–141
   photomicrographs, 139–141
Bryozoans, 244, 261
Bubble trains, 47
Bulk density, 9
Burning Springs anticline, 32
Butler Co.:
   Kansas, 75
   Kentucky, 58

Cadeville, photomicrographs, 237–239
Cadeville Sandstone, 234–240
Calcispheres, 70
Calcite, Rodney, 51
Calhoun field, 234–240

Caliche–Big Pine Key, Florida, 191
Cambrian, 50
Cambrian Reagan Sand, 253, 257
Cambridge arch, oil fields, 255
Cambro-Ordovician, 152
Canadian Shield, 3, 58
Capillary pressure curves, 90, 93, 99, 303, 304, 305
   definition of, 33
Carbonate cement, 9, 48
Carbonate content, 36
Carson Co., Texas, 130–137, 142
Catagenesis, 173
Catskill delta, 19
Catskill sea, 19
Cement rinds, 124
Central basin platform, 149, 152
   migration and emplacement of oil, 188
Central Stable Interior, 3
Chadron–Cambridge arch, 251, 253, 254, 255
Chalcedonic quartz, 295
Chalcedony, 292
Chamosite oolites, 102, 103, 104
Channel deposit, 40
Chemical weathering, 133
Chemung facies, 19
Cherokee Formation, 114
Cherokee shale, as source rock, 120
Chertified crinoid columnals, 126
Chert occurrence, 190, 213, 301, 370
Chert pebbles, 316
Chester series, 59
Chlorite, 330
Chuska Mountains, 325
Cincinnati arch, 6, 50
Clastic ratio map, 166
Clay matrix, 35
Clay minerals, 24
Clotted texture, 192
Coastal sabkhas, 277
Coastal tidal flats, 190
Coastal upwelling, 338
Collophane peloids, 292, 295
Colorado Plateau, 3
Competence of currents, 43, 44
Compositional variables, 9
Conodont color studies, 328, 333
Conodonts, 56, 70
Continental core, 3
Continental sabkhas, 277
Cook Inlet, 370, 377
Coral(s), 244, 246
Coral fragment, Matecombe Key, 127
Coral structure, Key Largo, 128
Cordilleran thrust belt, 3
Core photograph, San Andres, 176
Corpus Christi Bay, 230, 231
Correlation index:
   Admire Sand, El Dorado field, 83
   California crude oils, 365, 366
   definition of, 17
   Ellenburger, TXL field, 159
   Frio crude oils, 222, 226, 227, 228

Subject Index

Frontier crude oil, 275
James Lime, Fairway field, 244, 247
Panhandle crude oils, 148
Reagan Sand crude oil, 262
San Andres, Welch field, 173
Simpson Sand, El Dorado field, 105
Wilcox Sand, Oklahoma City field, 120
Cotton Valley group, 234
Cratonic sedimentation, 58
Cretaceous marine shales, 322
Crinoid columnals, 53, 261
Crinoid debris, 125, 126
Cristobalite-tridymite, 339, 345
Cross-bedding, 19, 60
Cross-sections:
  Amarillo uplift, 132
  axis of Bakersfield arch, 351
  diatomite mine, 344
  Lost Hills anticline, 340
  Phosphoria, 283
  Pleasant Valley field, 357
  Prudhoe Bay field, 315
  Wind River basin, 265
Cross-stratification, 239, 277
Crude oil maturation, 366
Cumulative production and inferred reserves:
  gas, 373
  oil, 372
Cuttings:
  extracts, 182
  samples, 182

Dawson Co., Texas, 170, 172
Deep-marine sandstone, 348
Deep water exploration, temperature limits, 370
Delaware basin, regional cross-section, 153
Delaware basin margins, 190
Deltaic depocenters, 200
Deltaic deposition, 321
Deltaic splay channel, 206
Depositional centers, 367
Des Moines Series, 252
Destin, Florida, 30
Detrital carbonate, 48
Dhahran, Saudi Arabia, 56
Diabase, 136
Diatomaceous mudstones, 338
Diatom blooms, 350
Diatomite oil mine, 342–345
Diatremes, 326
Dikes, 326
Dineh-Bi-Keyah field, 325–333
  photomicrographs, 330, 331
Diopside, 320, 327
Directions of sand transport, 66
Dissolution:
  dolomite rhomb centers, 361, 363, 364
  feldspars, 350, 352, 353
Disordered cristobalite, 292
Distal-margin facies, 350
Distal shelf, 19
Distillation curves, 98, 104, 105

Distillation data, 83
Distillation fraction, 18
Dolomite, fine grained, 144
Dolomite cement, 48
Dolomite content, 24
Dolomite rhombs:
  Big Injun, Walton field, 27
  Dundee, Rodney field, 52
  Ellenburger, TXL field, 159, 160
  Gatchell–McAdams, Kettleman Hills field, 363
  Hunton, Yeage field, 110
  Lansing–Kansas City, Victory field, 125
  Rogers City–Dundee, North Wise field, 69
  Viola, El Dorado field, 88
Dolomitization, 50, 89
Dolomitized oolite, 139, 140
Distributary, 206
Down-to-the-coast faults, 196, 199
Downwarps and faults, 367
Dravo process, 344–345
Dundee, grainsize, 56
Dundee Formation, photomicrographs, 54, 55
Dune fields, 277

Economical set, 9
Ector Co., Texas, 154
El Dorado:
  production, 75
  structure, 89, 90
  Viola Formation, 88–96
El Dorado field, Admire Sand, 75–87
Electric log:
  Cadeville Sand, Calhoun field, 236
  Hardinsburg Sandstone, 60
  James Lime, Fairway field, 244
  Nordegg, Paddle River field, 298
  productive sill, Dineh-Bi-Keyah field, 329
  Pyrite zone, Prudhoe Bay field, 321
  Sadlerochit Formation, Prudhoe Bay field, 317
  Spraberry Sand, Spraberry Trend, 163
Electric log(s), misleading interpretation, 355, 356
Electric log cross-section:
  Kettleman North Dome, 362
  Salt Creek field, 270
Elk Hills, 366
Ellenburger Formation, 153, 154–161
  photomicrographs, 159–160
Eolian Sand, 56
Erosion channels, 60
Etchegoin sand, Lost Hills, 365
Eustatic drops in sea level, 170
Everglades, Florida, 107
Exploration intensity, 370, 376
Exploration success ratio *vs.* yield, 377
Exploration yield, 370, 376
Extracratonic basin, 203
Extractable hydrocarbons, 285, 286
Extractor tanks, 344

Fairway field, 241–250
  photomicrographs, 245, 246
Fault, cut off, 155, 157

Fault displacement, 114
Feldspar, 9
Feldspar overgrowths, 223
Ferric hydroxide, gelatinous, 305
Finely crystalline dolomite, 180
Five-spots, 15
Flamingo, Florida, 107
Florida panhandle, clean quartz sand, 30
Flow paths, radioactive tracers, 248
Fluid transmissibility, 278
Fluvial, 59
Fluvially-dominated delta, Mississippi, 309
Folding, faulting, and truncation, 113, 114
Fracture(s):
   across-the-grain, 364
   fracture porosity, Oriskany Sand, Thornwood, 45
   Hardinsburg Sand, Silver City, 60
   Red Cave Sand, Panhandle field, 142
   significance, 369
   Spraberry Sand, Spraberry Trend, 163, 166, 167
   timing, Dineh-Bi-Keyah field, 328
   Viola limestone, El Dorado field, 89
Fracture channels, 186
Fractured chert, Prudhoe Bay, 316, 320
Fractured Monterey Chert, 337, 340, 345, 346
Fractured shale pool, Lost Hills field, 337
Fracture filling, 259
Fracture lining, 141
Fracture porosity, 45
Fracture systems, conduits for oil migration, 179, 194
Fracture zone, basement granite, 136
Frio Formation, 203–233
   electric log, May field, 213
   light mineral profiles, May field, 208–212
   mineralogy, 206
   photomicrographs, 223, 224, 230, 231
   rock consolidation stages, 214
Frontier Formation, 263–275
   photomicrographs, 274
Fusilinids, 125, 261

Gas chromatographic analyses, saturated hydrocarbons, Phosphoria, 287
Gas pay map, 143
Gas reserves, Pekisko unconformity fields, 291
Gas slugs, 247
Gas sweep configurations, 247
Gastropod fragments, 236
Gatchell–McAdams Sandstone, 355–368
Generalized cross section of Gulf of Mexico, 197
Genessee Formation, 19
Geologic map, Arizona, 326
Gilsonite, 244
Glauconite, 40, 103, 213, 306
Goldsmith field, 180–189
Grain size:
   Basal Belly River Sand, Pembina field, 304
   Big Injun Sand, Walton field, 37
   Bradford Third Sand, Bradford field, 12, 13
   Frio Sand, May field, 222, 226, 227
   Hardinsburg Sand, Silver City field, 64
   Oriskany Sand, Thornwood field, 44
   Sadlerochit sandstone, Prudhoe Bay field, 321
   Simpson Sand, El Dorado field, 99
   Spraberry Sand, Spraberry Trend, 165
   Wilcox Sand, Oklahoma City field, 119
Granite ridges, 75
Granite Wash:
   photomicrographs, 134–136
   sampling for core analysis, 133
Granite Wash Formation, 130–137
Granitic monadnocks, 121
Grawackes, 13
Greenbriar group, 32
Greta–Cavancahuac barrier, strand-plain, 228, 229
Guijarral Hills field, 355–359
   photomicrographs, 357, 358
Gulf Coast, 195–250
Gypsum deposited in well bores, 176, 177

Hample fractions, 18
Hardinsburg Sandstone, 58–66
   photomicrographs, 62, 63
Haskell Co., Kansas, 122, 123
Heavy minerals, 214, 215–219
Helderberg limestone, 41
Hematite oolites, 101, 102, 103, 104
Henderson Co., Texas, 241
Highly rounded quartz, 121
High paraffinicity, 155
Hinge line fracturing, 173
Horizontal fractures, 186, 188
Horton Anticline, 43
Hugoton embayment, 122
Huntersville Chert, 41
Hunton Formation, 109–112
Hunton, photomicrographs, 110, 111
Hutchinson Co., Texas, 138
Hydrocarbon-like organics, 182
Hydrocarbons vs. organic carbon, 286
Hydrochloric acid, 17
Hypersthene, 136

Illinois basin, 59
Ilmenite and magnetite, 220
Initial displacement pressure, 33
Injectivity, 15
Interfacial tension, 85
Intergranular porosity, 72, 236
Intermediate base crudes, 18
Intertidal environment, 172
Intracratonic basins, 58
Iron-stained clay matrix, 134, 135
Irreducible water saturation, 34, 35, 90
Isabella Co., Michigan, 67
Isobaric map, 248
Isopach map:
   Basal Belly River Sand, Pembina-Keystone field, 308
   Cadeville Sand, Calhoun field, 239
   Frio Sand, Seeligson field, 207
   Frontier Formation, Wyoming, 266
   James Limestone, Fairway field, 243
   Nordegg Member, Paddle River field, 298
   Sadlerochit Sandstone, Prudhoe Bay field, 318

Subject Index

Upper Spraberry unit, Midland basin, 164
Viola limestone, Oklahoma and Kansas, 96
Isopotential map, 167
Isoprenoids, 286
ITIO, No. 1 Mary Sudik, 114, 115, 117, 118

Jackson parish, 234
James limestone, 241–250
Jim Hogg Co., Texas, 199
Joints and fractures, 155

Kanawha Forest field, 23
Kaolinite, 60
    alteration of feldspars, 356
    interstitial, 355, 356, 357, 358
Kaolinite "books," 305
Kaolinite plus montmorillonite–illite and chlorite, 301, 306
Kaycee-Tisdale Mountain area, 271
Kern Co., California, 337, 342
Kettleman Hills field, 360–368
    photomicrographs, 363, 364
Key Largo limestone, 127, 128
King and Fresno Counties, California, 360
Kleberg Co., Texas, 203
Krypton 85, 247
Kuparuk Formation, 314

Labradorite, 136
Lansing–Kansas City, Butler Co., Kansas, correlation index curves, 104
Lansing–Kansas City Group, 122–129
Large bivalves (clams), 244
Late Devonian, 50
Leaching, 298
    mineralized fluids, 328
Leaching and fracturing, 109, 112
Lea Co., New Mexico, 190
Lensing and channeling, 19
Lewis Run, 9
Lisburne carbonates, 314
Lithofacies, maps, 25, 29, 31
Lithologic and electric log, Baily Flats, Wyoming, 272
Lithologic log, environments, 273
Littoral, 59
Local dolomitization, 67
Long axes of quartz grains, Paddle River field, 297
Long-range migration, 367
Loose sand, 115, 118
Lost Hills field, 331–341
Low-grade metamorphism, 135
Low-magnitude directional permeability, 277
Low-rank metamorphic rocks, 7
Low-wave-energy conditions, 192
Lurgi pilot plant, 345

Magnetite and titaniferous magnetite, 331
Major regional tectonic features, 61
Major sediment accumulations, map, 196
Mangrove islands, 108
Map and cross-section:
    Getty diatomite mine, 343
    Guijarral Hills in relation to folding, 356

location of Fairway field, 242
location of fractured shale pool, 337
major rock units and fault systems, California, 335
major sedimentary basins, California, 336
major structures, North Slope, 313
Rocky Mountain province, 252
San Joaquin basin, California, 349
Tisdale Mountain, Kaycee area, 271
western Canada, 301
Marathon thrust belt, 3
Marginal marine environment, 322
Matador arch, 149, 150
Maturation indexes, 172
Maximum heptane value maps, 182, 185, 187
May field, 203–228
McCracken Sandstone, 328
McKittrick field, 342–345
McKittrick thrust, 342, 344
Meade Peak Formation, 282, 283
Mechanical sieving, 13
Megaquartz filling, 294
Metamorphic heavy minerals, 220
Metaquartzite, 37
Methanol, 182
Miami oolite, 107
Micellar–polymer process, 85
Micellar solutions, 7, 13
Michigan basin stratigraphy, 68
Michigan basin structure, 69
Micrite, 70, 284
Micrite coatings, 193
Microcline, 134
Microconglomerates, 13
Microfractures, 155, 160
Microfracturing, 339, 340
Mid-continent, 75–148
    basement structure, 147
    paleogeography, 146
Migration, vertical, 366
Migration pathways, 366
Miliolids, 246
Mineral alteration, 19
Mineral composition:
    Dineh-Bi-Keyah field, 327
    Kettleman North Dome field, 362
    Rodney field, 51
    Thornwood field, 41, 44
    Walton field, 37
Mineralogical sorting, 19
Mineralogy, Frontier Formation, 271
Miscible enriched gas flood, 324
Mixed-layer montmorillonite/illite, 305
Mobeetie field, 133, 137
Mobility, 182
Mobility buffer, 13
Modal analysis, 297
    Sadlerochit Sand, 322
Model, submarine fan deposition, 354
Moldic porosity:
    Brown Dolomite Formation, 140
    Cadeville Sand, 236
    Hunton Formation, 109, 110

Moldic porosity (*Continued*)
  James Lime, 245
  Nordegg Member, 294
  Rogers City–Dundee formation, 70
  San Andres Formation, 174, 175, 176
  Yates Formation, 192
Monroe uplift, 234, 235
Monterey Chert, photomicrographs, 346
Monterey diatomite, 337–341
  photomicrographs, 339
Mount Scott, Wichita Wildlife Refuge, Lawton, Oklahoma, 143
Mud balls, 213
Mud mounds, 69
Mudmound reconstruction, 73
Multieffect evaporators, 344
Multiple regression, 9
Muscovite, 38

Naphthene-based crudes, 18
Naphthenic distillates, 18
Naturally fractured Monterey Chert, 345, 346
Needmore shale, 41
Nemaha anticline, 77, 89
Newburg Gas Sand, photomicrographs, 27, 28
Newburg Sand:
  grainsize, 29
  mineral composition, 24, 26
Non-reservoir dolomite, 277
Non-wetting phase, 341
Nordegg Member, Alberta, Canada, 289–299
  photomicrographs, 293–296
Norias Delta System, 205
Normal paraffin hydrocarbons, 18
North Slope, 311–324
North Wise field, 67–74
Norton Co., Kansas, 257

Offshore Louisiana, 370, 377
Offshore prograding sand sea, 56, 106
Oil-generative window, 367
Oil implacement, 179
Oil migration, 366
Oil pay map, 139
Oil saturation, 260
  post waterflood, 15
Oklahoma City field, 113–121
Onandaga Limestone, 41
Ooid shoal, 140
Oolite(s), leached, 124, 126
Oolite ghosts, 139, 140
Oolite reservoir, 138, 139, 140
Oolitic zone, 98, 101–103
Opal, chalcedony, 133
Opaline frustules, 338
Open marine deposits, 190
Ordovician sole fault, 43
Organic geochemical profiles, 191, 194
Organic-rich micro layers, 81
Oriskany, photomicrographs, 47, 48
Oriskany Sandstone, 41
Orthoclase feldspar, 214
Orthoquartzite:
  Hardinsburg Sandstone, 59

Newburg Gas Sand, 27
Oriskany Sandstone, 41
Tensleep Sandstone, 276, 277
Wilcox Sandstone, 115, 121
Ostracodes, 258
Ouachita thrust belt, 3
Outcrop map, Frontier Formation, 271
Outcrop sample map, Phosphoria Formation, 285
Overpressuring, 339, 340
Owl Creek Mountains, 283
Oxygenated organics, 182

Pacific Margin System, 3
Paddle River gas field, 289–299
  location, 290
Paleocurrents, 43, 45
Paleogeography, Texas coastal plain and shelf, Frio, 206
Paleotemperatures, 56
Palo Duro basin, 149, 150
Panhandle field, Texas, 130–148
Paralic environments, 193
Partial columnar section, Fairway field, 242
Pearsall Formation, 241, 242
Pebble bed, Frontier Formation, 267, 268
Pekisko Formation, 289, 290, 291, 292
Pelecypod fragments, 236
Pellets, 70, 140, 245, 284
Pelmicrite, 72
Peloids, spherical and ellipsoidal, 192
Pembina, basal Belly River Sand, electric log, sonic log, 308
Pembina oil field, 300–310
Pendular areas, 35
Penn Grade Micellar Displacement Project, 15
Permeability, 27, 28, 33, 50
Permeability–Poisson distribution, 89, 92
Permian basin, 150
Permian-Triassic sediments, 313, 314
Perthite feldspar, 134, 135
Perthite leucogranites, 136
Phosphatization, 292
Phosphoria, photomicrographs, 284
Phosphoria Formation, 282–288
Phreatic cement, 124
Phyllites, 7
Phylogopite, 327, 330, 331
Phytane, 286
Piedmont, 3
Pilot test, 15
Pinchout of Basal Belly River Sand, 308
Plagioclase feldspar, 134, 136, 213, 357
Playa lakes, 193
Pleistocene reef, Florida, 127, 128
Pocono interval, 32
Poikilotopic calcite,
  Frio Formation, Corpus Christi Bay field, 231
  Nordegg Member, Paddle River gas field, 293, 295
  Reagan Sand, Ray field, 257
  Sadlerochit Group, Prudhoe Bay field, 318, 319
  Simpson Sand, El Dorado field, 103
Point Arguello field, 345
Point count mineralogy, 2, 301, 303
Polarizing microscope, 2, 4
Polar solvent, 182

Subject Index

Polyacrylamides, 15, 85
Polymer flooding, 13
Procelanite, 341
Pore diameter vs. porosity, 178
Porosity, 27, 28, 50, 109, 112, 115, 118
Porosity and permeability, Fairway field, 249
Porosity-permeability profiles, 280
Porosity vs. permeability, 176
Post-depositional history, 91
Postsubsidence phase, 367
Powder River basin, 267
Primary production, 7
Pristane, 286
Probability, Wyoming-Utah-Idaho thrust belt, 370
Production from basement granite, 136
Production trend map, South Texas, 197
Promontories, erosional edge, 298
Proximal fans, Granite Wash, 137
Prudhoe Bay, photomicrographs, 318–320
Prudhoe Bay field, 313–324
Pseudohexagonal biotite, 206, 220, 221, 222
Pyrite:
    Cadeville Sand, Calhoun field, 237, 239
    Frio Formation, May field, 217, 218, 220, 222
    Hunton Formation, Yeage field, 111
    Sadlerochit Formation, Prudhoe Bay field, 319, 321
    Simpson Formation, El Dorado field, 102, 104
    Spraberry Formation, Spraberry Trend, 166
    Yates Formation, West Teas field, 193
Pyrite cubes and framboids, 174, 339
Pyroxene, augite, 136

Quartz, crenulated contacts, 358
Quartz arenites, 121
Quartz cement, 38, 41, 165, 214
Quartz cementation, 321
Quartz content, 52
Quartz overgrowths:
    Hardinsburg Sand, Silver City pool, 62
    Newburg Gas Sand, Kanawha Forest field, 27
    Oriskany Sand, Thornwood field, 47
    Simpson Sand, El Dorado field, 100–101
    Wilcox Sand, Oklahoma City field, 119

Radioactive isotopes, 247
Rapid subsidence, 367
Rate of deposition, Frontier Formation, 274
Rattlesnake Mountains, 284, 285
Ray field, 253–257
Reagan Co., Texas, 162
Reagan Sand, photomicrographs, 257
Recent coastal sediments, Northwest Gulf of Mexico, 232
Red beds, 19
Red Cave, photomicrographs, 144–145
Red Cave Formation, Panhandle field, 142–148
Reducing conditions, 166
Red Willow Co., Nebraska, 252
Reef, 53, 124
    on-growing structure, 241, 243
Reef detritus, Key Largo, 128
Regional structure, Dineh-Bi-Keyah, 327
Rehydrogenation, Lurgi, 345
Relative permeability curves, 281

Reserves, productive capacity:
    Fairway field, 249–250
    Prudhoe Bay field, 323
Reservoir age vs. oil and gas, 337
Reservoir pressure (original), 236
Reservoir quality factors, 225
Resistivity, 86
Resistivity log, lithology of Nordegg Member, 296
Restricted basin, 166
Restricted marine environment, 322
Retort shale, 282, 283
"Reverse drag," 196
Rhinestreet formation, 19
Riley Co., Kansas, 109
Rio Grande, deltaic plain, 207
Rio Grande embayment, 204
Roane Co., West Virginia, 32
Rock fragments, 301
Rocky Mountains, 251–310
Rodney oil field, 50–57
Rogers City–Dundee Formation, 67–74
    photomicrographs, 70–72
Romeo field, 199–202
Rudistids, 244

Sabine uplift, 234, 235
Sadlerochit Formation, 313–324
Salt Creek field, 263–275
Salt flowage, 243
Sampling interval, 182
San Andres, Goldsmith:
    photomicrographs, 183–184
    west-to-east structural cross-section, 182
San Andres, Welch, photomicrographs, 174–177
San Andres Formation, 170–189
Sanidine, 327
San Joaquin Valley, 337
Schematic Frio stratigraphic sections, 204
Schists, 7
Schuler Formation, 234–236
Secondary anhydrite, 172
Secondary leaching, 67
Secondary migration, 267
Secondary porosity, 350, 352, 353
Second Wall Creek Sand, 263–275
Section, pre-Mississippian, TXL field, 156
Serpulid worm tubes, 236, 237, 238
Shallow-marine shelf, 59
Shallow water marine environment, 190
Silica cement, 9
Silicious volcanics, 301
Sills, 326
Silver City pool, 58–66
Simpson Sand, El Dorado, 97–108
    Oklahoma City field, photomicrographs, 118–119
    photomicrographs, 100–103
Simpson Sandstone, 113–121
Slates, 7
Sleepy Hollow field, 252–258
Slugs, 14
Smectite clays, 341
Sodium hypochlorite, 17
Sorting, 24

Source beds, Frontier, 268
Source of oil, Panhandle field, 143
Source rock(s), 222, 366
   El Dorado, 105, 106
   Phosphoria Formation, 286
South-to-north cross-section, Dineh-Bi-Keyah, 328
South Padre Island, 227
Sparry calcite, 110, 236
Spastoliths, 103
Specific gravities, 18
Spherulitic structures, 293
Spicules, 292
Sponge spicules, 70
Spraberry, photomicrographs, 163, 165
Spraberry field, 162–169
Spraberry Formation, 162–169
Spraberry oil, correlation index, 167, 169
Spraberry Sand:
   mineral composition, 164
   source of oil, 166, 167
Stable interior, 3
Stable isotopes, 322
Stained calcite, 53
Statistical comparison, 9
Stevens Sandstone, 348–354, 365, 366
   photomicrographs, 352–353
Straight chain hydrocabons, 366
Stratified reservoir with cross flow, 53
Stratigraphic chart:
   Cadeville Sand, 235
   South Texas, 198
Stratigraphic column:
   Kansas, 78
   McKean Co., Pennsylvania, 9
   Panhandle–Hugoton field, Texas, Oklahoma, Kansas, 132
   Prudhoe Bay field, Alaska, 314
   San Joaquin Valley, California, 351
   West Texas, 151
   Wyoming, 264
Stratigraphic cross-section:
   Schuler Formation, 235
   west-to-east, Jim Hogg Co., Texas, 199
Stratigraphic entrapment, 236
Stratigraphic section:
   Nordegg Member, 291
   Oklahoma, 116
   Walton field, 34
Stratigraphic sequence, southwestern Alberta, 302
Stratigraphic terminology, Monterey Formation, 338
Stratigraphic traps, 267
Stratigraphy, Lansing–Kansas City, 123
Stromatolites, 73
Stromatoporids, 244
Structural shelf, 190
Structure:
   Fairway field, 242
   Kettleman Hills field, 361
   Oklahoma City field, 115, 117
   Oregon Basin field, 278
   Prudhoe Bay field, 316
   Salt Creek field, 269
   Southwest Powder River basin, 268

Structure contour map, 65
   TXL field, 157
   upper surface, Nordegg Member, 298
Stylolites, 69, 186, 188, 293
Subaerial environment, 190
Subaerial erosion, 298
Subaerial exposure, 170
Subgraywacke, 32
Submarine fans, 348
Subsurface reef complex, 241
Sum of heptane values map, 186, 187
Supratidal environment, 172
Supratidal flats, 277
Surface lineaments, 171
Surfactants, 14, 85
Sutured boundaries, 292, 364
Sweep efficiency, 13

Tectonic framework, 3
Tectonic map:
   Kansas, 77, 124
   Nebraska, 253
   South Texas, 197
   Texas, 150
   West Texas, southeastern New Mexico, 152
Temblor Formation, 360
Tensleep, photomicrographs, 279
Tensleep Formation, 276–281
Terrestrially derived organic matter, 322
Tertiary recovery, 3, 7, 13, 83
Textularids, 246
Textural variables, 9
Thermal alteration index, 341
Thermal cracking, 366
Thermally immature area, 286
Thermal maturation map, 328–332
Third Bradford Sand, photomicrographs, 10, 11
Thornwood gas field, 41
Thrust faults, 43
Tidal flat, 79
Titaniferous magnetite, 135
Toadlena anticline, 325, 327, 328
Total organic carbon, 222
Transgression, Jacksonian stage, 200
Transgressive–regressive cycles, 282
Tritiated hydrogen, 247
Tritiated methane, 247
Tulare Formation, 365
Turbidite deposits, 348, 350, 361, 362
Turbidite fans, 370
   southern San Joaquin Valley, California, 350
Two-dimensional chromatography, 182
TXL field, 154–161
   cross-sections, 158

Ultraviolet fluorescence, 182
Undiscovered recoverable crude oil, 374
Undiscovered recoverable natural gas, 375
Unweathered igneous reservoir rock, 325
Upper Mississippi Valley, 58
Upper San Andres Formation, 180

Subject Index

Valley and Ridge province, 45
Veriform microstructure, 71
Vertical fractures, 106, 186, 188, 278, 340
Victory field, 122–129
   photomicrographs, 125, 126, 127, 128
Viola, photomicrographs, 94, 95
Volcanic rock fragments, 206, 213
Vuggy dolomite, 89

Walton field, 32
   photomicrographs, 38, 39
Warfield anticline, 24
Waterflooding, 7, 133
Wave-cut benches, 25
Wave-dominated, Nile delta, 309
Wave-dominated delta, 274
Welch field, 170–179
   porosity pinchout, 172
   source rock, 173
Well-rounded quartz, 100–101, 103, 119
Wichita Mountains, 136
Wilcox sandstone, 113–121
West Coast, 334–368
West-to-east cross-section:
   Cambridge arch, 255
   Sleepy Hollow field, 256
West-to-east section, Nebraska, 254
West-to-east stratigraphic diagram, 266
West Elk Hills field, 348–354
West Texas field, 190–194
West Texas–eastern New Mexico, 149–161
Wolfcampian source rocks, 173
Worm burrows, 72
Wyoming, 263–288

Xanthan gum, 85
X-ray diffraction, clays, 2

Yates Formation, 190–194
   photomicrographs, 192–193
Yeage field, 109–112
Yegua, Romeo field, photomicrographs, 201–202
Yegua Formation, 199–202

Zebraic chalcedony, 141
Zoned crystals, 110, 112
Zoned dolomite rhombs, 295, 361, 363
Zoned gypsum crystals, 177
Zoned quartz overgrowths, 91